Systolic Geometry and Topology

Mathematical
Surveys
and
Monographs

Volume 137

Systolic Geometry and Topology

Mikhail G. Katz

**With an Appendix
by Jake P. Solomon**

American Mathematical Society

2000 *Mathematics Subject Classification.* Primary 53C23; Secondary 11R52, 16K20, 17B25, 28D20, 30F10, 37C35, 52C07, 53C20, 55M30, 57M27, 55R37, 57N65.

For additional information and updates on this book, visit
www.ams.org/bookpages/surv-137

Library of Congress Cataloging-in-Publication Data

Katz, Mikhail Gersh, 1958–
Systolic geometry and topology / Mikhail G. Katz ; with an appendix by Jake P. Solomon.
p. cm. — (Mathematical surveys and monographs, ISSN 0076-5376 ; v. 137)
Includes bibliographical references and index.
ISBN 978-0-8218-4177-8 (alk. paper)
1. Geometry, Algebraic. 2. Riemann surfaces. 3. Topology. 4. Inequalities (Mathematics) I. Title.

QA564.K368 2007
516.3′5—dc22 2007060668

∞ The paper used in this book is acid-free and falls within the guidelines
established to ensure permanence and durability.
Visit the AMS home page at `http://www.ams.org/`

10 9 8 7 6 5 4 3 2 1 12 11 10 09 08 07

Dedicated to the memory of my parents,
Tsvi Dovid ben Moshe and Chaya bas Binyomin

The photograph on the back cover of the book had been taken by the author's late mother. The author is married and lives in Bnei Braq, Israel. The author is blessed with seven children, and he strives to emulate the serene ways of his late father, in raising them.

Contents

Preface

This text is based upon a recent course at Bar Ilan University, as well as upon numerous collaborative efforts (see the Acknowledgments Chapter following this Preface). After dealing with classical geometric preliminaries, including the *theorema egregium* of Gauss, we present new geometric inequalities of systolic type on Riemann surfaces, as well as their higher dimensional generalisations. Thus, the text can be viewed as an expanded version of a part of the survey by C. Croke and the present author [**CrK03**], which itself dealt with a small part of the material in M. Gromov's seminal text *Filling Riemannian manifolds* [**Gro83**]. Most of the results presented here have been obtained over the past four years, a reflection of the rapid progress in the field since the publication of the monograph [**Gro99**] seven years ago. M. Berger's recent monograph [**Berg03**] provides a fascinating historical perspective unburdened by proofs, *cf.* Section 2.1.

Acknowledgments

Many people have contributed to the present work at various stages of its development.

The present text involves the work of a number of collaborators. In alphabetical order, they are V. Bangert, C. Croke, S. Ivanov, C. Lescop, Y. Rudyak, S. Sabourau, M. Schaps, S. Shnider, J. Solomon, U. Vishne, and S. Weinberger. The author warmly thanks all of them for many fruitful interactions and enriching discussions.

We are grateful to D. Kazhdan and M. Belolipetsky for helpful discussions of arithmetic surfaces, and to A. Reid for help with simplicial volume.

We express appreciation to L. Ambrosio, G. Dula, J. Lagarias, C. LeBrun, F. Morgan, and S. Weinberger for insightful comments.

We are grateful to H. Farkas, S. Krushkal, and S. Shnider for helpful discussions of hyperelliptic surfaces; to M. Dutour for a helpful discussion of the combinatorial arguments of Section 8.7; and to F. Auer for the Super Bowl figures of Section 8.5.

We are grateful to D. Ebin, E. Leichtnam, and J. Lafontaine for discussions concerning the smooth dependence on parameters in Moser's method, exploited in Proposition 16.13.2. We thank E. Kuwert for providing references concerning closed forms minimizing the L^p-norm in their cohomology class.

We are grateful to A. Marin for a proof of Proposition 18.2.6 on self-linking numbers.

We are grateful to B. White for a helpful discussion of the deformation theorem and the material of Section 18.2.

We express appreciation to P. Biran, S. Donaldson, C. LeBrun, and J. Solomon for insightful comments related to the surjectivity conjecture, and to R. Borcherds for help with the automorphism group of the Lorentzian lattice.

We are grateful to I. Babenko for detailed criticisms in the context of systolic category, to R. Brown for helpful comments on non-abelian cohomology, and to J. Oprea for a calculation of the rational category of an interesting homogeneous space in Proposition 14.4.2.

We are grateful to M. Kapovich for helpful comments concerning groups generated by reflections, related to the material of Appendix A.

We are grateful to R. Hain for clarifying the issue of the compatibility of two Massey product theories, *cf.* Theorem 14.2.2.

The author is grateful to Marcel Berger and to Louis de Branges for consenting to the inclusion of their comments respectively in Section 2.1 and Subsection 2.2.3.

We thank I. Chavel for some probing questions, and T. Sakai, J. Lafontaine, and N. Hitchin for reading an early version of the manuscript of the book, and making valuable suggestions.

Our thanks are due to Mrs. M. Beller for careful editing of portions of the text.

I thank my wife for unbounded patience, and for accomodating unreasonable work schedules inherent in a task such as writing a book.

I thank Hashem Yisborach for giving me the strength to lead this work to its completion.

Part 1

Systolic geometry in dimension 2

CHAPTER 1

Geometry and topology of systoles

1.1. From Loewner to Gromov via Berger

The *systole* of a compact metric space X is a metric invariant of X, defined to be the least length of a noncontractible loop in X (more detailed definitions appear in Chapter 5). We will denote it $\mathrm{sys}\pi_1$, following M. Gromov's notation [**Gro96**]. When X is a graph, the invariant is usually referred to as the *girth*, ever since the 1947 article [**Tu47**] by W. Tutte (1917-2002), see Subsection 2.4.2 below. Possibly inspired by Tutte's article, C. Loewner started thinking about systolic questions on surfaces in the late 1940's, resulting in a 1950 thesis by his student P.M. Pu.

1.1.1. Thom. This line of research was, apparently, given further impetus by a remark of the venerable René Thom, in a conversation with Marcel Berger in the library of Strasbourg University during the 1961-62 academic year, shortly after the publication of the papers of R. Accola [**Ac60**] and C. Blatter [**Bl61a**]. Referring to these systolic inequalities, Thom reportedly exclaimed: *"Mais c'est fondamental!"* [These results are of fundamental importance!] Subsequently, M. Berger popularized the subject in [**Berg65, Berg93, Berg03**]. For an intuitive description of the problem, see [**Berg70**]. M. Berger's personal account of systolic history appears in Section 2.1.

C. Loewner's seminal ideas in what is known today as systolic geometry have had a slow start, with only a pair of articles appearing in the 1950's: by J. Hersch [**Her55**] and P. Pu [**Pu52**]; three articles in the 1960's: by R. Accola [**Ac60**], M. Berger [**Berg65**], and C. Blatter [**Bl61a**]; and four in the 1970's: by Berger [**Berg72a, Berg72b**], P. Buser [**Bus78**], and I. Chavel [**Cha72**]. Over a dozen publications appear in the 1980's: by W. Abikoff [**Ab86**], C. Bavard [**Bav86, Bav88**], section in a book by Y. Burago and V. Zalgaller [**BuraZ80**, p. 43], C. Croke [**Cr88**], M. Gromov [**Gro81a, Gro83**], R. Harvey and H. B. Lawson [**HarL82**], J. Hebda [**Heb81, Heb86**], F. Jenni [**Je84**], S. Kodani [**Kod87**], T. Sakai [**Sak88**], and others.

The systolic geometry website currently lists over a hundred articles. The website at http://www.math.biu.ac.il/~katzmik/sgt.html is regularly updated, see Section 2.4.

Systolic geometry is a rapidly developing field, featuring a number of recent publications in leading journals. Recently, an intriguing connection has emerged with the Lusternik-Schnirelmann category, which may signal the genesis of a subject that could be called "systolic topology", alluded to in the book title.

1.1.2. Flavor. To give a preliminary idea of the flavor of the field, one could make the following observations.

The main thrust of Thom's remark to Berger quoted in Subsection 1.1.1 above, appears to be the following. Whenever one encounters an inequality relating geometric invariants, such a phenomenon in itself is interesting; all the more so when the inequality is sharp (*i.e.* optimal). The elegance of the inequalities of C. Loewner, P. Pu, and M. Gromov (see below) is indisputable, *cf.* [**Gro99**, Systolic reminiscences, p. 271].

In systolic questions about surfaces, integral-geometric identities play a particularly important role. Roughly speaking, there is an integral identity relating area on the one hand, and an average of energies of a suitable family of loops, on the other. By the Cauchy-Schwarz inequality, energy is an upper bound for length squared, hence one obtains an inequality between area and the square of the systole. Such an approach works both for the Loewner inequality for the torus,

$$\mathrm{sys}\pi_1(\mathbb{T}^2)^2 \leq \gamma_2 \,\mathrm{area}(\mathbb{T}^2), \tag{1.1.1}$$

where $\gamma_2 = \frac{2}{\sqrt{3}}$, *cf.* (5.4.1), and for Pu's inequality for the real projective plane,

$$\mathrm{sys}\pi_1(\mathbb{RP}^2)^2 \leq \tfrac{\pi}{2} \,\mathrm{area}(\mathbb{RP}^2), \tag{1.1.2}$$

cf. (5.2.2). Biographical notes on C. Loewner and P. Pu appear in Sections 2.2 and 2.3, respectively.

Similar inequalities exist for the Klein bottle [**Bav86**], as well as for the Möbius band [**Bl61b**]. A number of new inequalities of this type have recently been discovered, including universal volume lower bounds, *cf.* [**CrK03**].

1.1.3. Gromov's spectacular inequality. The inequality of M. Gromov [**Gro83**, (0.1), p. 3] for the homotopy 1-systole of an essential n-manifold M,

$$\mathrm{sys}\pi_1(M)^n \leq C_n \,\mathrm{vol}_n(M), \tag{1.1.3}$$

cf. (12.2.1), is the deepest result in the field. I would compare it to Gromov's theorem on almost flat manifolds [**Gro78**], which also originally appeared about a quarter century ago.

The difference between these two results of Gromov's is that in the latter case, P. Buser and H. Karcher almost immediately set to work, and produced a splendid monograph [**BusK81**] to explain the theorem. In the case of Gromov's systolic inequality, no such monograph has appeared, neither immediately afterwards, nor any time later. Gromov's 1999 book [**Gro99**] explains the general idea behind the proof, but does not present the full proof. To my knowledge, the original dense 35 page proof in [**Gro83**] is the only reference where the inequality is proved in full.

Gromov's proof of (1.1.3) involves a version of what is known in minimal surface theory as the monotonicity formula, but in a situation when the ambient space is a Banach space, combined with an application of the deformation theorem from geometric measure theory. Gromov's proof passes via a number of auxiliary inequalities (*e.g.* cone inequality, isoperimetric inequality, coarea formula), and involves additional invariants, such as diameter and filling volume. A presentation of a full proof would fall outside the scope of the present monograph, which focuses on the research of the past four years. A summary of a proof, based on recent results in geometric measure theory by S. Wenger [**Wen05**, Lemma 3.4], building upon earlier work by L. Ambrosio and B. Kirchheim [**AmK00**], appears in Section 12.2 below.

A recent application of (a generalisation of) Gromov's 1983 inequality, can be found in an article by G. Besson, G. Courtois, and S. Gallot [**BesCG03**], *cf.* Remark 12.2.5 below.

1.1.4. Gromov's optimal inequality for complex projective space. A significant difference between 1-systolic invariants (defined in terms of lengths of loops) and the higher, k-systolic invariants (defined in terms of areas of cycles, *etc.*) should be kept in mind. While a number of optimal systolic inequalities, involving the 1-systoles, have by now been obtained and will be detailed in this book, just about the only optimal inequality involving purely the higher k-systoles is M. Gromov's optimal stable 2-systolic inequality for complex projective space,

$$\mathrm{stsys}_2(\mathbb{CP}^n, \mathcal{G})^n \leq n! \, \mathrm{vol}_{2n}(\mathbb{CP}^n, \mathcal{G}), \tag{1.1.4}$$

cf. (13.2.4), where the optimal bound is attained by the Fubini-Study metric.

Just how exceptional inequality (1.1.4) is, only became clear recently. Namely, we discovered that, contrary to expectation, the symmetric metric on the quaternionic projective plane $\mathbb{HP}^2$ is *not* its systolically optimal metric (see [**BaKSS06**]), in contrast with the complex projective case, see Subsection 1.3.3 below for more details. As of this writing, no optimal inequality is available for the quaternionic or the Cayley two-point homogeneous spaces, *cf.* Section 13.3.

Similarly, just about the only nontrivial *lower* bound for a k-systole with $k \geq 2$, namely

$$\sup_{\mathcal{G}} \mathrm{confsys}_2 \left(\mathbb{CP}^2 \# n \overline{\mathbb{CP}}^2, \mathcal{G} \right)^2 \geq \sqrt{\tfrac{n}{2\pi e}}(1 + o(1)), \text{ for } n \to \infty, \tag{1.1.5}$$

cf. (20.1.2), results from recent work in gauge theory and J-holomorphic curves.

1.1.5. Schottky problem. Perhaps one of the most striking applications of systoles is in the context of the Schottky problem, by P. Buser and P. Sarnak [**BusS94**], who distinguished the Jacobians of Riemann surfaces among principally polarized abelian varieties. They showed that the locus of the Jacobians is small because their systoles are severely restricted (by the right-hand bound of (1.1.6) below) compared to the general case. Furthermore, they exhibited Riemann surfaces Σ_g of genus g, satisfying also the lower bound below:

$$C^{-1} \log g < \mathrm{confsys}_1(\Sigma_g)^2 < C \log g, \text{ for } g \to \infty \tag{1.1.6}$$

cf. (20.6.1).

1.1.6. Tools and techniques. Asking systolic questions often stimulates questions in related fields. Thus, a notion of "systolic category" of a manifold is defined and investigated in Chapter 12, where we exhibit a connection to the Lusternik-Schnirelmann (LS) category. Note that the systolic category (as well as the LS category) is, by definition, an integer (not to be confused with the foundational notion of "category theory"). Once the connection is established, the influence is mutual: known results about LS category stimulate systolic questions, and vice versa.

Asymptotic phenomena for the systole of surfaces of large genus have been shown to be related to interesting ergodic phenomena in Chapter 11, and to properties of congruence subgroups of arithmetic groups in [**KatzSV07, Vis07**].

The study of lower bounds for the conformal 2-systole of 4-manifolds has led to a simplified proof of the density of the image of the period map, *cf.* Appendix A.

To illustrate the variety of the tools involved, we mention a few connections to other areas:

- Banaszczyk's functional analytic results on the successive minima of a pair of dual lattices;
- dynamical systems notion of volume entropy and Katok's inequality;
- the topological techniques of Lusternik-Schnirelmann category;
- quaternion algebras and congruence subgroups of arithmetic Fuchsian and Kleinian groups;
- J-holomorphic curves and Seiberg-Witten invariants.

1.1.7. Statement of problem. Let M^n be a closed smooth n-dimensional manifold, and let $k > 0$ be an integer no greater than n. We study "curvature-free" inequalities satisfied by the various flavors of k-systolic invariants of M (detailed definitions appear in Chapters 5 and 12). Here "curvature-free" means that the coefficients in the inequality are independent of the metric, in the spirit of the Loewner inequality (5.4.1) for the two-torus.

In the first part of the book (Chapters 3 through 11), we start with a brief presentation of classical and semi-classical preliminaries, and continue by presenting systolic inequalities in dimension 2. In the second part (Chapters 12 through 20), we present higher dimensional generalisations. At the cost of a slight redundancy of definitions and basic inequalities, the individual chapters are fairly self-contained. Furthermore, in each chapter we tried to give an indication of the interrelationship of various results, with a further minor increase in redundancy.

Let us now describe the contents of each chapter in more detail.

1.2. Contents of Part 1

Following five chapters (3 through 7) which deal mostly with introductory material, Chapters 8, 9, and 10 explore systolic ramifications of hyperellipticity.

1.2.1. Filling area theorem. We prove M. Gromov's filling area conjecture in the hyperelliptic case (Chapter 8). These results first appeared in [**BaCIK05**]. In particular, we establish the conjecture for all genus 1 fillings of the circle, extending P. Pu's result in genus 0. We translate the problem into a question about closed ovalless real Riemann surfaces. The conjecture then results from a combination of two ingredients. On the one hand, we exploit an integral geometric result of J. Hersch [**Her55**]. The result permits an optimal comparison of the given metric, with orbifold metrics of constant positive curvature on real surfaces of even positive genus. Here the singular points are Weierstrass points. On the other hand, we exploit an analysis of the combinatorics on unions of closed curves, arising as geodesics of such orbifold metrics.

1.2.2. Loewner inequality and hyperellipticity. It turns out that the Loewner inequality for the torus is also satisfied by metrics, on higher genus surfaces Σ, whose underlying conformal structure is hyperelliptic (Chapter 9). These results first appeared in [**KatzS06a**]. In genus 2, we first construct the Loewner loops on the (mildly singular) companion tori, locally isometric to Σ away from Weierstrass points. The loops are then transplanted to Σ, and surgered to obtain a Loewner loop on Σ. In higher genus, we exploit M. Gromov's area estimates for ε-regular metrics on Σ.

1.2.3. Genus two, CAT(0) metrics, and Bolza surface. The optimal systolic ratio of the genus 2 surface is still unknown, inspite of sustained efforts of a number of researchers, *e.g.* [**Cal96, Bry96**]. Meanwhile, a precise answer can be obtained in the context of nonpositively curved metrics (Chapter 10). These results first appeared in [**KatzS06b**]. We prove an optimal systolic inequality for an arbitrary CAT(0) surface Σ_2 of genus 2, namely

$$\mathrm{sys}\pi_1(\Sigma_2)^2 \leq \tfrac{1}{3}\cot\tfrac{\pi}{8}\ \mathrm{area}(\Sigma_2), \tag{1.2.1}$$

cf. (10.1.1). We use a Voronoi cell technique, introduced by C. Bavard [**Bav92**] in the hyperbolic context. The optimal bound (1.2.1) is attained by a flat singular metric in the conformal class defined by the smooth completion of the affine complex algebraic curve $y^2 = x^5 - x$. Thus, among all CAT(0) metrics, the one with the best systolic ratio is composed of six flat regular octagons centered at the Weierstrass points of the Bolza surface.

1.2.4. Katok's inequality, volume entropy, and systole. Chapter 11 examines asymptotic bounds for the optimal systolic ratio of a surface whose genus tends to infinity. Asymptotic questions in systolic geometry can be studied via the volume entropy. These results first appeared in [**KatzS05**]. We find an upper bound for the entropy of a systolically extremal surface, in terms of its systole. We combine the upper bound with A. Katok's lower bound [**Kat83**] in terms of the volume, to obtain a simpler alternative proof of M. Gromov's asymptotic estimate for the optimal systolic ratio of surfaces of large genus. Furthermore, we improve the multiplicative constant in Gromov's estimate. Namely, we show the following bound:

$$\mathrm{sys}\pi_1(\Sigma_g)^2 \leq \frac{\log^2 g}{\pi g}\,\mathrm{area}(\Sigma_g)(1+o(1)) \quad \text{when } g \to \infty, \tag{1.2.2}$$

cf. (11.1.4). We show that every surface of genus at least 20 is Loewner.

Note that asymptotic *lower* bounds for the systole were studied in 1994 by P. Buser and P. Sarnak [**BusS94**]. They constructed hyperbolic surfaces of large genus, whose systole behaves logarithmically in the genus. The Fuchsian groups in their examples are principal congruence subgroups of a fixed arithmetic group with rational trace field. Their construction can be generalized to principal congruence subgroups of arbitrary arithmetic surfaces [**KatzSV07**], by exploiting a new trace estimate valid for an arbitrary ideal in a quaternion algebra. The estimate yields a particularly sharp bound for a principal congruence tower of Hurwitz surfaces (PCH), *i.e.* (2, 3, 7) triangle surfaces, namely

$$\mathrm{sys}\pi_1(\Sigma_{\mathrm{PCH}}) \geq \frac{4}{3}\log g(\Sigma_{\mathrm{PCH}}). \tag{1.2.3}$$

Similar results are obtained in [**KatzSV07**] for the systole of hyperbolic 3-manifolds, compared to their simplicial volume, *cf.* [**AdR00**].

1.3. Contents of Part 2

The second part of the book deals with higher dimensional generalisations, with regard to both the ambient manifold, and the systolic invariants themselves.

1.3.1. Systolic category and Lusternik-Schnirelmann category. We open the second part of the book, by definining the homotopy systole, homology systoles, stable systoles, and systolic category in arbitrary dimension (Chapter 12). Note that both systolic and LS categories are integers, not to be confused with the foundational notion of "category theory".

The invariant called *systolic category*, denoted $\mathrm{cat}_{\mathrm{sys}}(M)$, was first defined in [**KatzR06**]. It is a Riemannian analogue of the Lusternik-Schnirelmann category $\mathrm{cat}_{\mathrm{LS}}(M)$. The invariant $\mathrm{cat}_{\mathrm{sys}}(M)$ is defined in terms of the existence of systolic inequalities satisfied by every metric $\mathcal{G}$ on M, as initiated by C. Loewner and later developed by M. Gromov. Such an approach provides a topological framework in which systolic questions can be studied. We compare the two categories. In all our examples, the inequality

$$\mathrm{cat}_{\mathrm{sys}} M \leq \mathrm{cat}_{\mathrm{LS}} M \tag{1.3.1}$$

is satisfied, which typically turns out to be an equality, *e.g.* in dimensions 2 and 3. We show that a number of existing systolic inequalities can be reinterpreted as special cases of such equality. The comparison with the value of $\mathrm{cat}_{\mathrm{LS}}(M)$ leads us to prove or conjecture new systolic inequalities on M.

1.3.2. Optimal inequality for $\mathbb{CP}^n$, Massey products. Given a compact Riemannian manifold $(M, \mathcal{G})$, its real homology group $H_*(M, \mathbb{R})$ is naturally endowed with the stable norm. Briefly, if $h \in H_k(M, \mathbb{R})$ then the stable norm of h is the infimum of the Riemannian k-volumes of real cycles representing h. The stable k-systole, denoted $\mathrm{stsys}_k(\mathcal{G})$, is the minimum of the stable norm over nonzero elements in the lattice of integral classes in $H_k(M, \mathbb{R})$. In Chapter 13, we prove Gromov's optimal stable systolic inequality (1.1.4) for complex projective space. The proof relies on the Wirtinger inequality.

1.3.3. Exceptional Lie algebra E_7 and systoles. We undertake the study of optimal curvature-free inequalities of the type discovered by C. Loewner and M. Gromov, using a generalisation of the Wirtinger inequality for the comass, *cf.* [**BaKSS06**]. Using a model for the classifying space BS^3 built inductively out of BS^1, we prove that the symmetric metrics of certain 2-point homogeneous manifolds turn out not to be the systolically optimal metrics on those manifolds. Thus, we obtain the following surprising bound for the optimal systolic ratio of the quaternionic projective plane:

$$6 \leq \mathrm{SR}(\mathbb{HP}^2) \leq 14, \tag{1.3.2}$$

whereas its symmetric metric has a systolic ratio of $10/3$. We point out the unexpected role played by the exceptional Lie algebra E_7 in systolic geometry, via the calculation of Wirtinger constants, see Appendix B below.

1.3.4. Massey systoles. We show that, like the Lusternik-Schnirelmann category, systolic category is sensitive to Massey products (Chapter 14). For example, we show that every metric $\mathcal{G}$ on the 19-dimensional manifold $SU(6)/SU(3) \times SU(3)$ satisfies the bound

$$\mathrm{stsys}_4(\mathcal{G})^2\, \mathrm{stsys}_6(\mathcal{G})^2 \leq 19!\, \mathrm{IQ}(\mathcal{G})\, \mathrm{vol}_{19}(\mathcal{G}), \tag{1.3.3}$$

cf. (14.4.2), where IQ is a suitable isoperimetric quotient.

1.3.5. Systoles and cup-length. We study the constraints imposed by multiplicative relations in the cohomology ring of a manifold, upon its stable systoles (Chapter 15). These results first appeared in [**BaK03**]. Relying on results from the geometry of numbers due to W. Banaszczyk [**Bana93**], and extending work by M. Gromov [**Gro83**] and J. Hebda [**Heb86**], we prove metric-independent inequalities for products of stable systoles. Thus, we show that if the fundamental cohomology class of an orientable manifold M^n is a cup product of classes of dimensions $k_1, \ldots, k_m$, then every Riemannian metric $\mathcal{G}$ on M satisfies the inequality

$$\prod_{j=1}^{m} \operatorname{stsys}_{k_j}(\mathcal{G}) \leq C(n) \left(\prod_j b_{k_j}(M) \left(1 + \log b_{k_j}(M)\right) \right) \operatorname{vol}_n(\mathcal{G}), \tag{1.3.4}$$

cf. (15.2.1). Here the product can be as long as the real cup-length of M, which therefore constitutes a lower bound for the systolic category of M.

Chapters 16 and 17 contain two distinct higher-dimensional optimal generalisations of the Loewner inequality (1.1.1).

1.3.6. Systolic inequalities, dual-critical lattices, and volume preserving flows. We present a new optimal systolic inequality for a closed Riemannian manifold M^n, which generalizes a number of earlier inequalities, including those of C. Loewner and J. Hebda (Chapter 16). These results first appeared in [**BaK04**]. Namely, we show that

$$\operatorname{confsys}_1(M) \operatorname{sys}_{n-1}(M) \leq \gamma'_b \operatorname{vol}_n(M)^{\frac{n-1}{n}}, \tag{1.3.5}$$

cf. (16.2.1), where $b = b_1(M)$ is the first Betti number, while γ'_b is the Bergé-Martinet constant (5.3.3). We characterize the boundary case of equality in terms of the geometry of the Abel-Jacobi map, or more specifically the Lichnerowicz harmonic map $\mathcal{A}_M$, of M. A metric attaining the optimal bound (1.3.5) has the property that the map $\mathcal{A}_M$ is a Riemannian submersion with minimal fibers, onto a flat torus. Here the base of $\mathcal{A}_M$ possesses a group of deck transformations which is a dual-critical lattice. The latter arise in the context of an extremal problem for Euclidean lattices, studied by A.-M. Bergé and J. Martinet [**BeM89**]. Given a closed manifold M that admits a submersion F to its Jacobi torus $\mathbb{T}^{b_1(M)}$, we construct all metrics on M that realize equality in our inequality. While one can choose arbitrary metrics of fixed volume on the fibers of F, the horizontal space is chosen using a multi parameter version of J. Moser's method of constructing volume-preserving flows.

1.3.7. Abel-Jacobi maps and optimal inequalities. We generalize optimal inequalities due to C. Loewner and M. Gromov, by proving optimal lower bounds for the total volume in terms of the homotopy systole and the stable systole (Chapter 17). These results first appeared in [**IvK04**]. Namely, we show that every Riemannian n-manifold M with Betti number $b = b_1(M) > 0$ satisfies the inequality

$$\deg(\mathcal{A}_M) \operatorname{stsys}_1(M)^b \leq (\gamma_b)^{\frac{b}{2}} \operatorname{vol}_n(M). \tag{1.3.6}$$

cf. (17.2.3). Here γ_b is the Hermite constant, while $\mathcal{A}_M$ is the Abel-Jacobi map. Our main tool is the construction of an area-decreasing map to the Jacobi torus, streamlining and generalizing the construction of D. Burago and S. Ivanov [**BuraI95**]. It turns out that one can successfully combine this construction with the coarea formula, to yield new optimal inequalities. As a consequence, we obtain a lower bound for systolic category which is stronger than the real cup-length lower bound studied in Chapter 15.

We study a certain generalisation of the stable systole called the conformal systole in Chapters 18 and 20.

1.3.8. Conformal systolic inequalities. We prove certain optimal systolic inequalities for a closed Riemannian manifold $(M, \mathcal{G})$, depending on a pair of parameters, n and b, where n is the dimension of M, while b is its first Betti number (Chapter 18). These results first appeared in [**BaCIK07**]. The proof of the inequalities involves constructing Abel-Jacobi maps from M to its Jacobi torus $\mathbb{T}^b$, which are area-decreasing (on b-dimensional areas), with respect to suitable norms. These norms are the stable norm of $\mathcal{G}$, the conformally invariant norm, as well as other L^p-norms. Here we exploit L^p-minimizing differential 1-forms in cohomology classes. We characterize the case of equality in our optimal inequalities, in terms of the criticality of the lattice of deck transformations of $\mathbb{T}^b$, while the Abel-Jacobi map is a harmonic Riemannian submersion. That the resulting inequalities are actually nonvacuous follows from an isoperimetric inequality of Federer and Fleming, under the assumption of the nonvanishing of the homology class of the lift of the typical fiber of the Abel-Jacobi map to the maximal free abelian cover.

1.3.9. Conformal 2-systole and the surjectivity conjecture. The asymptotic behavior of the conformal 2-systole of four-manifolds is studied in Chapter 20. These results first appeared in [**Katz03**]. In 1994, P. Buser and P. Sarnak showed that the maximum, over the moduli space of Riemann surfaces of genus g, of the least conformal length of a nonseparating loop, is logarithmic in g. We present an application of (polynomially) dense Euclidean packings, to estimates for an analogous 2-dimensional conformal systolic invariant of a 4-manifold M with indefinite intersection form. The estimate turns out to be polynomial, rather than logarithmic, in $b_2(M)$:

$$\sup_{\mathcal{G}} \operatorname{confsys}_2\left(\mathbb{CP}^2 \# n\overline{\mathbb{CP}}^2, \mathcal{G}\right)^2 \geq \sqrt{\tfrac{n}{2\pi e}}(1 + o(1)), \text{ for } n \to \infty. \tag{1.3.7}$$

It relies on the results of T. Li and A. Liu [**Li01**] concerning the density of the image of of the period map, *cf.* Appendix A. These results allow one to insert suitable lattices with metric properties prescribed in advance, into the second de Rham cohomology group of M, as its integer lattice. The idea is to adapt the well-known Lorentzian construction of the Leech lattice, by replacing the Leech lattice by the Conway-Thompson unimodular lattices which define asymptotically dense packings. The final step can be described, in terms of the successive minima λ_i of a lattice, as deforming a λ_2-bound into a λ_1-bound.

1.3.10. Image of period map and J-holomorphic curves. J. Solomon wrote an appendix for the book (Apendix A), proving the density of the image of the period map for blow-ups of the complex projective plane, a result originally due to T. Li and A. Liu [**Li01**]. Such density is proved using the correspondence between Seiberg-Witten and Gromov-Witten invariants developed by C. Taubes, as well as recent results in gauge theory, symplectic geometry, and J-holomorphic curves.

To summarize the relation of this result to systoles, note that Chapter 20 proved a relative result, to the effect that *if* the image is dense, *then* one has a certain lower bound for the conformal 2-systole. Now that one has such density, the relative result becomes an absolute theorem. The theorem is the only non-trivial lower bound in the literature for k-systoles with $k \geq 2$, in contrast with the abundance of such bounds for $k = 1$, some of which have been described in the book.

The existence of metrics asserted by this theorem poses a challenge to the community of Riemannian geometers. The gauge-theoretic proof of the existence of such metrics gives no inkling of what they might look like, or how to construct them using traditional tools of Riemannian geometry.

We conclude with a final Appendix B containing a list of open problems in systolic geometry, topology, and arithmetic.

CHAPTER 2

Historical remarks

2.1. A la recherche des systoles, by Marcel Berger

This section contains Marcel Berger's personal story concerning systoles, as told by the master in the summer of 2005. We have limited editorial changes to a bare minimum, so as to preserve the flavor of the original account.

2.1.1. Thom on Blatter. "To the best and fairness of my knowledge, here is my recollection of my relations with systoles.

"For me things started in the very small [room of the] Strasbourg Mathematics Department Library, I think in the fall of 1961 or in the spring of 1962, I cannot remember [exactly]. [René] Thom was there, looking at the newly arrived journals, I was close to him in the small room. Of course he was my hero. Then he came upon the Blatter paper (Blatter 1961), and spoke to me, saying, 'You know Berger (still now most French mathematicians call between them or quoting their [last] names not by the first name), this paper is very interesting because it concerns the general problem to find universal (independent of the metric) relations between the volumes of various homology, etc. classes.'

"Coming from Thom, I considered that question as good (and natural also, which I like). I started looking at various generalizations. I was really getting nowhere. The only thing I did was to learn [P.M.] Pu's proof, and then to make propaganda for the topic in my Bombay lectures (Berger 1965).

2.1.2. Chern, Wirtinger inequality, and calibration. "I then remember talking about Thom's problem with [S.S.] Chern (I think in Berkeley where I was spending the summer quarter 1968 or the summer of 1969, I am not sure) about the question. He told me two things, the first was that his feeling was that such a universal inequality could be true in the middle dimension (for even dimensions). The second thing he did when thereafter I asked him about the complex projective space, since I was trying to look at Pu generalization to other projective spaces. He told me for the complex case that this followed from Wirtinger inequality (I did not know Wirtinger stuff at that time).

"I kept working with the topic, and I could really get nowhere. The only thing not trivial I could prove was that the inequality was OK for the quaternionic projective space and the Cayley plane (for their standard metric). To do this I had only to prove that the 4-form (respectively, the 8-form) was what is called today calibrating after (Harvey and Lawson 1982). The quaternionic case was easy, but I had a hard time to prove the Cayley case. To make a living, I published the two papers (Berger 1972a) and (Berger 1972), but you perfectly know that there is nothing in them, except propaganda for the subject, and the two calibrations. Then I [made] no contribution, except having kept making propaganda. You can look at the wonderful introduction book (Morgan 1988-2000), page 75, section 6.5: history of calibration. I kept making propaganda, in lectures, and by some writing (in part).

"Then came [M.] Gromov, and his followers; I still continue propaganda, but as you know the subject started soon to be popular.

2.1.3. Three notes. "The first is the explanation for the two 'funny' titles, in case you are not familiar with the famous French novel 'A la recherche du temps perdu', by Marcel Proust. The year 1971 was Proust centenary, so I used that pretext to imitate the first two sections of his novel which are called 'Du côté de chez Swann', and 'A l'ombre des jeunes filles en fleur'.

"The second thing is the origin of the word 'systole', in case I have not already told you. I was looking at [the] time for a word of the type 'iso-???-ic' both for the systoles and for the injectivity radius. I looked at Greek language dictionaries (French to Greek, basically there are no good ones) and found various wording. [Luckily], I was doing physical education together with a Greek literature colleague; he told me what I was proposing was 'low Greek', of course I explained to him, in ordinary words, what were a systole and the injectivity radius. The next week physical education session he came back with the two proposal: 'isoclysteric' and 'isosphincteric'. I was still young in the bad sense, say provocative, I was [amused]; but I told these two wording to Besse seminar next week, they were horrified and told me 'Marcel, you cannot use that'. So at the next session I asked him to find a less bad taste 'ic' stuffs. Next week he came back with "isosystolic" and 'isoembolic'. Seminar people were happy; you understand that in short I switched from proctology to cardiology. Both wording were immediately accepted by Gromov.

"The third thing is more important. In the same library, Thom asked me another question: 'Berger, I give you a compact manifold: does there exist a preferred metric on it?' This started my propaganda for Einstein manifolds. So that in conclusion, after my dissertation and the pinching stuff, all my interests were given to me by Thom."

2.1.4. Bibliography for this section. The above account refers to the following texts.

(1) Berger, M. (1965). Lecture on Geodesics in Riemannian Geometry, Tata Institute of Fundamental Research.
(2) Berger, M. (1972). A l'ombre de Loewner. Ann. Scient. Ec. Norm. Sup. 5: 241-260.
(3) Berger, M. (1972a). Du côté de chez Pu. Ann. Scient. Ec. Norm. Sup. 5: 1-44.
(4) Blatter, C. (1961). Über Extremallänge auf geschlossenen Flächen. Comment. Math. Helvetici 35: 153-168.
(5) Harvey, R. and Lawson, H.B. (1982). Calibrated geometries. Acta Mathematica 148: 48-157.
(6) Morgan, F. (1988-2000). Geometric measure theory: a beginner's guide, third edition in 2000, Academic Press.

2.2. Charles Loewner (1893-1968)

In this section, and in Section 2.3 below, we include biographical notes for the two pioneers of systolic geometry.

A considerable amount has been written about Loewner's Mathematics in the biographies available in the literature and on the internet (see also Subsection 2.2.12 below). On the human side, what one finds in the biographies is invariably brief. This side of him is perhaps as much his heritage as the Mathematics. To this day, some colleagues who knew him personally, regret that there are not more mathematicians like him nowadays. Others still remember fondly their contacts with Loewner, nearly half a century later. We aim to bring out this aspect of Loewner the man in the notes below.

2.2.1. Childhood. Originally named Karel (later Karl) Löwner, he was born into an Orthodox Jewish family in the village of Lány, near Prague, which was at the time the capital of Bohemia.

Karel's father, Sigmund (or Zygmund) Löwner, owned a general store in Lány which supplied the needs of all the villagers, as well as those of the local Count, whose castle dominated the village. Inspite of this, the family was on a strict budget, as the goods were sold at a very small markup. Thus caring for others and helping them, the elder Löwner planted the seeds of the *middah* (character trait) of *rachmanus* (see below) that would later blossom in his offspring. *Rachmanus* can be loosely translated as concern for others, pity, or lovingkindness; see gemara Talmud Babli, maseches Beitzah, page 32b; as well as Maseches Yevamos, page 79a. Karel lost his father close to bar mitzvah age, six years prior to the onset of World War 1. In accordance with Jewish law, he said the traditional Kaddish prayer for his father every day for a year following his loss.

In line with Sigmund's wishes, Karel studied in a Prague gymnasium where German was not only taught, but was also the language of instruction. The attendant infringement upon traditional Jewish learning, may have seemed a small price to pay for the entry ticket to Europe's most refined ivory towers. Sigmund's infatuation with all things German has been noted by biographers, *cf.* editor's introduction to [**Lo88**].

2.2.2. The Holocaust. Whether or not the infatuation went sour, a generation later, as many as four of Sigmund's children were destined to perish in the Holocaust. Meanwhile, his son Viktor (Karel's only sibling to survive the war) married out (well before the war), fabricated a genealogical tree (perhaps to save his life) purporting to prove that his father's family was not Jewish in the first place, and switched to an unrelated surname (derived from the name of his hometown).

A helpful website on mathematicians under the Third Reich is maintained by Prof. Thomas Huckle at Munich, detailing also Karel's narrow escape from the Nazis' clutches. See also Rachel Pomerantz's highly informative book *The world in flames. A short history of the Holocaust* [**Po93**] for background, root causes, and lessons of the Holocaust.

Karel Löwner held several positions in Germany in the 1920's, and later in Prague, before fleeing for his life from the Nazis. According to his *curriculum vitae* from 1939, his students in Europe were E. Lammel, F. Kraus, O. Dobsch, and Ch. Bers. The latter is Lipman Bers, destined to become his colleague, neighbor, and ultimately editor [**Lo88**].

2.2.3. Berlin. We will spare the reader the litany of mathematical celebrities who surrounded Löwner at Berlin University, typically found in internet McBiographies. A name conspicuous through its absence from such lists is that of Ludwig

Bieberbach (see below). It needs to be mentioned that among Löwner's colleagues in Berlin in the 1920's was Albert Einstein, who played a pivotal role in securing his first job in the US, see Subsection 2.2.7.

Bieberbach was a major influence on Löwner. He may have been the source of the invitation for Löwner to leave his central European university, where he defended his thesis under Georg Pick in 1917, and come to Berlin in 1922. Löwner's famous proof of the Bieberbach conjecture in the first highly nontrivial case, that of the third coefficient, was published the following year [**Lo23**]. The manuscript was sent to Bieberbach for confirmation when it was submitted for publication. In accepting the work, Bieberbach wrote in the margin that it was an outstanding contribution. Löwner's *Habilitation* was mainly reviewed by Bieberbach in 1923 [**Bie88**, pp. 203, 367].

2.2.4. Aryan *versus* Jew. Bieberbach's reputation as a mathematician is solid. Alas, he was also a Nazi sympathizer, capable of arriving to a university lecture while sporting a brownshirt uniform, of harrassing his Jewish colleagues during the 1930's, as well as of pursuing a sinister thesis concerning the existence of a dichotomy of Aryan *versus* Jewish Mathematics. Ostensibly, his dichotomy centers on purely didactic and pedagogical qualities; yet the Aryan aversion to the Jewish trait of *rachmanus* (pity; *cf.* Subsection 2.2.1) is well-known [**Po93**].

The disappointment with Bieberbach is particularly acute for those who wish to believe that toiling in Mathematical pursuits provides a vaccine of sorts against indecency [**Go83**, vol. 2, p. 267, lines 19-25].

As a reaction against Bieberbach's indecency (*cf.* Subsection 2.2.12), his name is routinely doctored out of lists of mathematical celebrities at Berlin University during Löwner's employment there. The name only appears as an afterthought at the end of Bers' introductory essay in [**Lo88**], in the context of a brief discussion of his conjecture.

2.2.5. Paradox. Yet, there is a bit of a paradox in attempting to obscure Bieberbach's mathematical influence on fellow mathematicians. The fact is that O. Teichmuller's name has been canonized in the name of the space covering the moduli space of Riemann surfaces, inspite of his Nazi sympathies (which were even more extreme than Bieberbach's). As it happens, Bers himself actually took part in such a canonisation, although the author of these lines personally heard him express regrets about his role in it, in a lecture at Columbia University in the 1980's.

Louis de Branges succeeded in proving Bieberbach's 1916 conjecture, over half a century later [**Bra85**]. He feels that Löwner would not have undertaken the extremely difficult project of carrying out the necessary estimates for the case of the third coefficent, were it not for the guiding and directing force of Bieberbach's intuition. De Branges further wrote in the fall of 2006:

> [Löwner] uses an original construction which was not appreciated by his colleagues, and indeed is not even cited in the 1978 book by H. Grunsky [**Grun78**]. No one seems to know how Löwner had the confidence to undertake the proof of what was generally considered to be a wild and improbable conjecture by a mathematician who was noted for imprecision in his publications.

An introduction to the proof of the conjecture may be found in [**Kor86**].

2.2.6. Emigration. In an effort to leave the continent and escape its Nazi menace, Löwner applied for university positions both in England and in America. In a testimony to a trait of decency and unselfishness that will have characterized an entire lifetime, Richard von Mises wrote from Istanbul, in a 1939 letter of recommendation for Löwner:

> During his activity at the University of Berlin, [Löwner] was, among all the instructors in Mathematics, the one who had the strongest influence upon the students, stimulated them to independent research, and helped them in his unselfish way. Much of the work involved in the published theses of his pupils is not only due to his influence, but can in a true sense be considered as his work.

See also Subsection 2.2.11 below.

2.2.7. America. Loewner's first job in the United States, already under the suitably Americanized spelling of the surname *sans umlaut*, was at the University of Louisville, Kentucky, arranged by John von Neumann in 1939. However, the deal was not clinched until Loewner's former colleague Einstein agreed to transfer an offprint of the original edition of his renowned work on relativity theory [**Ei13**] to a lawyer and collector at the University of Louisville named W. Bullitt. The autographed offprint is currently in the Bullitt collection of the University of Louisville [**Dav89**].

Loewner's phenomenal generosity immediately made itself felt in Louisville. Thus, he conceded to a request by undergraduate students to offer an advanced Mathematics course, which had to be taught without remuneration. The only available location turned out to be the local beer brewery. Loewner did in fact teach his course there, at a vigorous 7 a.m. time slot, before the arrival of the first shift of brewery workers.

2.2.8. Character. Loewner's daughter recalls that one of his main character strengths was the ability to listen to people and empathize with their viewpoint, even when disagreeing with them. This sometimes led people of many different persuasions to assume, by projection, that he was in agreement with them, whether or not this may have been the case. At the same time, he would always seek out a point of agreement, and back them up on it. This may have been one of the secrets of his popularity, and certainly helped navigate the political tensions of the postwar decade.

2.2.9. Syracuse. After a period of employment at Brown University during World War 2, Loewner obtained a permanent position at Syracuse University. One of his students here was P.M. Pu, see Section 2.3.

When a nephew, Paul Gráf, a Holocaust survivor, arrived from Europe in 1947, the Loewners not only welcomed him into their home, but adopted him.

For a time, it seemed as though Syracuse would become one of the leading research departments in Mathematics. However, certain tensions arose within the department, which may have been exacerbated by the parallel political tensions of the decade. Some of the faculty were called in for questioning by the House of Representatives' Committee on Un-American Activities (HUAC). Certain appointments were made against the wishes of the star mathematicians in the department.

In the end, many of the star faculty ended up leaving Syracuse. Loewner left for Stanford University, where he would stay for the rest of his life.

2.2.10. Identity. Though unobservant as an adult, Loewner "deeply felt his Jewish identity" [**Lo88**, p. ix, line 18]. "A person of Jewish faith", as Loewner is revealingly described by his colleague A. Gelbart [**Gelb53**, p. 1571] in his testimony before the HUAC, he was a decent man and *baal rachmanus* (someone possessing the trait of *rachmanus*, *cf.* Subsection 2.2.1), by all accounts.

A telling circumstance is that, inspite of professional links with leftist academics, Loewner himself was never cross-examined by the HUAC. This circumstance suggests that, contrary to an editorial claim in [**Lo88**, p. ix, line 20], he was a largely apolitical man, as well. Artfully planted in the midst of a rhetorical flourish, there is a claim in [**Lo88**, p. ix, line 20] that Loewner was "a man of the left". Yet, Loewner himself would have likely shunned the Procrustean bed of the left-right political dichotomy, his sympathy for H. Wallace's ideas, and his appeal for clemency for E. and J. Rosenberg, notwithstanding.

The aforementioned editorial claim in [**Lo88**, p. ix, line 20] appears to be a personal projection by an admiring student, see Subsection 2.2.8 above. Indeed, the teacher, unlike the student, never called himself a *menshevik*, *cf.* [**AlAR90**, p. 20], [**AlR87**]. Nor is there any evidence that Loewner shared Bers' sense of belonging to a strain of marxism euphemistically described as "the [J.] Martov tradition", in an interview Bers gave to Constance Reid, of *Hilbert–Courant* [**Re86**] fame, see [**AlAR90**, p. 21].

While the aforementioned editorial claim has been recycled by internet biographies (of both men), one of his colleagues at the time feels today that Loewner was never engaged politically.

2.2.11. Students. Loewner felt particular resposibility toward students who were not among his strongest ones. His personal investment in his students was proverbial; a colleague of his reports, only partly in jest, that "it became generally known that the quality of his students' dissertations tended to vary inversely as the quality of the student". Some of his students' theses were almost entirely Loewner's own work, in America as in Europe, see Subsection 2.2.6 above.

With consistent generosity, the Loewners always had Charles' graduate students for dinner on holidays, while his wife Elisabeth (née Alexander) was still alive. He lost her in 1956. Their daughter recalls that her father insisted on always keeping the only telephone in the house, in his bedroom, "so as to be able to discuss his students' and colleagues' problems whenever they called, even after he had gone to bed". When a Chinese student found himself without suitable housing, Loewner took him into his home in Syracuse for a while.

2.2.12. Mathematics. One of Loewner's central scientific contributions is his proof of the case $n = 3$ of the celebrated Bieberbach conjecture, exploiting what is known today as the Loewner differential equation. The latter turned out to be instrumental in the proof of the conjecture [**Bra85**], see Subsection 2.2.3. The stochastic Loewner equation provided the inspiration for the work of Wendelin Werner, a 2006 Fields medalist.

Two of Loewner's most prolific students, Adriano Garsia and Lipman Bers, have each on the order of a hundred publications, in widely divergent fields. Both have started their own schools, each numbering dozens of students, *cf.* [**Gar06**].

Loewner's seminal role in systolic geometry is discussed in Section 1.1 above.

As early as 1955, Bers urged Loewner to publish a book based on his course at the University of California at Berkeley, and even proposed a publisher. The fact that the book [**Lo71**] was only completed posthumously, is surely a reflection of Loewner's dedication to his students, whom he continued to supervise even after retiring officially from Stanford.

In addition to his published articles, invariably path-breaking, a lot of his work was published, with his encouragement, by his students, for which he never claimed credit. Thus, it would be difficult to gauge the full extent of his influence as one of last century's great mathematicians. But first and foremost, he was a decent man and a *baal rachmanus*, *cf.* Subsections 2.2.4 and 2.2.10.

2.2.13. Acknowledgments. The above notes drew upon interviews and/or email exchanges with Loewner's children, Mrs. Marian Tracy and Mr. Paul Loewner; with his colleagues, Prof. Robert Finn, Prof. Gene Golub, and Prof. Zvi Ziegler; with his colleague's son, Prof. Steve Gelbart; with Prof. Reinhard Siegmund-Schultze; with Mrs. Delinda Buie, curator, Ekstrom Library, University of Louisville; with Prof. Louis de Branges; and with Rodney Ross, Center for Legislative Archives (NARA), Washington, D.C. The author is grateful to all of them for invaluable help.

2.3. Pu, Pao Ming (1910-1988)

2.3.1. Thesis under Loewner. The first name is spelled Bao Ming in the recently modified system of transliteration. Pu received his Ph.D. at Syracuse University in 1950 under the supervision of C. Loewner, resulting in the publication of the seminal paper [**Pu52**] containing both inequalities (1.1.1) and (1.1.2). The listing at the website of the Mathematical Genealogy Project indicates that his first name, according to Syracuse University records, was *Frank*.

Following Chiang Kai-shek's ouster from the mainland in 1949, there was apparently a wave of recalls of Chinese academics working in the West. One such case was Wu, Wen Tsün, a student of Cartan's in Paris, who one day disappeared from Paris without saying a word to anyone.

2.3.2. Mainland. Pu may have also been forced to return to the mainland by the communist authorities. He was professor at Sichuan University, Chengdu. Perhaps because someone who had spent time in the West was automatically suspect, Pu was unable to advise students for most of his scientific career. According to the official obituary at [**Lu90**], he was finally granted the coveted status of *boshi sheng dao shi* (supervisor of graduate students) at the ripe age of 71, four years after the end of the infamous cultural revolution. Most of his later papers concern fuzzy topology.

2.4. A note to the reader

2.4.1. Website. The systolic geometry website currently lists over a hundred articles and over fifty contributors, and is regularly updated. One can visit the website at

http://www.math.biu.ac.il/~katzmik/sgt.html

(suggestions and corrections are welcome). Please email the author at the address katzmik@math.biu.ac.il with any comments, which will be used to improve a possible new edition of the book.

2.4.2. What is not in this book. One particularly significant omission is the vast literature, of over a thousand articles, on the girth of graphs, originating with Tutte's article [**Tu47**] (see the first paragraph of Section 1.1 above), and particularly with the influential "solution to advanced problem" he published under the charming pseudonym of Blanche Descartes [**UD54**].

The techniques used in this subfield of graph theory are somewhat disjoint from those used in the present book. A discussion of girth, Moore graphs, and related literature may be found in B. Bollobás [**Bol78**, Chapter 3]. See also A. Lubotzky [**Lub94**].

Similarly, no effort is made to account for the extensive literature on dense sphere packings and lower bounds for the Hermite constant. A related bibliography may be found in [**RT90**] and [**ConS99**].

CHAPTER 3

The *theorema egregium* of Gauss

The present Chapter is an introduction to Gaussian curvature, and is thus simultaneously elementary, and essential to the book.

It is true that some of the material in the book *formally* stands on its own even without understanding Gaussian curvature. Still, the material on CAT(0) metrics in genus 2 (Chapter 10) requires an understanding of negative Gaussian curvature, while Chapter 8 on the filling area theorem requires an understanding of constant positive curvature metrics.

To go beyond formalism, however, it would be very difficult to *understand* the geometric classification of surfaces without understanding, first, the intrinsic nature of Gaussian curvature, *i.e.* the *theorema egregium* of Gauss.

A beautiful historical account and an analysis of Gauss's proof of the *theorema egregium* may be found in [**Do79**].

Our formula (3.3.3) for Gaussian curvature is actually very similar to the traditional formula for the Riemann curvature tensor in terms of the Levi-Civita connection (note the antisymmetrisation in both formulas, and the corresponding two summands), without the burden of the connection formalism.

We present a compact formula for Gaussian curvature and its proof, along the lines of the argument in M. do Carmo's book [**Car76**]. The appeal of an old-fashioned, computational, coordinate notation proof is that it obviates the need for higher order objects such as connections, tensors, exponental map, *etc.*, and is, hopefully, directly accessible to a student not yet familiar with the subject, *cf.* Remark 3.4.4.

REMARK 3.0.1. Throughout this Chapter, we will make use of the Einstein summation convention, *i.e.* suppress the summation symbol $\sum$ when the summation index occurs simultaneously as a subscript and a superscript in a formula, *cf.* (3.2.1).

3.1. Intrinsic *vs* extrinsic properties

We would like to distinguish two types of properties of a surface in Euclidean space: intrinsic and extrinsic. Understanding the extrinsic/intrinsic dichotomy is equivalent to understanding the *theorema egregium* of Gauss. The *theorema egregium* is the key insight lying at the foundation of differential geometry as conceived by B. Riemann in his essay [**Ri1854**] presented before the Royal Scientific Society of Göttingen a century and a half ago.

The *theorema egregium* asserts that an infinitesimal invariant of a surface in Euclidean space, called Gaussian curvature, is an "intrinsic" invariant of the surface Σ. In other words, Gaussian curvature of Σ is independent of its isometric imbedding in Euclidean space. This theorem paves the way for an intrinsic definition of curvature in modern Riemannian geometry.

We will prove that Gaussian curvature K is an *intrinsic* invariant of a surface in Euclidean space, in the following precise sense: K can be expressed in terms of the coefficients of the *first* fundamental form (see Section 3.2) and their derivatives alone. *A priori* the possibility of thus expressing K is not obvious, as the naive definition of K involves the coefficients of the *second* fundamental form (alternatively, of the Weingarten map).

REMARK 3.1.1. The intrinsic nature of Gaussian curvature paves the way for a transition from classical differential geometry, to a more abstract approach of modern differential geometry. The distinction can be described roughly as follows. Classically, one studies surfaces in Euclidean space. Here the first fundamental form (g_{ij}) of the surface is the restriction of the Euclidean inner product. Meanwhile, abstractly, a surface comes equipped with a set of coefficients, which we deliberately denote by the same letters, (g_{ij}) in each coordinate patch, or equivalently, its element of length. One then proceeds to study its geometry without any reference to a Euclidean imbedding, *cf.* (4.1.2).

Such an approach was pioneered in higher dimensions in Riemann's essay. The essay contains a single formula [**Ri1854**, p. 292] (*cf.* [**Sp79**, p. 147]), namely the formula for the element of length of a surface of constant (Gaussian) curvature $K \equiv \alpha$:

$$\frac{1}{1+\frac{\alpha}{4}\sum x^2}\sqrt{\sum dx^2} \tag{3.1.1}$$

(today, of course, we would incorporate a summation index as part of the notation), *cf.* formulas (4.1.3), (4.5.1), and (4.5.2).

3.2. Preliminaries to the *theorema egregium*

Given a surface Σ in 3-space which is the graph of a function of two variables, consider a critical point $p \in \Sigma$.

DEFINITION 3.2.1. The *Gaussian curvature* of the surface at p is the determinant of the Hessian of the function, *i.e.* the determinant of the two-by-two matrix of its second derivatives.

The implicit function theorem allows us to view any point of a regular surface, as such a critical point, after a suitable rotation. We have thus given the simplest possible definition of Gaussian curvature at any point of a regular surface.

REMARK 3.2.2. The appeal if this definition is that it allows one immediately to grasp the basic distinction between negative *versus* positive curvature, in terms of the dichotomy "saddle point *versus* cup/cap".

It will be more convenient to use an alternative definition in terms of the Weingarten map, which is readily shown to be equivalent to Definition 3.2.1.

We denote the coordinates in $\mathbb{R}^2$ by (u^1, u^2). In $\mathbb{R}^3$, let $\langle\,,\rangle$ be the Euclidean inner product. Let $x = x(u^1, u^2) : \mathbb{R}^2 \to \mathbb{R}^3$ be a regular parametrized surface. Here "regular" means that the vector valued function x has differential dx of rank 2 at every point. Consider partial derivatives $x_i = \frac{\partial}{\partial u^i}(x)$, where $i = 1, 2$. Thus, vectors x_1 and x_2 form a basis for the tangent plane at every point. Similarly, let

$$x_{ij} = \frac{\partial^2 x}{\partial u^i \partial u^j} \in \mathbb{R}^3.$$

Let $n = n(u^1, u^2)$ be a unit normal to the surface at the point $x(u^1, u^2)$, so that $\langle n, x_i \rangle = 0$.

DEFINITION 3.2.3 (symbols g_{ij}, L_{ij}, Γ^k_{ij}). The first fundamental form (g_{ij}) is given in coordinates by $g_{ij} = \langle x_i, x_j \rangle$. The second fundamental form (L_{ij}) is given in coordinates by $L_{ij} = \langle n, x_{ij} \rangle$. The symbols Γ^k_{ij} are uniquely defined by the decomposition

$$\begin{aligned} x_{ij} &= \Gamma^k_{ij} x_k + L_{ij} n \\ &= \Gamma^1_{ij} x_1 + \Gamma^2_{ij} x_2 + L_{ij} n, \end{aligned} \tag{3.2.1}$$

where the repeated (upper and lower) index k implies summation, in accordance with the Einstein summation convention, *cf.* Remark 3.0.1.

DEFINITION 3.2.4 (symbols $L^i_{\ j}$). The Weingarten map $(L^i_{\ j})$ is an endomorphism of the tangent plane, namely $\mathbb{R}x_1 \oplus \mathbb{R}x_2$. It is uniquely defined by the decomposition

$$\begin{aligned} n_j &= L^i_{\ j} x_i \\ &= L^1_{\ j} x_1 + L^2_{\ j} x_2, \end{aligned}$$

where $n_j = \frac{\partial}{\partial u^j}(n)$.

DEFINITION 3.2.5. We will denote by square brackets [ ] the antisymmetrisation over the pair of indices found in between the brackets, *e.g.*

$$a_{[ij]} = \tfrac{1}{2}(a_{ij} - a_{ji}).$$

Note that $g_{[ij]} = 0$, $L_{[ij]} = 0$, and $\Gamma^k_{[ij]} = 0$. We will use the notation $\Gamma^k_{ij;\ell}$ for the ℓ-th partial derivative of the symbol Γ^k_{ij}.

PROPOSITION 3.2.6. *We have the following relation*

$$\Gamma^q_{i[j;k]} + \Gamma^m_{i[j}\Gamma^q_{k]m} = -L_{i[j}L^q_{\ k]}$$

for each set of indices i, j, k, q (with, as usual, an implied summation over the index m).

PROOF. Let us calculate the tangential component, with respect to the basis $\{x_1, x_2, n\}$, of the third partial derivative x_{ijk}. Recall that $n_k = L^p_{\ k} x_p$ and $x_{jk} = \Gamma^\ell_{jk} x_\ell + L_{jk} n$. Thus, we have

$$\begin{aligned} (x_{ij})_k &= \left(\Gamma^m_{ij} x_m + L_{ij} n\right)_k \\ &= \Gamma^m_{ij;k} x_m + \Gamma^m_{ij} x_{mk} + L_{ij} n_k + L_{ij;k} n \\ &= \Gamma^m_{ij;k} x_m + \Gamma^m_{ij}\left(\Gamma^p_{mk} x_p + L_{mk} n\right) + L_{ij}\left(L^p_{\ k} x_p\right) + L_{ij;k} n. \end{aligned}$$

Grouping together the tangential terms, we obtain

$$\begin{aligned} (x_{ij})_k &= \Gamma^m_{ij;k} x_m + \Gamma^m_{ij}\left(\Gamma^p_{mk} x_p\right) + L_{ij}\left(L^p_{\ k} x_p\right) + (\ldots)n \\ &= \left(\Gamma^q_{ij;k} + \Gamma^m_{ij}\Gamma^q_{mk} + L_{ij}L^q_{\ k}\right) x_q + (\ldots)n \\ &= \left(\Gamma^q_{ij;k} + \Gamma^m_{ij}\Gamma^q_{km} + L_{ij}L^q_{\ k}\right) x_q + (\ldots)n, \end{aligned}$$

since the symbols Γ^q_{km} are symmetric in the two subscripts. Now the symmetry in j, k (equality of mixed partials) implies the following identity: $(x_{i[j})_{k]} = 0$. Therefore

$$\begin{aligned} 0 &= (x_{i[j})_{k]} \\ &= \left(\Gamma^q_{i[j;k]} + \Gamma^m_{i[j}\Gamma^q_{k]m} + L_{i[j}L^q{}_{k]}\right) x_q + (\ldots)n, \end{aligned}$$

and therefore $\Gamma^q_{i[j;k]} + \Gamma^m_{i[j}\Gamma^q_{k]m} + L_{i[j}L^q{}_{k]} = 0$ for each $q = 1, 2$. □

3.3. The *theorema egregium* of Gauss

DEFINITION 3.3.1. The Gaussian curvature function $K = K(u^1, u^2)$ is the determinant of the Weingarten map:

$$K = \det(L^i{}_j) = 2L^1{}_{[1}L^2{}_{2]}. \tag{3.3.1}$$

THEOREM 3.3.2. *We have the following three equivalent formulas for Gaussian curvature K:*

(a) $K = \det(L^i{}_j) = 2L^1{}_{[1}L^2{}_{2]}$;
(b) $K = \frac{\det(L_{ij})}{\det(g_{ij})}$;
(c) $K = -\frac{2}{g_{11}}L_{1[1}L^2{}_{2]}$.

PROOF. The first formula is our definition of K. To prove the second formula, we use the relation $L_{ij} = -L^k{}_j g_{ki}$. By the multiplicativity of determinant with respect to matrix multiplication, we have

$$\det(L^i{}_j) = (-1)^2 \frac{\det(L_{ij})}{\det(g_{ij})}.$$

Let us now prove formula (c). The proof is a calculation. Note that by definition, $2L_{1[1}L^2{}_{2]} = L_{11}L^2{}_2 - L_{12}L^2{}_1$. Hence

$$\begin{aligned} -2L_{1[1}L^2{}_{2]} &= \left(L^1{}_1 g_{11} - L^2{}_1 g_{21}\right) L^2{}_2 - \left(L^1{}_2 g_{11} - L^2{}_2 g_{21}\right) L^2{}_1 \\ &= g_{11}L^1{}_1 L^2{}_2 - g_{21}L^2{}_1 L^2{}_2 - g_{11}L^1{}_2 L^2{}_1 + g_{21}L^2{}_2 L^2{}_1 \\ &= g_{11}\left(L^1{}_1 L^2{}_2 - L^1{}_2 L^2{}_1\right) \\ &= g_{11}\det(L^i{}_j) = g_{11}K. \end{aligned} \tag{3.3.2}$$

Note that we can either calculate using the formula $L_{ij} = -L^k{}_j g_{ki}$, or the formula $L_{ij} = -L^k{}_i g_{kj}$. Only the former one leads to the appropriate cancellations as in (3.3.2). □

THEOREM 3.3.3 (Theorema egregium). *The Gaussian curvature function $K = K(u^1, u^2)$ can be expressed in terms of the coefficients of the first fundamental form alone (and their first and second derivatives) as follows:*

$$K = \frac{2}{g_{11}}\left(\Gamma^2_{1[1;2]} + \Gamma^j_{1[1}\Gamma^2_{2]j}\right), \tag{3.3.3}$$

where the symbols Γ^k_{ij} can be expressed in terms of the derivatives of g_{ij} be the formula $\Gamma^k_{ij} = \frac{1}{2}(g_{i\ell;j} - g_{ij;\ell} + g_{j\ell;i})g^{\ell k}$, where (g^{ij}) is the inverse matrix of (g_{ij}).

	Weingarten map $(L^i{}_j)$	Gaussian curvature K	mean curvature H
plane	$\begin{pmatrix} 0 & 0 \\ 0 & 0 \end{pmatrix}$	0	0
cylinder	$\begin{pmatrix} 1 & 0 \\ 0 & 0 \end{pmatrix}$	0	$\frac{1}{2}$
invariance of curvature		yes	no

TABLE 3.3.1. Plane and cylinder have the same intrinsic geometry (K), but different extrinsic geometries (H)

REMARK 3.3.4. Unlike Gaussian curvature K, the mean curvature $H = \frac{1}{2}L^i{}_i$ cannot be expressed in terms of the g_{ij} and their derivatives. Indeed, the plane and the cylinder have parametrisations with identical g_{ij}, but with different mean curvature, *cf.* Table 3.3.1. To summarize, Gaussian curvature is an intrinsic invariant, while mean curvature an extrinsic invariant, of the surface.

PROOF OF *theorema egregium*. We present a streamlined version of do Carmo's proof [**Car76**, p. 233]. The proof is in 3 steps.

(1) We express the third partial derivative x_{ijk} in terms of the Γ's (intrinsic information) and the L's (extrinsic information).
(2) The equality of mixed partials yields an identification of a suitable expression in terms of the Γ's, with a certain combination of the L's.
(3) The combination of the L's is expressed in terms of Gaussian curvature.

The first two steps were carried out in Proposition 3.2.6. Choosing the valus $i = j = 1$ and $k = q = 2$ for the indices, we obtain

$$\begin{aligned}\Gamma^2_{1[1;2]} + \Gamma^m_{1[1}\Gamma^2_{2]m} &= -L_{1[1}L^2{}_{2]} \\ &= g_{1i}L^i{}_{[1}L^2{}_{2]} \\ &= g_{11}L^1{}_{[1}L^2{}_{2]}\end{aligned}$$

since the term $L^2{}_{[1}L^2{}_{2]} = 0$ vanishes. Applying Theorem 3.3.2(c), we obtain the desired formula for K and complete the proof of the *theorema egregium*. □

3.4. The Laplacian formula for Gaussian curvature

An (x, y) chart in which the metric becomes conformal (see Definition 4.1.5 of Chapter 4) to the standard flat metric $dx^2 + dy^2$, is referred to as *isothermal coordinates*. The existence of the latter is proved in [**Bes87**].

DEFINITION 3.4.1. The Laplace-Beltrami operator for a metric $\lambda\delta_{ij}$ in isothermal coordinates is

$$\Delta_{LB} = \frac{1}{\lambda}\left(\frac{\partial^2}{\partial(u^1)^2} + \frac{\partial^2}{\partial(u^2)^2}\right).$$

COROLLARY 3.4.2. *Given a metric $\lambda(x, y)(dx^2+dy^2)$ in isothermal coordinates, its Gaussian curvature is minus half the Laplace-Beltrami operator of the* log *of the*

Γ_{ij}^1	$j=1$	$j=2$
$i=1$	μ_1	μ_2
$i=2$	μ_2	$-\mu_1$

Γ_{ij}^2	$j=1$	$j=2$
$i=1$	$-\mu_2$	μ_1
$i=2$	μ_1	μ_2

TABLE 3.4.1. Symbols Γ_{ij}^k of a metric $e^{2\mu(u^1,u^2)}\delta_{ij}$

conformal factor λ*:*

$$K = -\frac{1}{2}\Delta_{LB}\log\lambda. \tag{3.4.1}$$

PROOF. Let $\lambda = e^{2\mu}$. We have from Table 3.4.1:

$$2\Gamma^2_{1[1;2]} = \Gamma^2_{11;2} - \Gamma^2_{12;1} = -\mu_{22} - \mu_{11}.$$

It remains to verify that the $\Gamma\Gamma$ term in the expression (3.3.3) for the Gaussian curvature vanishes:

$$\begin{aligned} 2\Gamma^j_{1[1}\Gamma^2_{2]j} &= 2\Gamma^1_{1[1}\Gamma^2_{2]1} + 2\Gamma^2_{1[1}\Gamma^2_{2]2} \\ &= \Gamma^1_{11}\Gamma^2_{21} - \Gamma^1_{12}\Gamma^2_{11} + \Gamma^2_{11}\Gamma^2_{22} - \Gamma^2_{12}\Gamma^2_{12} \\ &= \mu_1\mu_1 - \mu_2(-\mu_2) + (-\mu_2)\mu_2 - \mu_1\mu_1 \\ &= 0. \end{aligned}$$

Then from formula (3.3.3) we have

$$K = \frac{2}{\lambda}\Gamma^2_{1[1;2]} = -\frac{1}{\lambda}(\mu_{11}+\mu_{22}) = -\Delta_{LB}\mu.$$

Meanwhile, $\Delta_{LB}\log\lambda = \Delta_{LB}(2\mu) = 2\Delta_{LB}(\mu)$, proving the result. □

Setting $\lambda = f^2$, we restate the corollary as follows.

COROLLARY 3.4.3. *Given a metric* $f^2(x,y)(dx^2+dy^2)$ *in isothermal coordinates, its Gaussian curvature is minus the Laplace-Beltrami operator of the* log *of the conformal factor* f*:*

$$K = -\Delta_{LB}\log f. \tag{3.4.2}$$

This formula is exploited in (10.4.2). Either one of the formulas (3.3.3), (3.4.1), or (3.4.2) can serve as the intrinsic definition of Gaussian curvature, replacing the extrinsic definition (3.3.1), *cf.* Remark 3.1.1.

REMARK 3.4.4. For a reader familiar with elements of Riemannian geometry, it is worth mentioning that the Jacobi equation

$$y'' + Ky = 0$$

of a Jacobi field y on M (expressing an infinitesimal variation by geodesics) sheds light on the nature of curvature in a way that no mere formula for K could. However, to prove the Jacobi equation, one would need to have already an intrinsically well-defined quantity on the left hand side, $y''+Ky$, of the Jacobi equation. In particular, one would need an already intrinsic notion of curvature K. Thus, a proof of the *theorema egregium* necessarily precedes the deeper insights provided by the Jacobi equation.

Similarly, the Gaussian curvature at $p \in M$ is the first significant term in the asymptotic expansion of the length of a "small" circle of center p. This fact, too, sheds much light on the nature of Gaussian curvature. However, to define what one means by a "small" circle, requires introducing higher order notions such as the exponential map, which are usually understood at a later stage than the notion of Gaussian curvature, *cf.* [**Car76, Car92, Cha06, GaHL04**].

CHAPTER 4

Global geometry of surfaces

4.1. Metric preliminaries

DEFINITION 4.1.1. By a (2-dimensional) closed Riemannian manifold we mean a compact subset $\Sigma \subset \mathbb{R}^n$ whose every point has a (relative) open neighborhood diffeomorphic to $\mathbb{R}^2$.

One can then use the diffeomorphism as a parametrisation, to calculate the coefficients g_{ij} (see Definition 3.2.3) of the first fundamental form, as, for example, in Theorem 4.3.4 and Example 4.5.1. The collection of all such data is then denoted by the pair $(\Sigma, \mathcal{G})$, where $\mathcal{G}$ is referred to as "the metric".

REMARK 4.1.2. Differential geometers like the $(\Sigma, \mathcal{G})$ notation, because it helps separate the topology Σ from the geometry $\mathcal{G}$. Strictly speaking, the notation is redundant, since the object $\mathcal{G}$ already incorporates all the information, including the topology. However, geometers have found it useful to use $\mathcal{G}$ when one wants to emphasize the geometry, and Σ when one wants to emphasize the topology.

Note that, as far as the intrinsic geometry of a Riemannian manifold is concerned, the imbedding in $\mathbb{R}^n$ referred to in Definition 4.1.1 is irrelevant to a certain extent, all the more so since certain basic examples, such as flat tori, are difficult to imbed in a transparent way.

EXAMPLE 4.1.3. The round 2-sphere $S^2 \subset \mathbb{R}^3$ defined by the equation

$$x^2 + y^2 + z^2 - 1$$

is a closed Riemannian manifold. Indeed, consider the function $f(x, y)$ defined in the unit disk $x^2 + y^2 < 1$ by setting $f(x, y) = \sqrt{1 - x^2 - y^2}$. Define a coordinate chart

$$x_1(u^1, u^2) = (u^1, u^2, f(u^1, u^2)).$$

Thus, each point of the open northern hemisphere admits a neighborhood diffeomorphic to a ball (and hence to $\mathbb{R}^2$). To cover the southern hemisphere, use the chart $\underline{x}_2(u^1, u^2) = (u^1, u^2, -f(u^1, u^2))$. To cover the points on the equator, use in addition charts $\underline{x}_3(u^1, u^2) = (u^1, f(u^1, u^2), u^2)$, $\underline{x}_4(u^1, u^2) = (u^1, -f(u^1, u^2), u^2)$, as well as the charts $\underline{x}_5(u^1, u^2) = (f(u^1, u^2), u^1, u^2)$, $\underline{x}_6(u^1, u^2) = (f(u^1, u^2), u^1, u^2)$.

Recall that we are working with dual bases $\left(\frac{\partial}{\partial u^i}\right)$ for vectors, and (du^i) for covectors (*i.e.* elements of the dual space), such that

$$du^i \left(\frac{\partial}{\partial u^j} \right) = \delta^i_j, \tag{4.1.1}$$

where δ^i_j is the Kronecker delta. We can thus write the first fundamental form as follows:

$$\mathcal{G} = g_{ij}(u^1, u^2) du^i du^j \tag{4.1.2}$$

with the Einstein summation convention. We will work with such data independently of any Euclidean imbedding, as discussed in Remark 3.1.1. For example, if the metric coefficients form an identity matrix, we obtain

$$\mathcal{G} = \left(du^1\right)^2 + \left(du^2\right)^2, \tag{4.1.3}$$

where the interior superscript denotes an index, while exterior superscript denotes the squaring operation.

THEOREM 4.1.4. *Define the* area *of* $(\Sigma, \mathcal{G})$ *by means of the formula*

$$\text{area} = \sum_{\{U\}} \int_U \sqrt{\det(g_{ij})} du^1 du^2, \tag{4.1.4}$$

namely by choosing a partition $\{U\}$ *of* Σ *subordinate to a finite open cover as in Definition 4.1.1, performing a separate integration in each open set, and summing the resulting areas. Then the total area is independent of the partition.*

PROOF. Consider a change from a coordinate chart (u^i) to another coordinate chart, denoted (v^α). In the overlap of the two domains, the coordinates can be expressed in terms of each other, *e.g.* $v = v(u)$, and we have the 2 by 2 Jacobian matrix

$$\frac{\partial(v^1, v^2)}{\partial(u^1, u^2)},$$

which is the matrix of partial derivatives. Denote by

$$\text{Jac}(u) = \det\left(\frac{\partial(v^1, v^2)}{\partial(u^1, u^2)}\right).$$

It is shown in advanced calculus that for any function $f(v)$ in a domain D, one has

$$\int_D f(v) dv^1 dv^2 = \int_D g(u) \text{Jac}(u) du^1 du^2, \tag{4.1.5}$$

where $g(u) = f(v(u))$.

Denote by $\tilde{g}_{\alpha\beta}$ the metric coefficients with respect to the chart (v^α). Thus, in the case of a metric induced by a Euclidean imbedding defined by $x = x(u) = x(u^1, u^2)$, we obtain a new parametrisation $y(v) = x(u(v))$. Then

$$\begin{aligned}
\tilde{g}_{\alpha\beta} &= \left\langle \frac{\partial y}{\partial v^\alpha}, \frac{\partial y}{\partial v^\beta} \right\rangle \\
&= \left\langle \frac{\partial x}{\partial u^i} \frac{\partial u^i}{\partial v^\alpha}, \frac{\partial x}{\partial u^j} \frac{\partial u^j}{\partial v^\beta} \right\rangle \\
&= \frac{\partial u^i}{\partial v^\alpha} \frac{\partial u^j}{\partial v^\beta} \left\langle \frac{\partial x}{\partial u^i}, \frac{\partial x}{\partial u^j} \right\rangle \\
&= g_{ij} \frac{\partial u^i}{\partial v^\alpha} \frac{\partial u^j}{\partial v^\beta}.
\end{aligned}$$

Since determinant is multiplicative, we obtain

$$\det(\tilde{g}_{\alpha\beta}) = \det(g_{ij}) \det\left(\frac{\partial(u^1, u^2)}{\partial(v^1, v^2)}\right)^2.$$

Hence using equation (4.1.5), we can write

$$\begin{aligned}\det{}^{\frac{1}{2}}(\tilde{g}_{\alpha\beta})dv^1dv^2 &= \det{}^{\frac{1}{2}}(\tilde{g}_{\alpha\beta})\ \mathrm{Jac}(u)du^1du^2\\ &= \det{}^{\frac{1}{2}}(g_{ij})\ \det\left(\frac{\partial(u^1,u^2)}{\partial(v^1,v^2)}\right)\ \mathrm{Jac}(u)du^1du^2\\ &= \det{}^{\frac{1}{2}}(g_{ij})\ du^1du^2\end{aligned}$$

since inverse maps have reciprocal Jacobians by chain rule. Thus the integrand is unchanged and area is well defined. □

DEFINITION 4.1.5. Two metrics, $\mathcal{G} = g_{ij}du^idu^j$ and $h = h_{ij}du^idu^j$, on Σ are called *conformally equivalent*, or *conformal* for short, if there exists a function $f = f(u^1, u^2) > 0$ such that

$$\mathcal{G} = f^2h,$$

in other words,

$$g_{ij} = f^2h_{ij} \quad \forall i, j. \tag{4.1.6}$$

DEFINITION 4.1.6. The function f above is called the *conformal factor* (note that sometimes it is more convenient to refer, instead, to the function $\lambda = f^2$ as the conformal factor).

REMARK 4.1.7. Note that the length of every vector at a point (u^1, u^2) is multiplied precisely by $f(u^1, u^2)$. More speficially, a vector $v = v^i\frac{\partial}{\partial u^i}$ which is a unit vector for the metric h, is "stretched" by a factor of f, *i.e.* its length with respect to $\mathcal{G}$ equals f. Indeed, the new length of v is

$$\begin{aligned}\sqrt{\mathcal{G}(v,v)} &= \mathcal{G}\left(v^i\frac{\partial}{\partial u^i}, v^j\frac{\partial}{\partial u^j}\right)^{\frac{1}{2}}\\ &= \left(\mathcal{G}\left(\frac{\partial}{\partial u^i}, \frac{\partial}{\partial u^j}\right)v^iv^j\right)^{\frac{1}{2}}\\ &= \sqrt{g_{ij}v^iv^j}\\ &= \sqrt{f^2h_{ij}v^iv^j}\\ &= f\sqrt{h_{ij}v^iv^j}\\ &= f.\end{aligned}$$

DEFINITION 4.1.8. An equivalence class of metrics on Σ conformal to each other is called a *conformal structure* on Σ.

THEOREM 4.1.9 (Riemann mapping/uniformisation). *Every metric on a connected surface is conformally equivalent to a metric of constant Gaussian curvature.*

From the complex analytic viewpoint, the uniformisation theorem states that every Riemann surface is covered by either the sphere, the plane, or the upper half-plane. Thus no notion of curvature is needed for the statement of the uniformisation theorem. However, from the differential geometric point of view, what is relevant is that every conformal class of metrics contains a metric of constant Gaussian curvature. See [**Ab81**] for a lively account of the history of the uniformisation theorem.

4.2. Geodesic equation and closed geodesics

The author has found that a number of mathematicians from fields other than differential geometry, while interested in systoles, need to be reminded what the definition of a geodesic is.

Perhaps the simplest possible definition of a geodesic β on a surface in 3-space is in terms of the orthogonality of its second derivative β'' to the surface. The nonlinear second order ordinary differential equation defining a geodesic is, of course, the "true" if complicated definition. We will now prove the equivalence of the two definitions. Consider a plane curve

$$\underset{s}{\mathbb{R}} \underset{\alpha}{\longrightarrow} \underset{(u^1,u^2)}{\mathbb{R}^2}$$

where $\alpha = (\alpha^1(s), \alpha^2(s))$. Let $x : \mathbb{R}^2 \to \mathbb{R}^3$ be a regular parametrisation of a surface in 3-space. Then the composition

$$\underset{s}{\mathbb{R}} \underset{\alpha}{\longrightarrow} \underset{(u^1,u^2)}{\mathbb{R}^2} \underset{x}{\longrightarrow} \mathbb{R}^3$$

yields a curve

$$\beta = x \circ \alpha.$$

DEFINITION 4.2.1. A curve $\beta = x \circ \alpha$ is a geodesic on the surface x if one of the following two equivalent conditions is satisfied:

(a) we have for each $k = 1, 2$,

$$(\alpha^k)'' + \Gamma^k_{ij}(\alpha^i)'(\alpha^j)' = 0 \quad \text{where} \quad {}' = \frac{d}{ds}, \tag{4.2.1}$$

meaning that

$$(\forall k) \quad \frac{d^2\alpha^k}{ds^2} + \Gamma^k_{ij}\frac{d\alpha^i}{ds}\frac{d\alpha^j}{ds} = 0;$$

(b) the vector β'' is perpendicular to the surface and one has

$$\beta'' = L_{ij}{\alpha^i}'{\alpha^j}'n. \tag{4.2.2}$$

To prove the equivalence, we write $\beta = x \circ \alpha$, then $\beta' = x_i {\alpha^i}'$ by chain rule. Furthermore,

$$\beta'' = \frac{d}{ds}(x_i \circ \alpha){\alpha^i}' + x_i{\alpha^i}'' = x_{ij}{\alpha^j}'{\alpha^i}' + x_k{\alpha^k}''.$$

Since $x_{ij} = \Gamma^k_{ij}x_k + L_{ij}n$ holds, we have

$$\beta'' - L_{ij}{\alpha^i}'{\alpha^j}'n = x_k\left({\alpha^k}'' + \Gamma^k_{ij}{\alpha^i}'{\alpha^j}'\right).$$

DEFINITION 4.2.2. A *closed geodesic* in a Riemannian 2-manifold Σ is defined equivalently as

(1) a periodic curve $\beta : \mathbb{R} \to \Sigma$ satisfying the geodesic equation ${\alpha^k}'' + \Gamma^k_{ij}{\alpha^i}'{\alpha^j}' = 0$ in every chart $\underline{x} : \mathbb{R}^2 \to \Sigma$, where, as usual, $\beta = \underline{x} \circ \alpha$ and $\alpha(s) = (\alpha^1(s), \alpha^2(s))$ where s is arclength. Namely, there exists a period $T > 0$ such that $\beta(s+T) = \beta(s)$ for all s.

(2) A unit speed map from a circle $\mathbb{R}/L_T \to \Sigma$ satisfying the geodesic equation at each point, where $L_T = T\mathbb{Z} \subset \mathbb{R}$ is the rank one lattice generated by $T > 0$.

DEFINITION 4.2.3. The *length* $L(\beta)$ of a path $\beta : [a,b] \to \Sigma$ is calculated using the formula

$$L(\beta) = \int_a^b \|\beta'(t)\| dt,$$

where $\|v\| = \sqrt{g_{ij}v^i v^j}$ whenever $v = v^i \underline{x}_i$. The energy is defined by $E(\beta) = \int_a^b \|\beta'(t)\|^2 dt$.

A closed geodesic as in Definition 4.2.2, item 2 has length T.

REMARK 4.2.4. The geodesic equation (4.2.1) is the Euler-Lagrange equation of the first variation of arc length. Therefore when a path minimizes arc length among all neighboring paths connecting two fixed points, it must be a geodesic. A corresponding statement is valid for closed loops, *cf.* proof of Theorem 4.2.5. See also the introduction to Chapter 5. In Sections 4.3 and 4.4 we will give a complete description of the geodesics for the constant curvature sphere, as well as for flat tori.

THEOREM 4.2.5. *Every free homotopy class of loops in a closed manifold contains a closed geodesic.*

PROOF. We sketch a proof for the benefit of a curious reader, who can also check that the construction is independent of the choices involved. The relevant topological notions are defined in Section 4.6 and [**Hat02**]. A free homotopy class α of a manifold M corresponds to a conjugacy class $g_\alpha \subset \pi_1(M)$. Pick an element $g \in g_\alpha$. Thus g acts on the universal cover $\tilde{M}$ of M. Let $f_g : M \to \mathbb{R}$ be the displacement function of g, *i.e.*

$$f_g(x) = d(\tilde{x}, g.\tilde{x}).$$

Let $x_0 \in M$ be a minimum of f_g. A first variation argument shows that any length-minimizing path between $\tilde{x}_0$ and $g.x_0$ descends to a closed geodesic in M representing α, *cf.* [**Car92, Cha06, GaHL04**]. □

4.3. Surfaces of constant curvature

By the uniformisation theorem 4.1.9, all surfaces fall into three types, according to whether they are conformally equivalent to metrics that are:

(1) flat (*i.e.* have zero Gaussian curvature $K \equiv 0$);
(2) spherical ($K \equiv +1$);
(3) hyperbolic ($K \equiv -1$).

For closed surfaces, the sign of the Gaussian curvature K is that of its Euler characteristic, *cf.* formula (4.6.1).

THEOREM 4.3.1 (Constant positive curvature). *There are only two surfaces, up to isometry, of constant Gaussian curvature $K = +1$. They are the round sphere S^2 of Example 4.1.3; and the real projective plane, denoted $\mathbb{RP}^2$.*

The latter can be defined as follows. Let $m : S^2 \to S^2$ denote the antipodal map of the sphere, *i.e.* the restriction of the map $v \mapsto -v$ in $\mathbb{R}^3$. Then m is an involution, *i.e.* each orbit of the $\mathbb{Z}_2$ action on S^2 defined by f consists of a pair of antipodal points, $\{\pm p\} \subset S^2$. On the set theoretic level, the real projective plane $\mathbb{RP}^2$ is the set of such orbits, *i.e.* the quotient of S^2 by the $\mathbb{Z}_2$ action. Denote by $Q : S^2 \to \mathbb{RP}^2$ the quotient map. The smooth structure and metric on $\mathbb{RP}^2$

are induced from S^2 in the following sense. Let $\underline{x} : \mathbb{R}^2 \to S^2$ be a chart on S^2 not containing any pair of antipodal points. Let g_{ij} be the metric coefficients with respect to this chart.

Let $\underline{y} = m \circ \underline{x} = -\underline{x}$ denote the "opposite" chart, and denote by h_{ij} its metric coefficients. Then

$$h_{ij} = \langle \frac{\partial \underline{y}}{\partial u^i}, \frac{\partial \underline{y}}{\partial u^j} \rangle = \langle -\frac{\partial \underline{x}}{\partial u^i}, -\frac{\partial \underline{x}}{\partial u^j} \rangle = \langle \frac{\partial \underline{x}}{\partial u^i}, \frac{\partial \underline{x}}{\partial u^j} \rangle = g_{ij}. \qquad (4.3.1)$$

Thus the opposite chart defines the identical metric coefficients. The composition $Q \circ \underline{x}$ is a chart on $\mathbb{RP}^2$, and the same functions g_{ij} form the metric coefficients for $\mathbb{RP}^2$ relative to this chart.

We can summarize the preceeding discussion by means of the following definition.

DEFINITION 4.3.2. The real projective plane $\mathbb{RP}^2$ is defined in the following two equivalent ways:

(1) the quotient of the round sphere S^2 by (the restriction to S^2 of) the antipodal map $v \mapsto -v$ in $\mathbb{R}^3$. In other words, a typical point of $\mathbb{RP}^2$ can be thought of as a pair of opposite points of the round sphere.
(2) the northern hemisphere of S^2, with opposite points of the equator identified.

The smooth structure of $\mathbb{RP}^2$ is induced from the round sphere. Since the antipodal map preserves the metric coefficients by the calculation (4.3.1), the metric structure of constant Gaussian curvature $K = +1$ descends to $\mathbb{RP}^2$, as well.

DEFINITION 4.3.3. A loop $\alpha : S^1 \to X$ of a space X is called *simple* if the map α is one-to-one, *cf.* Definition 4.6.1.

THEOREM 4.3.4. *All simple closed geodesics on S^2 have length 2π. The simple closed geodesics on $\mathbb{RP}^2$ have length π. The closed geodesics in both cases are defined by the great circles on the sphere.*

PROOF. We calculate the length of the equator of S^2 parametrized by $x(\theta, \phi) = (\sin\phi\cos\theta, \sin\phi\sin\theta, \cos\phi)$ in spherical coordinates (θ, ϕ). The equator is the curve $x \circ \alpha$ where $\alpha(s) = (s, \frac{\pi}{2})$ with $s \in [0, 2\pi]$. Recall that the metric coefficients are given by $g_{11}(\theta, \phi) = \sin^2\phi$, while $g_{22} = 1$ and $g_{12} = 0$. Thus

$$\|\beta'(s)\| = g_{ij}\left(s, \tfrac{\pi}{2}\right) \alpha^{i'} \alpha^{j'} = \left(\sin\tfrac{\pi}{2}\right)\left(\tfrac{d\theta}{ds}\right)^2 = 1.$$

Thus the length of β is

$$\int_0^{2\pi} \|\beta'(s)\| ds = \int_0^{2\pi} 1 ds = 2\pi.$$

A geodesic on $\mathbb{RP}^2$ is twice as short as on S^2, since the antipodal points are identified, and therefore the geodesic "closes up" sooner than (*i.e.* twice as fast as) on the sphere. For example, a longitude of S^2 is not a closed curve, but it descends to a closed curve on $\mathbb{RP}^2$, since its endpoints (north and south poles) are antipodal, and are therefore identified with each other. □

4.4. Flat surfaces

Let B be a finite-dimensional Banach space, *i.e.* a vector space together with a norm $\| \ \|$. Let $L \subset (B, \| \ \|)$ be a lattice of maximal rank, *i.e.* satisfying $\operatorname{rank}(L) = \dim(B)$. We define the notion of successive minima of L as follows, *cf.* [**GruL87**, p. 58].

DEFINITION 4.4.1. For each $k = 1, 2, \ldots, \operatorname{rank}(L)$, define the *k-th successive minimum* of the lattice L by

$$\lambda_k(L, \| \ \|) = \inf \left\{ \lambda \in \mathbb{R} \;\middle|\; \begin{array}{l} \exists \text{ lin. indep. } v_1, \ldots, v_k \in L \\ \text{with } \|v_i\| \leq \lambda \text{ for all } i \end{array} \right\}. \tag{4.4.1}$$

Thus the first successive minimum, $\lambda_1(L, \| \ \|)$ is the least length of a nonzero vector in L.

A metric is called *flat* if its Gaussian curvature K vanishes at every point. A closed surface of constant Gaussian curvature $K = 0$ is topologically either a torus $\mathbb{T}^2$ or a Klein bottle. Let us give a precise description in the former case.

EXAMPLE 4.4.2 (Flat tori). Every flat torus is isometric to a quotient $\mathbb{T}^2 = \mathbb{R}^2/L$ where L is a lattice, *cf.* [**Lo71**, Theorem 38.2]. In other words, a point of the torus is a coset of the additive action of the lattice in $\mathbb{R}^2$. The smooth structure is inherited from $\mathbb{R}^2$. Meanwhile, the additive action of the lattice is isometric. Indeed, we have $\operatorname{dist}(p, q) = \|q - p\|$, while for any $\ell \in L$, we have

$$\operatorname{dist}(p + \ell, q + \ell) = \|q + \ell - (p + \ell)\| = \|q - p\| = \operatorname{dist}(p, q).$$

Therefore the flat metric on $\mathbb{R}^2$ descends to $\mathbb{T}^2$.

Note that locally, all flat tori are indistinguishable from the flat plane itself. However, their global geometry depends on the metric invariants of the lattice, *e.g.* its successive minima, *cf.* Definition 4.4.1. Thus, we have the following.

THEOREM 4.4.3. *The least length of a nontrivial closed geodesic on a flat torus $\mathbb{T}^2 = \mathbb{R}^2/L$ equals the first successive minimum $\lambda_1(L)$.*

PROOF. The geodesics on the torus are the projections of straight lines in $\mathbb{R}^2$. In order for a straight line to close up, it must pass through a pair of points x and $x + \ell$ where $\ell \in L$. The length of the corresponding closed geodesic on $\mathbb{T}^2$ is precisely $\|\ell\|$, where $\| \ \|$ is the Euclidean norm. By choosing a shortest element in the lattice, we obtain a shortest closed geodesic on the corresponding torus. □

4.5. Hyperbolic surfaces

Most closed surfaces admit neither flat metrics nor metrics of positive curvature, but rather hyperbolic metrics. A hyperbolic surface is a surface equipped with a metric of constant Gaussian curvature $K = -1$. This case is far richer than the other two.

EXAMPLE 4.5.1. The pseudosphere (so called because its Gaussian curvature is constant, and equals -1) is the surface of revolution $(f(\phi)\cos\theta, f(\phi)\sin\theta, g(\phi))$ in $\mathbb{R}^3$ defined by the functions $f(\varphi) = e^\varphi$ and $g(\varphi) = \displaystyle\int_0^\varphi \left(1 - e^{2\psi}\right)^{1/2} d\psi$, where φ

ranges through the interval $-\infty < \varphi \le 0$. The usual formulas $g_{11} = f^2$ as well as $g_{22} = \left(\frac{df}{d\varphi}\right)^2 + \left(\frac{dg}{d\varphi}\right)^2$ yield in our case $g_{11} = e^{2\varphi}$, while

$$\begin{aligned} g_{22} &= (e^{\varphi})^2 + \left(\sqrt{1-e^{2\varphi}}\right)^2 \\ &= e^{2\varphi} + 1 - e^{2\varphi} \\ &= 1. \end{aligned}$$

Thus $(g_{ij}) = \begin{pmatrix} e^{2\varphi} & 0 \\ 0 & 1 \end{pmatrix}$. The pseudosphere has constant Gaussian curvature -1, but it is not a closed surface (as it is unbounded in $\mathbb{R}^3$).

The metric

$$\mathcal{G}_{\mathcal{H}^2} = \frac{1}{y^2}(dx^2 + dy^2) \tag{4.5.1}$$

in the upperhalf plane

$$\mathcal{H}^2 = \{(x, y) \mid y > 0\}$$

has constant Gaussian curvature $K = -1$. In coordinates (u^1, u^2), we can write it, a bit awkwardly, as

$$\mathcal{G}_{\mathcal{H}^2} = \frac{1}{(u^2)^2}\left(\left(du^1\right)^2 + \left(du^2\right)^2\right).$$

The Riemannian manifold $(\mathcal{H}^2, \mathcal{G}_{\mathcal{H}^2})$ is referred to as the *Poincaré upperhalf plane.* Its significance is that every closed hyperbolic surface Σ is isometric to the quotient of the Poincaré upperhalf plane by the action of a suitable group Γ:

$$\Sigma = \mathcal{H}^2/\Gamma.$$

Here the nonabelian group Γ is a discrete subgroup $\Gamma \subset PSL(2, \mathbb{R})$, where a matrix $A = \begin{pmatrix} a & b \\ c & d \end{pmatrix}$ acts on $\mathcal{H}^2 = \{z \in \mathbb{C} | \Im(z) > 0\}$ by $z \mapsto \frac{az+b}{cz+d}$, *cf.* [**Kato92**] where the following theorem is proved.

THEOREM 4.5.2. *Every geodesic in the Poincaré upperhalf plane is either a vertical ray, or a semicircle perpendicular to the x-axis.*

DEFINITION 4.5.3. A projection $(\mathbb{R}^2, \mathcal{G}) \to \mathbb{R}$ to the u^1-axis is called a *Riemannian submersion* if the metric coefficients in the horizontal direction are independent of u^2, or more precisely,

$$\mathcal{G} = h(u^1)\left(du^1\right)^2 + k(u^1, u^2)\left(du^2\right)^2, \tag{4.5.2}$$

for suitable functions h and k (note that the matrix of metric coefficients is diagonal but not necessarily scalar).

Being a Riemannian submersion is, of course, a local property, and we stated it for Euclidean space for convenience. We will use it mostly in the context of maps between compact manifolds.

EXAMPLE 4.5.4. The metric (4.5.1) of the Poincaré upperhalf plane admits a Riemannian submersion to (the positive ray of) the y-axis but not to the x-axis, as the conformal factor depends on y.

4.6. Topological preliminaries

We would like to provide a self-contained explanation of the topological ingredient which is necessary so as to understand the Loewner inequality, *i.e.* essentially the notion of a noncontractible loop and the fundamental group of a topological space X. See [**Hat02**, Chapter 1] for a more detailed account.

DEFINITION 4.6.1. A *loop* in X can be defined in one of two equivalent ways:
(1) a continuous map $\beta : [a,b] \to X$ satisfying $\beta(a) = \beta(b)$;
(2) a continuous map $\lambda : S^1 \to X$ from the circle S^1 to X.

LEMMA 4.6.2. *The two definitions of a loop are equivalent.*

PROOF. Consider the unique increasing linear function $f : [a,b] \to [0, 2\pi]$ which is one-to-one and onto. Thus, $f(t) = \frac{2\pi(t-a)}{b-a}$. Given a map $\lambda(e^{is}) : S^1 \to X$, we associate to it a map $\beta(t) = \lambda\left(e^{if(t)}\right)$, and vice versa. □

DEFINITION 4.6.3. A loop $S^1 \to X$ is said to be *contractible* if the map of the circle can be extended to a continuous map of the unit disk $\mathbb{D} \to X$, where $S^1 = \partial\mathbb{D}$.

DEFINITION 4.6.4. A space X is called *simply connected* if every loop in X is contractible.

THEOREM 4.6.5. *The sphere $S^n \subset \mathbb{R}^{n+1}$ which is the solution set of $x_0^2 + \ldots + x_n^2 = 1$ is simply connected for $n \geq 2$. The circle S^1 is not simply connected.*

One can refine the notion of simple connectivity by introducing a group, denoted $\pi_1(X) = \pi_1(X, x_0)$, and called the fundamental group of X relative to a fixed "base" point $x_0 \in X$.

A *based loop* is a loop $\alpha : [0,1] \to X$ satisfying the condition $\alpha(0) = \alpha(1) = x_0$. In terms of the second item of Definition 4.6.1, we choose a fixed point $s_0 \in S^1$. For example, we can choose $s_0 = e^{i0} = 1$ for the usual unit circle $S^1 \subset \mathbb{C}$, and require that $\alpha(s_0) = x_0$. Then the group $\pi_1(X)$ is the quotient of the space of all based loops modulo the equivalence equivalence relation defined by homotopies fixing the basepoint, *cf* Definition 4.6.6.

The equivalence class of the identity element is precisely the set of contractible loops based at x_0. The equivalence relation can be described as follows for a pair of loops in terms of the second item of Definition 4.6.1.

DEFINITION 4.6.6. Two based loops $\alpha, \beta : S^1 \to X$ are equivalent, or *homotopic*, if there is a continuous map of the cylinder $S^1 \times [c,d] \to X$ whose restriction to $S^1 \times \{c\}$ is α, whose restriction to $S^1 \times \{d\}$ is β, while the map is constant on the segment $\{s_0\} \times [c,d]$ of the cylinder, *i.e.* the basepoint does not move during the homotopy.

DEFINITION 4.6.7. An equivalence class of based loops is called a *based homotopy class*. Removing the basepoint restriction (as well as the constancy condition of Definition 4.6.6), we obtain a larger class called a *free homotopy class (of loops)*.

Composition of two loops is defined most conveniently in terms of item 1 of Definition 4.6.1, by concatenating their domains. Namely, the product of a pair of loops, $\alpha : [-1,0] \to X$ and $\beta : [0,+1] \to X$, is a loop $\alpha.\beta : [-1,1] \to X$, which coincides with α and β in their domains of definition. The product loop $\alpha.\beta$ is continuous since $\alpha(0) = \beta(0) = x_0$. Then Theorem 4.6.5 can be refined as follows.

THEOREM 4.6.8. *We have* $\pi_1(S^1) = \mathbb{Z}$*, while* $\pi_1(S^n)$ *is the trivial group for all* $n \geq 2$.

DEFINITION 4.6.9. The 2-torus $\mathbb{T}^2$ is defined to be the following Cartesian product: $\mathbb{T}^2 = S^1 \times S^1$, and can thus be realized as a subset

$$\mathbb{T}^2 = S^1 \times S^1 \subset \mathbb{R}^2 \times \mathbb{R}^2 = \mathbb{R}^4.$$

We have $\pi_1(\mathbb{T}^2) = \mathbb{Z}^2$. The familiar doughnut picture realizes $\mathbb{T}^2$ as a subset in Euclidean space $\mathbb{R}^3$.

DEFINITION 4.6.10. A 2-dimensional closed Riemannian manifold (*i.e.* surface) Σ is called *orientable* if it can be realized by a subset of $\mathbb{R}^3$.

Every imbedded closed surface in 3-space admits a continuous choice of a unit normal vector at every point. Note that no such choice is possible for an imbedding of the Mobius band, see [**Ar83**] for more details on orientability and imbeddings.

THEOREM 4.6.11 (Gauss-Bonnet theorem). *Every closed surface* Σ *satisfies the identity*

$$\int_\Sigma K(u^1, u^2)\sqrt{\det(g_{ij})}du^1du^2 = 2\pi\chi(\Sigma), \tag{4.6.1}$$

where K *is the Gaussian curvature function on* Σ*, whereas* $\chi(\Sigma)$ *is its Euler characteristic (here the integral is understood, as usual, in the sense of Theorem 4.1.4).*

See [**Hat02**] for a definition of the Euler characteristic. The latter is even for closed orientable surfaces, and the integer $g = g(\Sigma) \geq 0$ defined by

$$\chi(\Sigma) = 2 - 2g$$

is called the *genus* of Σ. We have $g(S^2) = 0$, while $g(\mathbb{T}^2) = 1$. In general, the genus can be understood intuitively as the number of "holes", *i.e.* "handles", in a familiar 3-dimensional picture of a pretzel. We see from formula (4.6.1) that the only compact orientable surface admitting flat metrics is the 2-torus. See [**Ar83**] for a friendly topological introduction to surfaces.

CHAPTER 5

Inequalities of Loewner and Pu

5.1. Definition of systole

The notions of contractible loops and simply connected spaces were reviewed in Section 4.6. We will denote by $\mathrm{sys}\pi_1(\mathcal{G})$, the infimum of lengths, referred to as the "systole" of $\mathcal{G}$, of a noncontractible loop β in a compact, non-simply-connected Riemannian manifold $(M, \mathcal{G})$:

$$\mathrm{sys}\pi_1(\mathcal{G}) = \inf_{\beta} \mathrm{length}(\beta), \tag{5.1.1}$$

where the infimum is over all noncontractible loops β in M. In graph theory, a similar invariant is known as the *girth* [**Tu47**], see the discussion in Subsection 2.4.2. The relation of the notion of systole expressed by (5.1.1) to the systolic arrays of [**Ku78**] is yet to be clarified. See Section 2.1 for a further discussion of the term, by its promulgator.

It can be shown that for a compact Riemannian manifold, the infimum is always attained, *cf.* Theorem 4.2.5. A loop realizing the minimum is necessarily a simple closed geodesic.

In systolic questions about surfaces, integral-geometric identities play a particularly important role. Roughly speaking, there is an integral identity relating area on the one hand, and an average of energies of a suitable family of loops, on the other. By the Cauchy-Schwarz inequality, there is an inequality relating energy and length squared, hence one obtains an inequality between area and the square of the systole.

Such an approach works both for the Loewner inequality (5.4.1) and Pu's inequality (5.2.2) (biographical notes on C. Loewner and P. Pu appear in Sections 2.2 and 2.3, respectively). One can prove an inequality for the Mobius band this way, as well [**Bl61b**].

5.2. Isoperimetric inequality and Pu's inequality

Pu's inequality can be thought of as an "opposite" isoperimetric inequality, in the following precise sense.

The classical isoperimetric inequality in the plane is a relation between two metric invariants: length L of a simple closed curve in the plane, and area A of the region bounded by the curve. Namely, every simple closed curve in the plane satisfies the inequality

$$\frac{A}{\pi} \le \left(\frac{L}{2\pi}\right)^2 .$$

This classical *isoperimetric inequality* is sharp, insofar as equality is attained only by a round circle.

In the 1950's, Charles Loewner's student P. M. Pu [**Pu52**] proved the following theorem. Let $\mathbb{RP}^2$ be the real projective plane endowed with an arbitrary metric, *i.e.* an imbedding in some $\mathbb{R}^n$. Then

$$\left(\frac{L}{\pi}\right)^2 \le \frac{A}{2\pi}, \tag{5.2.1}$$

where A is its total area and L is the length of its shortest non-contractible loop. This *isosystolic inequality*, or simply *systolic inequality* for short, is also sharp, to the extent that equality is attained only for a metric of constant Gaussian curvature, namely antipodal quotient of a round sphere, *cf.* Section 4.3. In our systolic notation (5.1.1), Pu's inequality takes the following form:

$$\operatorname{sys}\pi_1(\mathcal{G})^2 \le \tfrac{\pi}{2} \operatorname{area}(\mathcal{G}), \tag{5.2.2}$$

for every metric $\mathcal{G}$ on $\mathbb{RP}^2$. See Theorem 4.3.4 for a discussion of the constant. The inequality is proved in Section 6.5. Pu's inequality can be generalized as follows. We will say that a surface is *aspherical* if it is not a 2-sphere.

THEOREM 5.2.1. *Every aspherical surface $(\Sigma, \mathcal{G})$ satisfies the optimal bound (5.2.2), attained precisely when, on the one hand, the surface Σ is a real projective plane, and on the other, the metric $\mathcal{G}$ is of constant Gaussian curvature.*

The extension to aspherical surfaces follows from Gromov's inequality (5.2.3) below (by comparing the numerical values of the two constants). Namely, every aspherical compact surface $(\Sigma, \mathcal{G})$ admits a metric ball

$$B = B_p\left(\tfrac{1}{2}\operatorname{sys}\pi_1(\mathcal{G})\right) \subset \Sigma$$

of radius $\frac{1}{2}\operatorname{sys}\pi_1(\mathcal{G})$ which satisfies [**Gro83**, Corollary 5.2.B]

$$\operatorname{sys}\pi_1(\mathcal{G})^2 \le \frac{4}{3} \operatorname{area}(B). \tag{5.2.3}$$

We will need the following area bound in Chapter 11, *cf.* [**BuraZ80**, Chapter 1, Theorem 5.3.1, p. 43].

PROPOSITION 5.2.2. *Whenever a point $x \in \Sigma$ lies on a two-sided loop which is minimizing in its free homotopy class, cf. Definition 4.6.7, the metric ball $B_x(r) \subset \Sigma$ of radius $r \le \frac{1}{2}\operatorname{sys}\pi_1(\mathcal{G})$ satisfies the estimate*

$$\operatorname{area}(B_x(r)) \ge 2r^2. \tag{5.2.4}$$

PROOF. More generally, for an arbitrary 2-complex X, one can argue as follows. Let $\gamma(s)$ be an arclength parametrisation of the loop, with $\gamma(0) = x$. If points $\gamma(r)$ and $\gamma(-r)$ lie in distinct connected components of the r-level curve, then it is easy to see that the element $[\gamma] \in \pi_1(X, \gamma(0))$ splits off as a generator of an infinite cyclic factor in a free product decomposition of the fundamental group. If $X = \Sigma$ is a closed surface, its fundamental group does not admit such a decomposition. Hence $\gamma(r)$ and $\gamma(-r)$ must lie in a common connected component, $C \subset \Sigma$, of the level curve. Let $\delta_\pm \subset C$ be the two imbedded arcs connecting $\gamma(r)$ to $\gamma(-r)$. An easy argument shows that $\operatorname{length}(\delta_+ \cup \delta_-) \ge 2r$. We now perform an integration to deduce the bound (5.2.4) from the coarea formula, see [**KatzRS06**] for more details. □

A (different) generalisation of Pu's inequality for the relative systole of a surface of negative Euler characteristic appears in Theorem 8.2.3.

5.3. Hermite and Bergé-Martinet constants

Let $b \in \mathbb{N}$. The Hermite constant γ_b is defined in one of the following two equivalent ways:

(1) γ_b is the *square* of the biggest first successive minimum, *cf.* Definition 4.4.1, among all lattices of unit covolume;

(2) γ_b is defined by the formula

$$\sqrt{\gamma_b} = \sup \left\{ \frac{\lambda_1(L)}{\operatorname{vol}(\mathbb{R}^b/L)^{1/b}} \,\middle|\, L \subseteq (\mathbb{R}^b, \|\ \|) \right\}, \tag{5.3.1}$$

where the supremum is extended over all lattices L in $\mathbb{R}^b$ with a Euclidean norm $\|\ \|$.

A lattice realizing the supremum is called a *critical* lattice. A critical lattice may be thought of as the one realizing the densest packing in $\mathbb{R}^b$ when we place balls of radius $\frac{1}{2}\lambda_1(L)$ at the points of L.

The existence of the Hermite constant, as well as the existence of critical lattices, are both nontrivial results [**Cas71**].

LEMMA 5.3.1. *When $b = 2$, we have the following value for the Hermite constant: $\gamma_2 = \frac{2}{\sqrt{3}} = 1.1547\ldots$. The corresponding critical lattice is homothetic to the $\mathbb{Z}$-span of the cube roots of unity in $\mathbb{C}$.*

PROOF. Consider a lattice $L \subset \mathbb{C} = \mathbb{R}^2$. Clearly, multiplying L by nonzero complex numbers does not change the value of the quotient

$$\frac{\lambda_1(L)^2}{\operatorname{area}(\mathbb{C}/L)}.$$

Choose a "shortest" vector $z \in L$, *i.e.* we have $|z| = \lambda_1(L)$. By replacing L by the lattice $z^{-1}L$, we may assume that the complex number $+1 \in \mathbb{C}$ is a shortest element in the lattice. We will denote the new lattice by the same letter L, so that now $\lambda_1(L) = 1$. Now complete the element $+1 \in L$ to a $\mathbb{Z}$-basis $\{\tau, +1\}$ for L. Thus $|\tau| \geq \lambda_1(L) = 1$. Consider the real part $\Re(\tau)$. We can adjust the basis by adding a suitable integer to τ, so as to satisfy the condition $-\frac{1}{2} \leq \Re(\tau) \leq \frac{1}{2}$. Then the second basis vector τ lies in the closure of the standard fundamental domain

$$D = \left\{ z \in \mathbb{C} \,\middle|\, |z| > 1,\ |\Re(z)| < \tfrac{1}{2},\ \Im(z) > 0 \right\} \tag{5.3.2}$$

for the action of the group $\mathrm{PSL}(2, \mathbb{Z})$ in the upperhalf plane of $\mathbb{C}$, see [**Serr73**, p. 78]. The imaginary part satisfies $\Im(\tau) \geq \frac{\sqrt{3}}{2}$, with equality possible in the following two cases: $\tau = e^{i\frac{\pi}{3}}$ or $\tau = e^{i\frac{2\pi}{3}}$. Moreover, if $\tau = r\exp(i\theta)$, then

$$\sin\theta = \frac{\Im(\tau)}{|\tau|} \geq \frac{\sqrt{3}}{2|\tau|}.$$

Finally, we calculate the area of the parallelogram in $\mathbb{C}$ spanned by τ and $+1$, and write

$$\frac{\lambda_1(L)^2}{\operatorname{area}(\mathbb{C}/L)} = \frac{1}{|\tau|\sin\theta} \leq \frac{2}{\sqrt{3}}$$

to conclude the proof. □

EXAMPLE 5.3.2. In dimensions $b \geq 3$, the Hermite constants are harder to compute, but explicit values (as well as the associated critical lattices) are known for small dimensions (≤ 8), *e.g.* $\gamma_3 = 2^{\frac{1}{3}} = 1.2599\ldots$, while $\gamma_4 = \sqrt{2} = 1.4142\ldots$. Note that γ_n is asymptotically linear in n, *cf.* (5.3.5).

A related constant γ'_b is defined as follows, *cf.* [**BeM89**].

DEFINITION 5.3.3. The Bergé-Martinet constant γ'_b is defined by setting

$$\gamma'_b = \sup\left\{\lambda_1(L)\lambda_1(L^*)\,\middle|\, L \subseteq (\mathbb{R}^b, \|\ \|)\right\}, \tag{5.3.3}$$

where the supremum is extended over all lattices L in $\mathbb{R}^b$.

Here L^* is the lattice dual to L. If L is the $\mathbb{Z}$-span of vectors (x_i), then L^* is the $\mathbb{Z}$-span of a dual basis (y_j) satisfying $\langle x_i, y_j\rangle = \delta_{ij}$, *cf.* relation (4.1.1).

Thus, the constant γ'_b is bounded above by the Hermite constant γ_b of (5.3.1). We have $\gamma'_1 = 1$, while for $b \geq 2$ we have the following inequality:

$$\gamma'_b \leq \gamma_b \leq \frac{2}{3}b \quad \text{for all } b \geq 2. \tag{5.3.4}$$

Moreover, one has the following asymptotic estimates:

$$\frac{b}{2\pi e}(1+o(1)) \leq \gamma'_b \leq \frac{b}{\pi e}(1+o(1)) \quad \text{for } b \to \infty, \tag{5.3.5}$$

cf. [**LaLS90**, pp. 334, 337]. Note that the lower bound of (5.3.5) for the Hermite constant and the Bergé-Martinet constant is nonconstructive, but see [**RT90**] and [**ConS99**].

DEFINITION 5.3.4. A lattice L realizing the supremum in (5.3.3) is called *dual-critical.*

REMARK 5.3.5. The constants γ'_b and the dual-critical lattices in $\mathbb{R}^b$ are explicitly known for $b \leq 4$, *cf.* [**BeM89**, Proposition 2.13]. In particular, we have $\gamma'_1 = 1$, $\gamma'_2 = \frac{2}{\sqrt{3}}$.

EXAMPLE 5.3.6. In dimension 3, the value of the Bergé-Martinet constant, $\gamma'_3 = \sqrt{\frac{3}{2}} = 1.2247\ldots$, is slightly below the Hermite constant $\gamma_3 = 2^{\frac{1}{3}} = 1.2599\ldots$. It is attained by the face-centered cubic lattice, which is not isodual [**MilnH73**, p. 31], [**BeM89**, Proposition 2.13(iii)], [**ConS94**].

5.4. The Loewner inequality

Historically, the first lower bound for the volume of a Riemannian manifold in terms of a systole is due to Charles Loewner. In 1949, Loewner proved the first systolic inequality, in a course on Riemannian geometry at Syracuse University, *cf.* [**Pu52**]. Namely, he showed the following result, whose proof appears in Section 6.2.

THEOREM 5.4.1 (C. Loewner). *Every Riemannian metric $\mathcal{G}$ on the torus $\mathbb{T}^2$ satisfies the inequality*

$$\mathrm{sys}\pi_1(\mathcal{G})^2 \leq \gamma_2 \ \mathrm{area}(\mathcal{G}), \tag{5.4.1}$$

where $\gamma_2 = \frac{2}{\sqrt{3}}$ is the Hermite constant (5.3.1). *A metric attaining the optimal bound* (5.4.1) *is necessarily flat, and is homothetic to the quotient of $\mathbb{C}$ by the lattice spanned by the cube roots of unity, cf. Lemma 5.3.1.*

In the next chapter, we will prove both the Loewner inequality, and Pu's inequality.

CHAPTER 6

Systolic applications of integral geometry

6.1. An integral-geometric identity

Let $\mathbb{T}^2$ be a torus with an arbitrary metric. Let $\mathbb{T}_0 = \mathbb{R}^2/L$ be the flat torus conformally equivalent to $\mathbb{T}^2$. Let $\ell_0 = \ell_0(x)$ be a simple closed geodesic of $\mathbb{T}_0$. Thus ℓ_0 is the projection of a line $\tilde{\ell}_0 \subset \mathbb{R}^2$. Let $\tilde{\ell}_y \subset \mathbb{R}^2$ be the line parallel to $\tilde{\ell}_0$ at distance $y > 0$ from ℓ_0 (here we must "choose sides", *e.g.* by orienting $\tilde{\ell}_0$ and requiring $\tilde{\ell}_y$ to lie to the left of $\tilde{\ell}_0$). Denote by $\ell_y \subset \mathbb{T}_0$ the closed geodesic loop defined by the image of $\tilde{\ell}_y$. Let $y_0 > 0$ be the *smallest* number such that $\ell_{y_0} = \ell_0$, *i.e.* the lines $\tilde{\ell}_{y_0}$ and $\tilde{\ell}_0$ both project to ℓ_0.

Note that the loops in the family $\{\ell_y\} \subset \mathbb{T}^2$ are not necessarily geodesics with respect to the metric of $\mathbb{T}^2$. On the other hand, the family satisfies the following identity.

LEMMA 6.1.1 (An elementary integral-geometric identity). *The metric on $\mathbb{T}^2$ satisfies the following identity:*

$$\operatorname{area}(\mathbb{T}^2) = \int_0^{y_0} E(\ell_y)dy, \tag{6.1.1}$$

where E is the energy of a loop with respect to the metric of $\mathbb{T}^2$, see Definition 4.2.3.

PROOF. Denote by f^2 the conformal factor of $\mathbb{T}^2$ with respect to the flat metric $\mathbb{T}_0$. Thus the metric on $\mathbb{T}^2$ can be written as

$$f^2(x,y)(dx^2 + dy^2).$$

By Fubini's theorem applied to a rectangle with sides $\operatorname{length}_{\mathbb{T}_0}(\ell_0)$ and y_0, combined with Remark 4.1.7, we obtain

$$\begin{aligned} \operatorname{area}(\mathbb{T}^2) &= \int_{\mathbb{T}_0} f^2 dx dy \\ &= \int_0^{y_0} \left(\int_{\ell_y} f^2(x,y)dx \right) dy \\ &= \int_0^{y_0} E(\ell_y)dy, \end{aligned}$$

proving the lemma. □

REMARK 6.1.2. The identity (6.1.1) can be thought of as the simplest integral-geometric identity, *cf.* equation (6.4.3).

6.2. Two proofs of the Loewner inequality

We give a slightly modified version of M. Gromov's proof [**Gro96**], using conformal representation and the Cauchy-Schwarz inequality, of the Loewner inequality (5.4.1) for the 2-torus, see also [**CrK03**]. We present the following slight generalisation: there exists a pair of closed geodesics on $(\mathbb{T}^2, \mathcal{G})$, of respective lengths λ_1 and λ_2, such that

$$\lambda_1 \lambda_2 \leq \gamma_2 \operatorname{area}(\mathcal{G}), \tag{6.2.1}$$

and whose homotopy classes form a generating set for $\pi_1(\mathbb{T}^2) = \mathbb{Z} \times \mathbb{Z}$.

PROOF. The proof relies on the conformal representation

$$\phi : \mathbb{T}_0 \to (\mathbb{T}^2, \mathcal{G}),$$

where $\mathbb{T}_0$ is flat, *cf.* uniformisation theorem 4.1.9. Here the map ϕ may be chosen in such a way that $(\mathbb{T}^2, \mathcal{G})$ and $\mathbb{T}_0$ have the same area. Let f be the conformal factor, so that

$$\mathcal{G} = f^2 \left((du^1)^2 + (du^2)^2 \right)$$

as in formula (4.1.6), where $(du^1)^2 + (du^2)^2$ (locally) is the flat metric.

Let ℓ_0 be any closed geodesic in $\mathbb{T}_0$. Let $\{\ell_s\}$ be the family of geodesics parallel to ℓ_0. Parametrize the family $\{\ell_s\}$ by a circle S^1 of length σ, so that

$$\sigma \ell_0 = \operatorname{area}(\mathbb{T}_0).$$

Thus $\mathbb{T}_0 \to S^1$ is a Riemannian submersion, *cf.* Definition 4.5.3. Then

$$\operatorname{area}(\mathbb{T}^2) = \int_{\mathbb{T}_0} f^2.$$

By Fubini's theorem, we have $\operatorname{area}(\mathbb{T}^2) = \int_{S^1} ds \int_{\ell_s} f^2 dt$. Therefore by the Cauchy-Schwarz inequality,

$$\operatorname{area}(\mathbb{T}^2) \geq \int_{S^1} ds \frac{\left(\int_{\ell_s} f dt \right)^2}{\ell_0} = \frac{1}{\ell_0} \int_{S^1} ds \left(\operatorname{length} \phi(\ell_s) \right)^2.$$

Hence there is an s_0 such that $\operatorname{area}(\mathbb{T}^2) \geq \frac{\sigma}{\ell_0} \operatorname{length}\phi(\ell_{s_0})^2$, so that

$$\operatorname{length}\phi(\ell_{s_0}) \leq \ell_0.$$

This reduces the proof to the flat case. Given a lattice in $\mathbb{C}$, we choose a shortest lattice vector λ_1, as well as a shortest one λ_2 not proportional to λ_1. The inequality now follows from Lemma 5.3.1. In the boundary case of equality, one can exploit the equality in the Cauchy-Schwarz inequality to prove that the conformal factor must be constant. □

ALTERNATIVE PROOF. Let ℓ_0 be any simple closed geodesic in $\mathbb{T}_0$. Since the desired inequality (6.2.1) is scale-invariant, we can assume that the loop has unit length:

$$\operatorname{length}_{\mathbb{T}_0}(\ell_0) = 1,$$

and, moreover, that the corresponding covering transformation of the universal cover $\mathbb{C} = \tilde{\mathbb{T}}_0$ is translation by the element $1 \in \mathbb{C}$. We complete 1 to a basis $\{\tau, 1\}$ for the lattice of covering transformations of $\mathbb{T}_0$. Note that the rectangle defined by

$$\{z = x + iy \in \mathbb{C} \mid 0 < x < 1,\ 0 < y < \Im(\tau)]\}$$

is a fundamental domain for $\mathbb{T}_0$, so that area$(\mathbb{T}_0) = \Im(\tau)$. The maps

$$\ell_y(x) = x + iy, \; x \in [0,1]$$

parametrize the family of geodesics parallel to ℓ_0 on $\mathbb{T}_0$. Recall that the metric of the torus $\mathbb{T}$ is $f^2(dx^2 + dy^2)$.

LEMMA 6.2.1. *We have the following relation between the length and energy of a loop:*

$$\text{length}(\ell_y)^2 \le E(\ell_y).$$

PROOF. By the Cauchy-Schwarz inequality,

$$\int_0^1 f^2(x,y)dx \ge \left(\int_0^1 f(x,y)dx\right)^2,$$

proving the lemma. □

Now by Lemma 6.1.1 and Lemma 6.2.1, we have

$$\text{area}(\mathbb{T}^2) \ge \int_0^{\Im(\tau)} \left(\text{length}(\ell_y)\right)^2 dy.$$

Hence there is a y_0 such that

$$\text{area}(\mathbb{T}^2) \ge \Im(\tau)\text{length}(\ell_{y_0})^2,$$

so that length$(\ell_{y_0}) \le 1$. This reduces the proof to the flat case. Given a lattice in $\mathbb{C}$, we choose a shortest lattice vector λ_1, as well as a shortest one λ_2 not proportional to λ_1. The inequality now follows from Lemma 5.3.1. □

REMARK 6.2.2. Define a conformal invariant called the *capacity* of an annulus as follows. Consider a right circular cylinder

$$\zeta_\kappa = \mathbb{R}/\mathbb{Z} \times [0, \kappa]$$

based on a unit circle $\mathbb{R}/\mathbb{Z}$. Its capacity $C(\zeta_\kappa)$ is defined to be its height, $C(\zeta_\kappa) = \kappa$. Recall that every annular region in the plane is conformally equivalent to such a cylinder, and therefore we have defined a conformal invariant of an arbitrary annular region. Every annular region R satisfies the inequality area$(R) \ge C(R)\,\text{sys}\pi_1(R)^2$. Meanwhile, if we cut a flat torus along a shortest loop, we obtain an annular region R with capacity at least $C(R) \ge \gamma_2^{-1} = \frac{\sqrt{3}}{2}$. This provides an alternative proof of the Loewner theorem. In fact, we have the following identity:

$$\text{confsys}_1(\mathcal{G})^2 C(\mathcal{G}) = 1, \tag{6.2.2}$$

where confsys is the conformally invariant generalisation of the homology systole, while $C(\mathcal{G})$ is the largest capacity of a cylinder obtained by cutting open the underlying conformal structure on the torus.

QUESTION 6.2.3. Is the Loewner inequality (5.4.1) satisfied by *every* orientable nonsimply connected compact surface? Inspite of its elementary nature, and considerable research devoted to the area, the question is still open. Recently the case of genus 2 was settled (Chapter 9), as well as genus $g \ge 20$ (Chapter 11).

6.3. Hopf fibration and the Hamilton quaternions

The circle action in $\mathbb{C}^2$ restricts to the unit sphere $S^3 \subset \mathbb{C}^2$, which therefore admits a fixed point free circle action. Namely, the circle $S^1 = \{e^{i\theta} : \theta \in \mathbb{R}\} \subset \mathbb{C}$ acts on (z_1, z_2) by

$$e^{i\theta}.(z_1, z_2) = (e^{i\theta} z_1, e^{i\theta} z_2).$$

It is possible to show that the quotient manifold (space of orbits) S^3/S^1 is the 2-sphere S^2 (from the projective viewpoint, the quotient is the complex projective line $\mathbb{CP}^1$). Moreover, the quotient map

$$S^3 \to S^2, \tag{6.3.1}$$

called the *Hopf fibration*, is a Riemannian submersion, for which the natural metric on the base is a metric of constant Gaussian curvature $+4$ rather than $+1$. The latter is explained by the fact that the maximal distance between a pair of S^1 orbits is $\frac{\pi}{2}$ rather than π, in view of the fact that a pair of antipodal points of S^3 lies in a common orbit.

Furthermore, the sphere S^3 can be thought of as the Lie group formed of the unit (Hamilton) quaternions in $\mathbb{H} = \mathbb{C}^2$. The Lie group $SO(3)$ can be identified with $S^3/\{\pm 1\}$. Therefore the fibration (6.3.1) descends to a fibration

$$SO(3) \to S^2, \tag{6.3.2}$$

exploited in the next section.

REMARK 6.3.1. The group $G = SO(3)$ can be identified with the unit tangent bundle of S^2, as follows. Choose a fixed unit tangent vector v to S^2, and send an element $g \in G$, acting on S^2, to the image of v under $dg : TS^2 \to TS^2$.

REMARK 6.3.2. Let us elaborate a bit further on the quaternionic viewpoint, with an eye to aplying it in Section 6.4. A choice of a purely imaginary Hamilton quaternion in $\mathbb{H} = \mathbb{R}1 + \mathbb{R}i + \mathbb{R}j + \mathbb{R}k$, *i.e.* a linear combination of i, j, and k, specifies a complex structure on $\mathbb{H}$. Such a complex structure leads to a fibration of the unit sphere, as in (6.3.1). Choosing a pair of purely imaginary quaternions which are orthogonal, results in a pair of circle fibrations of S^3 such that a fiber of one fibration is perpendicular to a fiber of the other fibration whenever two such fibers meet. The construction of two "perpendicular" fibrations also descends to the quotient by the natural $\mathbb{Z}_2$ action on S^3.

6.4. Double fibration of $SO(3)$ and integral geometry on S^2

In this section, we present a self-contained account of an integral-geometric identity, used in the proof of Pu's inequality (5.2.2) in Section 6.5, as well as in our argument in Lemma 8.5.1. The identity has its origin in results of P. Funk [**Fu16**] determining a symmetric function on the two-sphere from its great circle integrals; see [**Hel99**, Proposition 2.2, p. 59], as well as Preface therein.

The sphere S^2 is the homogeneous space of the Lie group $SO(3)$, *cf.* (6.3.2), so that $S^2 = SO(3)/SO(2)$, *cf.* [**Ar83**, p. 82]. Denote by $SO(2)_\sigma$ the fiber over (stabilizer of) a typical point $\sigma \in S^2$. The projection

$$p : SO(3) \to S^2 \tag{6.4.1}$$

is a Riemannian submersion for the standard metric of constant sectional curvature $\frac{1}{4}$ on $SO(3)$. The total space $SO(3)$ admits another Riemannian submersion,

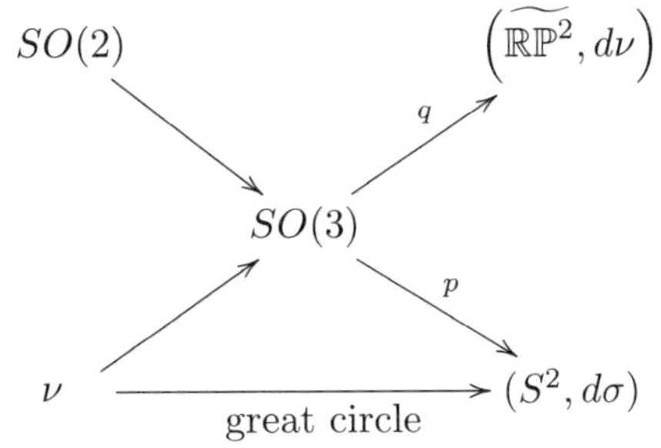

FIGURE 6.4.1. Integral geometry on S^2, *cf.* (6.4.1) and (6.4.2)

which we denote

$$q : SO(3) \to \widetilde{\mathbb{RP}}^2, \tag{6.4.2}$$

whose typical fiber ν is an orbit of the geodesic flow on $SO(3)$ when the latter is viewed as the unit tangent bundle of S^2, *cf.* Remark 6.3.1. Here $\widetilde{\mathbb{RP}}^2$ denotes the double cover of $\mathbb{RP}^2$, homeomorphic to the 2-sphere, *cf.* Definition 4.3.2. A fiber of fibration q is the collection of unit vectors tangent to a given directed closed geodesic (great circle) on S^2. Thus, each orbit ν projects under p to a great circle on the sphere. We think of the base space $\widetilde{\mathbb{RP}}^2$ of q as the configuration space of oriented great circles on the sphere, with measure $d\nu$. While fibration q may seem more "mysterious" than fibration p, the two are actually equivalent from the quaternionic point of view, *cf.* Remark 6.3.2.

The diagram of Figure 6.4.1 illustrates the maps defined so far.

Denote by $E_{\mathcal{G}}(\nu)$ the energy, and by $L_{\mathcal{G}}(\nu)$ the length, of a curve ν with respect to a (possibly singular) metric $\mathcal{G} = f^2\mathcal{G}_0$.

THEOREM 6.4.1. *We have the following integral-geometric identity:*

$$\operatorname{area}(\mathcal{G}) = \frac{1}{2\pi}\int_{\widetilde{\mathbb{RP}}^2} E_{\mathcal{G}}(\nu)d\nu.$$

PROOF. We apply Fubini's theorem [**Ru87**, p. 164] twice, to both q and p, to obtain

$$\begin{aligned}\int_{\widetilde{\mathbb{RP}}^2} E_{\mathcal{G}}(\nu)d\nu &= \int_{\widetilde{\mathbb{RP}}^2} d\nu\left(\int_\nu f^2\circ p\circ\nu(t)dt\right)\\ &= \int_{SO(3)} f^2\circ p\\ &= \int_{S^2} f^2\, d\sigma\left(\int_{SO(2)_\sigma} 1\right)\\ &= 2\pi\ \operatorname{area}(\mathcal{G}),\end{aligned} \tag{6.4.3}$$

completing the proof of the theorem. □

PROPOSITION 6.4.2. *Let f be a square integrable function on S^2, which is positive and continuous except possibly for a finite number of points where f either vanishes or has a singularity of type $\frac{1}{\sqrt{r}}$, cf. Remark 8.4.4. Then there is a great* *circle ν such that the following inequality is satisfied:*

$$\left(\int_\nu f(\nu(t))dt\right)^2 \leq \pi \int_{S^2} f^2(\sigma)d\sigma, \tag{6.4.4}$$

where t is the arclength parameter, and $d\sigma$ is the Riemannian measure of the metric $\mathcal{G}_0$ of the standard unit sphere. In other words, there is a great circle of length L in the metric $f^2\mathcal{G}_0$, so that $L^2 \leq \pi A$, where A is the Riemannian surface area of the metric $f^2\mathcal{G}_0$. In the boundary case of equality in (6.4.4), *the function f must be constant.*

PROOF. The proof is an averaging argument and shows that the average length of great circles is short. Comparing length and energy yields the inequality

$$\frac{\left(\int_{\widetilde{\mathbb{RP}^2}} L_{\mathcal{G}}(\nu)^2\right)}{2\pi} \leq \int_{\widetilde{\mathbb{RP}^2}} E_{\mathcal{G}}(\nu)d\nu. \tag{6.4.5}$$

The formula now follows from Theorem 6.4.1, since $\text{area}(\widetilde{\mathbb{R}P^2}) = 4\pi$.

In the boundary case of equality in (6.4.4), one must have equality also in inequality (6.4.5) relating length and energy. It follows that the conformal factor f must be constant along every great circle. Hence f is constant everywhere on S^2. □

6.5. Proof of Pu's inequality

A metric $\mathcal{G}$ on $\mathbb{RP}^2$ lifts to a centrally symmetric metric $\tilde{\mathcal{G}}$ on S^2. Applying Proposition 6.4.2 to $\tilde{\mathcal{G}}$, we obtain a great circle of $\tilde{\mathcal{G}}$-length L satisfying $L^2 \leq \pi A$, where A is the Riemannian surface area of $\tilde{\mathcal{G}}$. Thus we have $\left(\frac{L}{2}\right)^2 \leq \frac{\pi}{2}\left(\frac{A}{2}\right)$, where $\text{sys}\pi_1(\mathcal{G}) \leq \frac{L}{2}$ while $\text{area}(\mathcal{G}) = \frac{A}{2}$, proving inequality (5.2.2). See **[Iv02]** for an alternative proof.

6.6. A table of optimal systolic ratios of surfaces

Denote by $\text{SR}(\Sigma)$ the supremum of the systolic ratios,

$$\text{SR}(\Sigma) = \sup_{\mathcal{G}} \frac{\text{sys}\pi_1(\mathcal{G})^2}{\text{area}(\mathcal{G})},$$

ranging over all metrics $\mathcal{G}$ on a surface Σ. The known values of the optimal systolic ratio are tabulated in Figure 6.6.1. It is interesting to note that the optimal ratio for the Klein bottle $\mathbb{RP}^2\#\mathbb{RP}^2$ is achieved by a singular metric, described in the references listed in the table.

	SR(Σ)	numerical value	where to find it
$\Sigma = \mathbb{RP}^2$	$= \frac{\pi}{2}$ **[Pu52]**	≈ 1.5707	Section 6.5
infinite $\pi_1(\Sigma)$	$< \frac{4}{3}$ **[Gro83]**	$< 1.3333\ldots$	(5.2.3)
$\Sigma = \mathbb{T}^2$	$= \frac{2}{\sqrt{3}}$ (Loewner)	≈ 1.1547	(5.4.1)
$\Sigma = \mathbb{RP}^2 \# \mathbb{RP}^2$	$= \frac{\pi}{2^{3/2}}$	≈ 1.1107	**[Bav86, Bav06, Sak88]**
Σ of genus 2	$> \frac{1}{3}\left(\sqrt{2}+1\right)$	> 0.8047	(10.1.1)
Σ of genus 3	$\geq \frac{8}{7\sqrt{3}}$	> 0.6598	**[Cal96]**

FIGURE 6.6.1. Values for optimal systolic ratio SR of surface Σ

CHAPTER 7

A primer on surfaces

In this Chapter, we collect some classical facts on Riemann surfaces. More specifically, we deal with hyperelliptic surfaces, real surfaces, and Katok's optimal bound for the entropy of a surface.

7.1. Hyperelliptic involution

Let Σ be an orientable closed Riemann surface which is not a sphere. By a Riemann surface, we mean a surface equipped with a fixed conformal structure, *cf.* Definition 4.1.8, while all maps are angle-preserving.

DEFINITION 7.1.1. A *hyperelliptic involution* of a Riemann surface Σ of genus g is a holomorphic (conformal) map, $J : \Sigma \to \Sigma$, satisfying $J^2 = 1$, with $2g + 2$ fixed points.

DEFINITION 7.1.2. A surface Σ admitting a hyperelliptic involution will be called a *hyperelliptic surface.*

REMARK 7.1.3. The involution J can be identified with the nontrivial element in the center of the (finite) automorphism group of Σ (*cf.* [**FarK92**, p. 108]) when it exists, and then such a J is unique, *cf.* [**Mir95**, p.204].

It is known that the quotient of Σ by the involution J produces a conformal branched 2-fold covering

$$Q : \Sigma \to S^2 \tag{7.1.1}$$

of the sphere S^2.

DEFINITION 7.1.4. The $2g + 2$ fixed points of J are called *Weierstrass points.* Their images in S^2 under the ramified double cover Q of formula (7.1.1) will be referred to as *ramification points.*

A notion of a Weierstrass point exists on any Riemann surface, but will only be used in the present text in the hyperelliptic case.

EXAMPLE 7.1.5. In the case $g = 2$, topologically the situation can be described as follows. A simple way of representing the figure 8 contour in the (x, y) plane is by the reducible curve

$$(((x-1)^2 + y^2) - 1)(((x+1)^2 + y^2) - 1) = 0 \tag{7.1.2}$$

(or, alternatively, by the lemniscate $r^2 = \cos 2\theta$ in polar coordinates, *i.e.* the locus of the equation $(x^2 + y^2)^2 = x^2 - y^2$).

Now think of the figure 8 curve of (7.1.2) as a subset of $\mathbb{R}^3$. The boundary of its tubular neighborhood in $\mathbb{R}^3$ is a genus 2 surface. Rotation by π around the x-axis has six fixed points on the surface, namely, a pair of fixed points near each of the points -2, 0, and $+2$ on the x-axis. The quotient by the rotation can be seen to be homeomorphic to the sphere.

A similar example can be repackaged in a metrically more precise way as follows.

EXAMPLE 7.1.6. We start with a round metric on $\mathbb{RP}^2$. Now attach a small handle. The orientable double cover Σ of the resulting surface can be thought of as the unit sphere in $\mathbb{R}^3$, with two little handles attached at north and south poles, *i.e.* at the two points where the sphere meets the z-axis. Then one can think of the hyperelliptic involution J as the rotation of Σ by π around the z-axis. The six fixed points are the six points of intersection of Σ with the z-axis. Furthermore, there is an orientation reversing involution τ on Σ, given by the restriction to Σ of the antipodal map in $\mathbb{R}^3$. The composition $\tau \circ J$ is the reflection fixing the xy-plane, in view of the following matrix identity:

$$\begin{pmatrix} -1 & 0 & 0 \\ 0 & -1 & 0 \\ 0 & 0 & -1 \end{pmatrix} \begin{pmatrix} -1 & 0 & 0 \\ 0 & -1 & 0 \\ 0 & 0 & 1 \end{pmatrix} = \begin{pmatrix} 1 & 0 & 0 \\ 0 & 1 & 0 \\ 0 & 0 & -1 \end{pmatrix}. \tag{7.1.3}$$

Meanwhile, the induced orientation reversing involution τ_0 on S^2 can just as well be thought of as the reflection in the xy-plane. This is because, at the level of the 2-sphere, it is "the same thing as" the composition $\tau \circ J$. Thus the fixed circle of τ_0 is precisely the equator, *cf.* formula (7.3.3). Then one gets a quotient metric on S^2 which is roughly that of the western hemisphere, with the boundary longitude folded in two, *cf.* Proposition 8.4.2(2). The metric has little bulges along the z-axis at north and south poles, which are leftovers of the small handle.

7.2. Hyperelliptic surfaces

For a treatment of hyperelliptic surfaces, see [**Mir95**, p. 60-61]. By [**Mir95**, Proposition 4.11, p. 92], the affine part of a hyperelliptic surface Σ is defined by a suitable equation of the form

$$w^2 = f(z) \tag{7.2.1}$$

in $\mathbb{C}^2$, where f is a polynomial. On such an affine part, the map J is given by $J(z, w) = (z, -w)$, while the hyperelliptic quotient map $Q : \Sigma \to S^2$ is represented by the projection onto the z-coordinate in $\mathbb{C}^2$.

A slight technical problem here is that the map

$$\Sigma \to \mathbb{CP}^2, \tag{7.2.2}$$

whose image is the compactification of the solution set of (7.2.1), is not an imbedding. Indeed, there is only one point at infinity, given in homogeneous coordinates by $[0 : w : 0]$. This point is a singularity. A way of desingularizing it using an explicit change of coordinates at infinity is presented in [**Mir95**, p. 60-61]. The resulting smooth surface is unique [**DaS98**, Theorem, p. 100].

REMARK 7.2.1. To explain what happens "geometrically", note that there are two points on our affine surface "above infinity". This means that for a large circle $|z| = r$, there are two circles above it satisfying equation (7.2.1) where f has even degree $2g + 2$ (for a Weierstrass point we would only have one circle). To see this, consider $z = re^{ia}$. As the argument a varies from 0 to 2π, the argument of $f(z)$ will change by $(2g + 2)2\pi$. Thus, if $(re^{ia}, w(a))$ represents a continuous curve on our surface, then the argument of w changes by $(2g + 2)\pi$, and hence we end up where we started, and not at $-w$ (as would be the case were the polynomial of odd degree). Thus there are *two* circles on the surface over the circle $|z| = r$. We conclude that to obtain a smooth compact surface, we will need to add two points at infinity, *cf.* discussion around [**FarK92**, formula (7.4.1), p. 102].

Thus, the affine part of Σ, defined by equation (7.2.1), is a Riemann surface with a pair of punctures p_1 and p_2. A neighborhood of each p_i is conformally equivalent to a punctured disk. By replacing each punctured disk by a full one, we obtain the desired compact Riemann surface Σ. The point at infinity $[0 : w : 0] \in \mathbb{CP}^2$ is the image of both p_i under the map (7.2.2).

7.3. Ovalless surfaces

Denote by Σ^ι the fixed point set of an involution ι of a Riemann surface Σ. Let Σ be a hyperelliptic surface of even genus g. Let $J : \Sigma \to \Sigma$ be the hyperelliptic involution, *cf.* Definition 7.1.1. Let $\tau : \Sigma \to \Sigma$ be a fixed point free, antiholomorphic involution.

LEMMA 7.3.1. *The involution τ commutes with J and descends to S^2. The induced involution $\tau_0 : S^2 \to S^2$ is an inversion in a circle $C_0 = Q(\Sigma^{\tau\circ J})$. The set of ramification points is invariant under the action of τ_0 on S^2.*

PROOF. By the uniqueness of J, *cf.* Remark 7.1.3, we have the commutation relation

$$\tau \circ J = J \circ \tau, \tag{7.3.1}$$

cf. relation (7.1.3). Therefore τ descends to an involution τ_0 on the sphere. There are two possibilities, namely, τ is conjugate, in the group of fractional linear transformations, either to the map $z \mapsto \bar{z}$, or to the map $z \mapsto -\frac{1}{\bar{z}}$. In the latter case, τ is conjugate to the antipodal map of S^2.

In the case of even genus, there is an odd number of Weierstrass points in a hemisphere. Hence the inverse image of a great circle is a connected loop. Thus we get an action of $\mathbb{Z}_2 \times \mathbb{Z}_2$ on a loop, resulting in a contradiction.

In more detail, the set of the $2g+2$ ramification points on S^2 is centrally symmetric. Since there is an odd number, $g+1$, of ramification points in a hemisphere, a generic great circle $A \subset S^2$ has the property that its inverse image $Q^{-1}(A) \subset \Sigma$ is connected. Thus both involutions τ and J, as well as $\tau \circ J$, act fixed point freely on the loop $Q^{-1}(A) \subset \Sigma$, which is impossible. Therefore τ_0 must fix a point. It follows that τ_0 is an inversion in a circle. □

Suppose a hyperelliptic Riemann surface Σ admits an antiholomorphic involution τ. In the literature, the components of the fixed point set Σ^τ of τ are sometimes referred to as "ovals". When τ is fixed point free, we introduce the following terminology.

DEFINITION 7.3.2. A hyperelliptic surface (Σ, J) of even positive genus $g > 0$ is called *ovalless real* if one of the following equivalent conditions is satisfied:

(1) Σ admits an imaginary reflection, *i.e.* a fixed point free, antiholomorphic involution τ;
(2) the affine part of Σ is the locus in $\mathbb{C}^2$ of the equation

$$w^2 = -P(z), \tag{7.3.2}$$

where P is a monic polynomial, of degree $2g+2$, with real coefficients, no real roots, and with distinct roots.

LEMMA 7.3.3. *The two ovalless reality conditions of Definition 7.3.2 are equivalent.*

PROOF. A related result appears in [**GroH81**, p. 170, Proposition 6.1(2)]. To prove the implication (2) $\Longrightarrow$ (1), note that complex conjugation leaves the equation invariant, and therefore it also leaves invariant the locus of (7.3.2). A fixed point must be real, but P is positive hence (7.3.2) has no real solutions. There is no real solution at infinity, either, as there are two points at infinity which are not Weierstrass points, since P is of even degree, as discussed in Remark 7.2.1. The desired imaginary reflection τ switches the two points at infinity, and, on the affine part of the Riemann surface, coincides with complex conjugation $(z,w) \mapsto (\bar{z}, \bar{w})$ in $\mathbb{C}^2$.

To prove the implication (1) $\Longrightarrow$ (2), note that by Lemma 7.3.1, the induced involution τ_0 on $S^2 = \Sigma/J$ may be thought of as complex conjugation, by choosing the fixed circle of τ_0 to be the circle

$$\mathbb{R} \cup \{\infty\} \subset \mathbb{C} \cup \{\infty\} = S^2. \tag{7.3.3}$$

Since the surface is hyperelliptic, it is the smooth completion of the locus in $\mathbb{C}^2$ of *some* equation of the form (7.3.2), *cf.* (7.2.1). Here P is of degree $2g+2$ with distinct roots, but otherwise to be determined. The set of roots of P is the set of (the z-coordinates of) the Weierstrass points. Hence the set of roots must be invariant under τ_0. Thus the roots of the polynomial either come in conjugate pairs, or else are real. Therefore P has real coefficients. Furthermore, the leading cofficient of P may be absorbed into the w-coordinate by extracting a square root. Here we may have to rotate w by i, but at any rate the coefficients of P remain real, and thus P can be assumed monic.

If P had a real root, there would be a ramification point fixed by τ_0. But then the corresponding Weierstrass point must be fixed by τ, as well! This contradicts the fixed point freeness of τ. Thus all roots of P must come in conjugate pairs. □

7.4. Katok's entropy inequality

Let $(\Sigma, \mathcal{G})$ be a closed surface with a Riemannian metric. Denote by $(\tilde{\Sigma}, \tilde{\mathcal{G}})$ the universal Riemannian cover of $(\Sigma, \mathcal{G})$. Choose a point $\tilde{x}_0 \in \tilde{\Sigma}$.

DEFINITION 7.4.1. The volume entropy (or asymptotic volume) $h(M, \mathcal{G})$ of a surface $(\Sigma, \mathcal{G})$ is defined by setting

$$h(\Sigma, \mathcal{G}) = \lim_{R \to +\infty} \frac{\log\left(\mathrm{vol}_{\tilde{\mathcal{G}}} B(\tilde{x}_0, R)\right)}{R}, \tag{7.4.1}$$

where $\mathrm{vol}_{\tilde{\mathcal{G}}} B(\tilde{x}_0, R)$ is the volume (area) of the ball of radius R centered at $\tilde{x}_0 \in \tilde{\Sigma}$.

Since Σ is compact, the limit in (7.4.1) exists, and does not depend on the point $\tilde{x}_0 \in \tilde{\Sigma}$ [**Man79**]. This asymptotic invariant describes the exponential growth rate of the volume in the universal cover.

DEFINITION 7.4.2. The minimal volume entropy, MinEnt, of Σ is the infimum of the volume entropy of metrics of unit volume on Σ, or equivalently

$$\mathrm{MinEnt}(\Sigma) = \inf_{\mathcal{G}} h(\Sigma, \mathcal{G}) \, \mathrm{vol}(\Sigma, \mathcal{G})^{\frac{1}{2}} \tag{7.4.2}$$

where $\mathcal{G}$ runs over the space of all metrics on Σ. For an n-dimensional manifold in place of Σ, one defines MinEnt similarly, by replacing the exponent of vol by $\frac{1}{n}$.

The classical result of A. Katok [**Kat83**] states that every metric $\mathcal{G}$ on a closed surface Σ with negative Euler characteristic $\chi(\Sigma)$ satisfies the optimal inequality

$$h(\mathcal{G})^2 \geq \frac{2\pi|\chi(\Sigma)|}{\operatorname{area}(\mathcal{G})}. \tag{7.4.3}$$

Inequality (7.4.3) also holds for hom ent$(\mathcal{G})$ [**Kat83**], as well as the topological entropy, since the volume entropy bounds from below the topological entropy (see [**Man79**]). We recall the following well-known fact, *cf.* [**KatH95**, Proposition 9.6.6, p. 374].

LEMMA 7.4.3. *Let $(M, \mathcal{G})$ be a closed Riemannian manifold. Then,*

$$h(M, \mathcal{G}) = \lim_{T\to+\infty} \frac{\log(P'(T))}{T} \tag{7.4.4}$$

where $P'(T)$ is the number of homotopy classes of loops based at some fixed point x_0 which can be represented by loops of length at most T.

PROOF. Let $x_0 \in M$ and choose a lift $\tilde{x}_0 \in \tilde{M}$. The group

$$\Gamma := \pi_1(M, x_0)$$

acts on $\tilde{M}$ by isometries. The orbit of $\tilde{x}_0$ under Γ is denoted $\Gamma.\tilde{x}_0$. Consider a fundamental domain Δ for the action of Γ, containing $\tilde{x}_0$. Denote by D the diameter of Δ. The cardinal of $\Gamma.\tilde{x}_0 \cap B(\tilde{x}_0, R)$ is bounded from above by the number of translated fundamental domains $\gamma.\Delta$, where $\gamma \in \Gamma$, contained in $B(\tilde{x}_0, R + D)$. It is also bounded from below by the number of translated fundamental domains $\gamma.\Delta$ contained in $B(\tilde{x}_0, R)$. Therefore, we have

$$\frac{\operatorname{vol}(B(\tilde{x}_0, R))}{\operatorname{vol}(M, \mathcal{G})} \leq \operatorname{card}(\Gamma.\tilde{x}_0 \cap B(\tilde{x}_0, R)) \leq \frac{\operatorname{vol}(B(x_0, R + D))}{\operatorname{vol}(M, \mathcal{G})}. \tag{7.4.5}$$

Take the log of these terms and divide by R. The lower bound becomes

$$\begin{aligned} &\frac{1}{R}\log\left(\frac{\operatorname{vol}(B(R))}{\operatorname{vol}(\mathcal{G})}\right) = \\ &= \frac{1}{R}\log(\operatorname{vol}(B(R))) - \frac{1}{R}\log(\operatorname{vol}(\mathcal{G})), \end{aligned} \tag{7.4.6}$$

and the upper bound becomes

$$\begin{aligned} &\frac{1}{R}\log\left(\frac{\operatorname{vol}(B(R+D))}{\operatorname{vol}(\mathcal{G})}\right) = \\ &= \frac{R+D}{R}\frac{1}{R+D}\log(\operatorname{vol}(B(R+D))) - \frac{1}{R}\log(\operatorname{vol}(\mathcal{G})). \end{aligned} \tag{7.4.7}$$

Hence both bounds tend to $h(\mathcal{G})$ when R goes to infinity. Therefore,

$$h(\mathcal{G}) = \lim_{R\to+\infty} \frac{1}{R}\log(\operatorname{card}(\Gamma.\tilde{x}_0 \cap B(\tilde{x}_0, R))). \tag{7.4.8}$$

This yields the result since $P'(R) = \operatorname{card}(\Gamma.\tilde{x}_0 \cap B(\tilde{x}_0, R))$. □

CHAPTER 8

Filling area theorem for hyperelliptic surfaces

8.1. To fill a circle: an introduction

Consider a compact, connected, orientable manifold N of dimension $n \geq 1$ with a distance function d. Here d is not necessarily Riemannian. The notion of the Filling Volume, $\mathrm{FillVol}(N^n, d)$, of N was introduced in [**Gro83**]. We won't present the general definition here, but see Section 12.2. An alternative definition results from the observation that, as shown in [**Gro83**], when $n \geq 2$, we have

$$\mathrm{FillVol}(N^n, d) = \inf_{\mathcal{G}} \mathrm{vol}(X^{n+1}, \mathcal{G}) \tag{8.1.1}$$

where X is any fixed manifold such that $\partial X = N$. Here one can even take a cylinder $X = N \times [0, \infty)$. The infimum is taken over all complete Riemannian metrics $\mathcal{G}$ on X for which the boundary distance function is bounded below by d, *i.e.* the length of the shortest path in X between boundary points p and q is at least $d(p, q)$.

In the case $n = 1$, the topology of the filling X^2 can affect the infimum in (8.1.1), as is shown by example in [**Gro83**, Counterexamples 2.2.B].

The precise value of the filling volume is not known for any Riemannian metric. The following values for the canonical metrics on the spheres (of sectional curvature $+1$) are conjectured in [**Gro83**], immediately following Proposition 2.2.A.

CONJECTURE 8.1.1. $\mathrm{FillVol}(S^n, can) = \frac{1}{2}\omega_{n+1}$, *where* ω_{n+1} *represents the volume of the unit* $(n+1)$*-sphere.*

This conjecture is still open in all dimensions. The case $n = 1$ is known as the *filling area conjecture.* It can be broken up into separate problems depending on the genus of the filling X used in the infimum (8.1.1). The filling area of the circle of length 2π with respect to simply connected fillings (*i.e.* by a disk) is indeed 2π, by Pu's inequality (5.2.2), applied to the real projective plane obtained by identifying all pairs of opposite points of the circle. Here one may need to smooth the resulting metric, with an arbitrarily small change in the two invariants involved.

In Corollary 8.1.6 below, we prove the corresponding result when the filling is by a surface of genus one. These results originally appeared in [**BaCIK05**].

REMARK 8.1.2. Consider a filling of the circle by an orientable surface Σ'' of positive genus. Thus $\partial\Sigma'' = S^1$. We apply the same technique of glueing antipodal points of the boundary together to form a nonorientable surface Σ' (whose metric might again need to be smoothed as before). We are thus reduced to proving a conjectured relative version of Pu's inequality (5.2.2), see discussion around formula (8.2.3). This conjecture is most easily stated by passing to the orientable double cover Σ of Σ' with deck transformation $\tau : \Sigma \to \Sigma$, as follows.

CONJECTURE 8.1.3. *Let Σ be an orientable surface of even genus with a Riemannian metric $\mathcal{G}$ that admits an orientation reversing, isometric involution τ. Then there is a point $p \in \Sigma$ with*

$$\frac{\mathrm{dist}_{\mathcal{G}}(p, \tau(p))^2}{\mathrm{area}(\mathcal{G})} \leq \frac{\pi}{4}.$$

REMARK 8.1.4. The analogous conjecture is false in odd genus, as pointed out by H. Parlier [**Par04**].

The original proof of Pu's inequality employed conformal techniques (namely, uniformisation), combined with an application of an integral-geometric identity, as in Section 6.5. For another proof, see [**Iv02**]. It seems reasonable also to try to apply conformal techniques to the proof of the conjectured relative version of Pu's inequality. This works well in the hyperelliptic case, *cf.* Definition 7.1.1.

Our result about hyperelliptic surfaces is the following. Given a circle $C_0 \subset S^2$ and a point $w_0 \in S^2 \setminus C_0$, we can consider the orbifold metric with equator C_0 and poles at w_0 and at the image of w_0 under inversion in C_0, *cf.* Section 8.4. Given a hyperelliptic surface $Q : \Sigma \to S^2$ together with a circle $C \subset \Sigma$ double covering a circle $C_0 \subset S^2$, and a point $w \in \Sigma \setminus C$, denote by

$$AF(C, w) \tag{8.1.2}$$

the pullback to Σ of the orbifold metric for $C_0 = Q(C)$ and $w_0 = Q(w)$. Given an involution $\iota : \Sigma \to \Sigma$ fixing a circle, denote by $\Sigma^\iota \subset \Sigma$ its fixed circle.

A closed oriented Riemann surface Σ is called *ovalless real* if it admits a fixed point free, antiholomorphic involution τ, see Section 7.3 for a more detailed discussion.

THEOREM 8.1.5 (Optimal displacement bound). *Let (Σ, τ, J) be an ovalless real hyperelliptic surface of even genus g, and $\mathcal{G}$ a Riemannian metric in its conformal class. Then there is a point $p \in \Sigma$ with*

$$\frac{\mathrm{dist}_{\mathcal{G}}(p, \tau(p))^2}{\mathrm{area}(\mathcal{G})} \leq \frac{\pi}{4}. \tag{8.1.3}$$

Specifically, there exists a curve joining p and $\tau(p)$, of length at most $\left(\frac{\pi}{4}\,\mathrm{area}(\mathcal{G})\right)^{1/2}$, which consists of arcs of at most $g+1$ special curves γ_i. Each of the γ_i is a geodesic of the singular constant (positive) curvature metric

$$AF(\Sigma^{\tau \circ J}, w_i),$$

where w_i is one of the Weierstrass points, while the fixed point set $\Sigma^{\tau\circ J}$ of the involution $\tau \circ J$ is a circle.

Recall that every genus 2 surface is hyperelliptic [**FarK92**, Proposition III.7.2, page 100]. We therefore obtain the following corollary.

COROLLARY 8.1.6. *All orientable genus* 1 *fillings of the circle satisfy the filling area conjecture.*

Let us briefly review related results. For a precise calculation of a related invariant called the *filling radius*, see [**Katz83, KatzL05**] and Definition 12.2.3 below. In Chapter 9 we show that all hyperelliptic surfaces satisfy the Loewner inequality (5.4.1). Furthermore, in Chapter 11 we prove that all surfaces of genus at least 20 satisfy (5.4.1). An optimal inequality for CAT(0) metrics on the genus 2 surface is proved in Chapter 10.

The relation of our theorem to Pu's inequality is discussed in Section 8.2. The two-step proof of Theorem 8.1.5 is sketched in Section 8.3. An orbifold metric on the sphere that plays a key role in the proof is described in Section 8.4. The key integral-geometric ingredient of the proof appears in Section 8.5. The two steps appear, respectively, in Section 8.6 and Section 8.7.

8.2. Relative Pu's way

While the main focus of the present Chapter is on Riemann surfaces, we briefly discuss a link, *via* a generalisation of Pu's inequality, to systolic bounds in higher-dimensional manifolds. Namely, in Chapter 18, we study certain optimal systolic inequalities for a Riemannian manifold $(M, \mathcal{G})$, depending on a pair of parameters, n and b. Here n is the dimension of M, while b is its first Betti number. The proof of the inequalities involves streamlining and generalizing the techniques pioneered in [**BuraI94, BuraI95**] and [**Gro99**, pp. 259-260] (*cf.* [**CrK03**, inequality (5.14)]), resulting in a construction of suitable Abel-Jacobi maps, *cf.* (16.5.3), from M to its Jacobi torus $\mathbb{T}^b$. The maps we construct are area-decreasing, with respect to suitable norms on the torus.

The study of the metrics attaining the optimal bound in the particular case when $n = b + 2$ depends on the filling area conjecture and/or a suitable generalisation of Pu's inequality (5.2.2), as we now discuss.

DEFINITION 8.2.1. Let Σ' be a nonorientable closed surface. Let

$$\phi : \pi_1(\Sigma') \to \mathbb{Z}_2$$

be an epimorphism, corresponding to a map $\hat{\phi} : \Sigma' \to \mathbb{RP}^2$ of absolute degree $+1$, *cf.* [**BroBS01**]. We define the "1-systole relative to ϕ", denoted $\operatorname{sys}_1(\Sigma', \phi, \mathcal{G})$, of a metric $\mathcal{G}$ on Σ', by minimizing length over loops γ which are not in the kernel of ϕ:

$$\operatorname{sys}_1(\Sigma', \phi, \mathcal{G}) = \min\left\{\operatorname{length}(\gamma) \mid \phi([\gamma]) \neq 0 \in \mathbb{Z}_2\right\} \tag{8.2.1}$$

The following question can be thought of as a relative version of Pu's inequality, *cf.* [**CrK03**, conjecture 2.7].

QUESTION 8.2.2. Does every nonorientable surface Σ' and continuous map $\hat{\phi} : \Sigma' \to \mathbb{RP}^2$ of absolute degree one, satisfy the inequality

$$\operatorname{sys}_1(\Sigma', \phi, \mathcal{G})^2 \leq \tfrac{\pi}{2} \operatorname{area}(\mathcal{G}) \tag{8.2.2}$$

for all metrics $\mathcal{G}$ on Σ'?

We let

$$\sigma_2 = \sup_{(\Sigma', \phi, \mathcal{G})} \frac{\operatorname{sys}_1(\Sigma', \phi, \mathcal{G})^2}{\operatorname{area}(\mathcal{G})}, \tag{8.2.3}$$

where the supremum is over all triples $(\Sigma', \phi, \mathcal{G})$ as in Definition 8.2.1. Thus we ask whether $\sigma_2 = \frac{\pi}{2}$. The answer is affirmative in the class of metrics $\mathcal{G}$ whose underlying conformal structure (on the associated orientable surface) is hyperelliptic, *cf.* Theorem 8.1.5.

Consider a genus one filling Σ'' of the circle S^1, the associated nonorientable surface Σ', and its orientable double cover Σ, as in Remark 8.1.2. Then Σ' is diffeomorphic to the connected sum $3\mathbb{RP}^2$ of 3 copies of the real projective plane.

A path joining a pair of opposite points of $S^1 = \partial\Sigma''$ corresponds to a 1-sided loop in Σ', and thus defines an element of $\pi_1(\Sigma')$ outside the subgroup $\pi_1(\Sigma) \subset \pi_1(\Sigma')$. We can therefore restate Corollary 8.1.6 as follows.

THEOREM 8.2.3. *Consider the homomorphism* $\phi : \pi_1(3\mathbb{RP}^2) \to \mathbb{Z}_2$ *whose kernel is the fundamental group of the orientable double cover of* $3\mathbb{RP}^2$. *Then the relative systole* $\operatorname{sys}_1(3\mathbb{RP}^2, \phi, \mathcal{G})$ *satisfies relative Pu's inequality* (8.2.2).

An analogue for higher dimensional manifolds of the relative 1-systolic ratio was studied in [**Bab93, Bab04, Ba06**].

It is shown in Chapter 17 that the quantity σ_2 of (8.2.3) satisfies $\sigma_2 \in [\frac{\pi}{2}, 2]$, *cf.* Proposition 17.3.1. Furthermore, a theorem is proved in Chapter 17, which in the case of the manifold $\mathbb{RP}^2 \times \mathbb{T}^b$ states that every metric $\mathcal{G}$ on $\mathbb{RP}^2 \times \mathbb{T}^b$ satisfies the following inequality:

$$\operatorname{stsys}_1(\mathcal{G})^b \operatorname{sys}\pi_1(\mathcal{G})^2 \leq \sigma_2 \gamma_b^{\frac{b}{2}} \operatorname{vol}_{b+2}(\mathcal{G}), \tag{8.2.4}$$

where σ_2 is the optimal systolic ratio from (8.2.3).

Answering Question 8.2.2 in the affirmative would allow one to characterize the metrics which attain the optimal bound (8.2.4). Our Theorem 8.1.5 can be thought of as a partial result in this direction. Calculating σ_2 depends on calculating the filling area of the Riemannian circle. Removing the hyperellipticity assumption of Theorem 8.1.5 would amount to proving the filling area conjecture [**Gro83**], as explained in Section 8.1.

8.3. Outline of proof of optimal displacement bound

We will prove Theorem 8.1.5 in a way modeled on Pu's proof, *cf.* Section 6.5. The idea is to obtain estimates for an arbitrary metric in the conformal class of a hyperelliptic surface, by applying an integral-geometric argument to the quotient metric (by the hyperelliptic involution) on the 2-sphere. We start with a metric in the conformal class of an ovalless real hyperelliptic surface (Σ, τ, J) of even genus g and area A, and then proceed as follows.

Step 1: In Lemma 8.6.1, we will show that it is sufficient to assume that the commuting involutions J and τ are isometries of our metric. Then there is an induced metric $\mathcal{G}_1 = f^2(x)\mathcal{G}_0$ of area $A/2$ on S^2 (here $f(x)$ is a nonnegative square integrable function) which pulls back (away from the Weierstrass points) to the J-invariant metric on Σ. The involution τ induces an orientation reversing isometry, denoted τ_0, on S^2 which can be assumed to be the map that leaves the equator fixed and exchanges the northern and southern hemispheres. Clearly the involution τ_0 maps ramification points to ramification points, and hence there are $g+1$ of them in each hemisphere.

Step 2: There is a loop γ on S^2 of $\mathcal{G}_1$-length $\leq \sqrt{\frac{\pi}{4}A}$ through a point p on the equator (hence p is fixed by τ_0) such that a lift of γ starting at a preimage, $\tilde{p}$, of p back on Σ is not closed.

We call a loop whose $\mathcal{G}_1$-length $\leq \sqrt{\frac{\pi}{4}A}$ an *optimally short loop.* The loop we find lies in a hemisphere. In fact, we find a loop with a lift that is not closed, *i.e.* the sum of the winding numbers about the ramification points in the hemisphere is odd.

Note that this proves the theorem, since the endpoints of such a lift would have to be $\tilde{p}$ and $\tau(\tilde{p})$. INdeed, the fact that p is fixed by τ_0 means that τ (which has no fixed points) must exchange the two (distinct) preimages of p.

8.4. Near optimal surfaces and the football

It is often helpful, in the proof of sharp inequalities, to keep in mind how the argument applies to spaces that are optimal or near optimal. We present such a discusson below, and also find some optimally short loops on such optimal surfaces.

Denote by $\bar{\mathcal{G}}_0$ the round metric of Gaussian curvature +1 on the 2-sphere. We view S^2 as a compactification $\mathbb{C} \cup \{\infty\}$ of $\mathbb{C}$, with the usual coordinates (x, y), where $x = \Re(z)$ and $y = \Im(z)$, for $z \in \mathbb{C}$.

THEOREM 8.4.1. *The metric $\mathcal{G}_0$ on $\mathbb{C}$ obtained as the pullback of the round metric of the sphere by the stereographic projection, can be written as follows:*

$$\mathcal{G}_0 = \frac{4}{(1+x^2+y^2)^2}\left(dx^2+dy^2\right). \tag{8.4.1}$$

PROOF. See [**Ri1854**, p. 292]. □

A special role in our argument is played by the simply connected, compact, 2-dimensional orbifold, denoted $AF(\pi, \pi)$, which we will refer to henceforth as an (American) football, see Figure 8.5.1, *cf.* [**Bor94**]. It is the compact orbifold of constant curvature with a pair of conical singularities, each with total angle π. Similarly to Riemann's formula (8.4.1), we would like to express the metric in terms of a (possibly singular) conformal factor on $\mathbb{C}$. We will denote the metric $\mathcal{G}_{AF}$.

PROPOSITION 8.4.2. *The football admits the following two equivalent descriptions:*

1. *The orbifold $AF(\pi, \pi)$ is obtained from a round hemisphere by folding its boundary in two, i.e. using an identification on the boundary circle which is defined by a reflection of the circle with 2 antipodal fixed points.*
2. *The rotation by π around the z-axis in $\mathbb{R}^3$ gives rise to a ramified double cover $D : (S^2, \bar{\mathcal{G}}_0) \to AF(\pi, \pi)$, which is a local isometry away from the poles.*

The proof is obvious. Note that, if the singular points are viewed as north and south poles, it is helpful to think of the hemisphere in terms of the *western* hemisphere.

The singular metric on the orbifold can be expressed in terms of a singular conformal factor as follows, *cf.* Remark 10.1.2.

PROPOSITION 8.4.3. *The orbifold metric $\mathcal{G}_{AF}$ can be expressed in terms of the standard metric by means of the following conformal factor:*

$$\mathcal{G}_{AF} = \frac{1+\mathcal{O}(r)}{4r}\mathcal{G}_0, \tag{8.4.2}$$

where $\mathcal{O}(r) \to 0$ as $r \to 0$, while $r = |z| = \sqrt{x^2+y^2}$. Thus we have a homeomorphism $\phi : AF(\pi, \pi) \to S^2$ which is a conformal diffeomorphism away from the two conical singularities of $AF(\pi, \pi)$.

PROOF. We work with the first definition of Proposition 8.4.2. We represent a hemisphere defining the orbifold as the upperhalf plane $\mathcal{H}^2 \subset \mathbb{C}$. Then the orbifold can be viewed as a quotient of $(\mathcal{H}^2, \mathcal{G}_0)$, where the positive and negative rays of the x-axis are identified. Denote by w the complex coordinate in $\mathcal{H}^2$.

By the uniqueness of the underlying conformal structure, there exists a conformal map $c : (\mathcal{H}^2, w) \to (S^2, z)$, identifying the orbifold with the 2-sphere, with

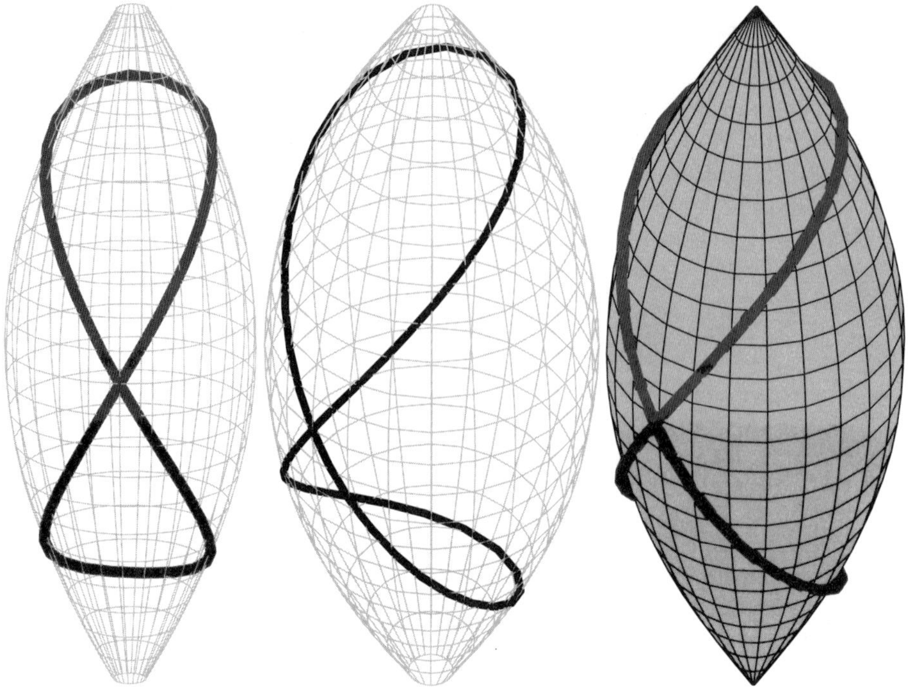

FIGURE 8.5.1. Figure eight geodesics on a football, *cf.* (8.4.2)

complex coordinate z. In our coordinates, we can represent c by the map $z = w^2$. Here the singular points of the orbifold correspond to 0 and ∞. Since $\frac{dc}{dw} = 2w$, the metric $\mathcal{G}_0$ pulls back under c to the quadratic differential $c^*(\mathcal{G}_0)(w) = \mu(|w|^2)\mathcal{G}_0$, where

$$\mu(t) = 4t\frac{(1+t)^2}{(1+t^2)^2} = 4t(1+\mathcal{O}(t)).$$

Note that $\mathcal{G}_0$ restricted to $\mathcal{H}^2$ is precisely the orbifold metric $c^*(\mathcal{G}_{AF})$ we are looking for. Thus

$$\begin{aligned} c^*(\mathcal{G}_{AF}) &= \frac{1}{\mu(|w|^2)} c^*(\mathcal{G}_0) \\ &= \frac{1}{\mu(|z|)} c^*(\mathcal{G}_0) \\ &= c^*\left(\frac{1+\mathcal{O}(|z|)}{4|z|}(\mathcal{G}_0)\right), \end{aligned}$$

as required. □

REMARK 8.4.4. The function $\frac{1}{\sqrt{r}}$ is locally square integrable in $\mathbb{C}$, which is important for our applications, *cf.* Proposition 6.4.2.

The football $AF(\pi, \pi)$ arises in the typical example where the associated systolic ratio is close to being optimal, see Example 7.1.6.

8.5. Finding a short figure eight geodesic

In our case, the optimal metric is not round, so it is natural to consider geodesics on the football $AF(\pi, \pi)$ instead of the round sphere. The branched double cover $D : S^2 \to AF(\pi, \pi)$ of Proposition 8.4.2(2) pulls back $\mathcal{G}_{AF}$ to the standard metric $\bar{\mathcal{G}}_0$ on S^2. This allows us easily to see the shape of the $\mathcal{G}_{AF}$-geodesics. They are of figure eight type with their point of self intersection on the equator (except the ones through the poles), see Figure 8.5.1. Note also that the loop in each of the northern and southern hemispheres winds around the pole precisely once.

Starting with a square integrable function $f(\sigma)$ on $AF(\pi, \pi)$, we can pull it back to S^2 and apply Proposition 6.4.2 to obtain the following bound.

LEMMA 8.5.1. *There is a $\mathcal{G}_{AF}$-geodesic γ of figure 8 type, satisfying the following estimate:*

$$\left(\int_\gamma f(\gamma(t))dt\right)^2 \leq \pi \int_{S^2} f^2(\sigma)d\sigma = 2\pi \operatorname{area}(f^2\mathcal{G}_{AF}).$$

The factor of 2 comes from the fact that the area of the pull back of the (possibly singular) metric $f^2\mathcal{G}_{AF}$ in S^2 (under the double cover $S^2 \to AF(\pi, \pi)$) has twice the area of the original metric. Now Lemma 8.5.1 implies the following. Assume that our function f is symmetric with respect to the inversion in the equator. Then with respect to the metric $f^2\mathcal{G}_{AF}$, the length L of each of the two hoops of the figure eight curve, satisfies the desired inequality $L \leq \sqrt{\frac{\pi}{2}A(f^2\mathcal{G}_{AF})}$. This inequality is sharp, since $A(f^2\mathcal{G}_{AF})$ is half the area A of the original hyperelliptic surface.

REMARK 8.5.2. An estimate equivalent to the above was proved half a century ago by J. Hersch [**Her55**, p. 412].

8.6. Proof of circle filling: Step 1

The lemma below allows us to assume that the hyperelliptic involution J and the anticonformal involution τ are isometries of the metric.

LEMMA 8.6.1. *Let F denote an angle preserving involution of Σ, and let $\bar{\mathcal{G}} = \frac{1}{2}\{\mathcal{G} + F^*\mathcal{G}\}$ be the averaged metric. Then for all $p, q \in \Sigma$ we have*

$$\frac{\operatorname{dist}_{\bar{\mathcal{G}}}(p,q)^2}{\operatorname{area}(\bar{\mathcal{G}})} \geq \frac{\left(\frac{1}{2}\{\operatorname{dist}_{\mathcal{G}}(p,q) + \operatorname{dist}_{\mathcal{G}}(F(p), F(q))\}\right)^2}{\operatorname{area}(\mathcal{G})}. \tag{8.6.1}$$

and hence if Theorem 8.1.5 holds for $\bar{\mathcal{G}}$ then it holds for $\mathcal{G}$, as well.

REMARK 8.6.2. Applying Lemma 8.6.1 to $F = \tau$ and $q = \tau(p)$, we obtain a τ-invariant metric $\mathcal{G}^-$, conformal to $\mathcal{G}$ in the sense of Definition 4.1.5, such that

$$\frac{\operatorname{dist}_{\mathcal{G}^-}(p,q)^2}{\operatorname{area}(\mathcal{G}^-)} \geq \frac{\operatorname{dist}_{\mathcal{G}}(p,q)^2}{\operatorname{area}(\mathcal{G})}.$$

If we apply Lemma 8.6.1 to $\mathcal{G}^-$ and $F = J$, $q = \tau(p)$, we obtain a J-invariant metric $\bar{\mathcal{G}}$, also conformal to $\mathcal{G}$. Now inequality (8.6.1) implies that if Theorem 8.1.5 holds for $\bar{\mathcal{G}}$, then it holds for $\mathcal{G}$ as well. Moreover, since J and τ commute, $\bar{\mathcal{G}}$ is also τ-invariant.

PROOF. Since F is an angle-preserving diffeomorphism, we have $F^*\mathcal{G} = f^2\mathcal{G}$ for some smooth positive function f. Then $\bar{\mathcal{G}} = \frac{1}{2}(f^2 + 1)\mathcal{G}$. (Note that a similar proof works also with $\bar{\mathcal{G}} = \left(\frac{1}{2}(f+1)\right)^2 \mathcal{G}$ as the average.) Let dx represent the volume element of $\mathcal{G}$. We have

$$\begin{aligned}\operatorname{area}(\bar{\mathcal{G}}) &= \int \frac{1}{2}(f^2(x)+1)dx \\ &= \frac{1}{2}\left(\int f^2(x)dx + \int 1dx\right) \\ &= \frac{1}{2}(\operatorname{area}(F^*(\mathcal{G})) + \operatorname{area}(\mathcal{G})) \\ &= \operatorname{area}(\mathcal{G}),\end{aligned}$$

and thus the denominator is unchanged. As for the numerator, start with any curve γ from p to q, then $F(\gamma)$ is a curve from $F(p)$ to the point $F(q)$. The length element for $\bar{\mathcal{G}}$ is $\sqrt{\frac{f^2+1}{2}} \geq \frac{f+1}{2}$, so that length of the curve γ for the averaged metric is bounded below by the average of the length of the curves γ and $F \circ \gamma$ for the original metric. Thus taking γ to be a $\bar{\mathcal{G}}$ minimizing path, we obtain

$$\begin{aligned}\operatorname{dist}_{\bar{\mathcal{G}}}(p,q) &= L_{\bar{\mathcal{G}}}(\gamma) \\ &\geq \frac{1}{2}(L_{\mathcal{G}}(\gamma) + L_{\mathcal{G}}(F\circ\gamma)) \\ &\geq \frac{1}{2}\{\operatorname{dist}_{\mathcal{G}}(p,q) + \operatorname{dist}_{\mathcal{G}}(F(p),(F(q)))\},\end{aligned}$$

proving inequality (8.6.1). □

Now if inequality (8.1.3) holds for $\bar{\mathcal{G}}$ then there is a point p such that $\frac{\pi}{4} \geq \frac{\operatorname{dist}_{\bar{\mathcal{G}}}(p,\tau(p))^2}{\operatorname{area}(\bar{\mathcal{G}})}$. Hence either $\frac{\pi}{4} \geq \frac{\operatorname{dist}_{\mathcal{G}}(p,\tau(p))^2}{\operatorname{area}(\mathcal{G})}$, or else we have

$$\frac{\pi}{4} \geq \frac{\operatorname{dist}_{\mathcal{G}}(F(p),\tau(F(p)))^2}{\operatorname{area}(\bar{\mathcal{G}})},$$

and the inequality follows for $\mathcal{G}$, as well. From now on we will assume that J and τ are isometries.

8.7. Proof of circle filling: Step 2

Let $C_0 = Q(\Sigma^{\tau\circ J}) \subset S^2$ be the fixed circle of the involution τ_0. Let $H \subset S^2$ be a connected component of $S^2 \setminus C_0$. Since the set of Weierstrass points is invariant under τ, there is an odd number of ramification points, namely $g+1$, in the "hemisphere" H.

We need to find an optimally short loop, *i.e.* a loop satisfying

$$L \leq \sqrt{\tfrac{\pi}{4}A},$$

based at a point on the "equator" C_0, such that the path lifting the loop to Σ does not close up. The path would then connect a pair of points of Σ which are an orbit of τ.

LEMMA 8.7.1. *For every ramification point $w \in H$, there is an optimally short loop (half of a figure 8 loop) in H with unit winding number around w.*

PROOF. By using a Möbius transformation if necessary, we can assume that C_0 is a great circle and our ramification point $w \in S^2$ is the polar point of the great

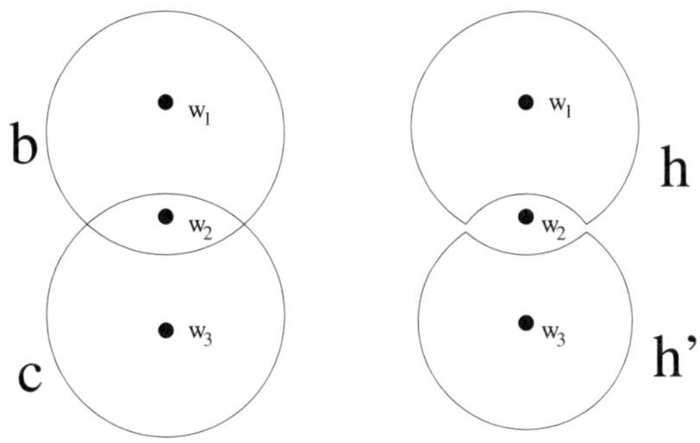

FIGURE 8.7.1. Loops b and c are even, while h, h' are odd

circle C_0, *i.e.* the "center" of the hemisphere H (note that we may need to take a different Möbius transformation for each w).

Recall that S^2 carries a metric induced from Σ. We then can use a conformal diffeomorphism from $AF(\pi, \pi)$ to S^2 which takes the conical singularities to the ramification points, *cf.* Proposition 8.4.2, to pull the metric back to $AF(\pi, \pi)$. Namely, we compare our surface Σ to $AF(\Sigma^{\tau\circ J}, w)$, *cf.* (8.1.2). Here the surface $AF(\Sigma^{\tau\circ J}, w)$ double covers the football $AF(\pi, \pi)$, while the circle $\Sigma^{\tau\circ J}$ double covers C_0 under Q. Recall that the conformal factor of our metric may be assumed invariant under τ.

Hence Lemma 8.5.1 produces an optimally short loop (half of a figure 8 curve) with unit winding number around w, based at a point of C_0.

By placing ourselves in a generic situation, we can assume that the loops we work with do not pass through ramification points, and thus the winding number is well defined. Namely, we assume that the geodesic we pick on $AF(\pi, \pi)$ is generic. This worsens the resulting estimate by an arbitrarily small $\epsilon > 0$. But if we prove that the optimal systolic ratio is within ϵ of $\pi/2$ for arbitrary ϵ, then it certainly follows that it equals $\pi/2$. □

If a loop produced by Lemma 8.7.1 encircles no other ramification points, then its lift to the ovalless real hyperelliptic Riemann surface (Σ, τ, J) connects a pair of points which form an orbit of τ, and we are done with Step 2. Otherwise we argue as follows.

Fix a finite set S of points in the hemisphere. In the end, S will be the set of ramification points, but the argument involves subsets, too. The loops involved in the argument are not assumed simple. Given a loop b, denote by $W_S(b)$ the set of points of S for which the winding number of b is odd. We will say that a set is odd if it contains an odd number of points.

PROPOSITION 8.7.2. *Let b and c be smooth optimally short loops in a disk, such that the set $W_S(c) \setminus W_S(b)$ is odd. Then there is an optimally short loop h (with the same basepoint as either b or c) such that $W_S(h)$ is odd.*

PROOF. The idea is to surger the loops b, c so as to obtain an odd loop h, *cf.* Figure 8.7.1. Consider a ramified double cover X of the disk, with ramification points at S. Given a loop, the number of ramification points with odd winding number is even if and only if the lift of the loop to the cover X, closes up.

Choose a fixed lift of the loop b to X, as well as a fixed lift of the loop c. Suppose both lifts are closed.

DEFINITION 8.7.3. The points of the intersection $b \cap c$ fall into two types: "crossroads", where their lifts intersect, as well; and "bridges", where the lifts do not intersect, *i.e.* lie in different sheets of the double cover X.

LEMMA 8.7.4. *The existence of a pair of intersection points of different types is sufficient to ensure the existence of an optimally short loop.*

PROOF. Indeed, if p and q are such points, consider the arc of b between p and q not containing the basepoint, and the similar arc of c. Interchanging these arcs produces a new pair of loops of the same combined length, whose lifts do not close up. Hence one of them is optimally short, as required. □

We continue with the proof of Proposition 8.7.2. We need the fact that a self-intersection of one of the loops remains a self-intersection after lifting. Suppose it does not. Then we have a sub-loop, not containing the basepoint, whose lift is not closed. Remove this sub-loop from the original loop. Then the remaining loop has the desired property: it is optimally short, and its lift to X is not closed.

To find a pair of intersection points of different types, to which we can apply Lemma 8.7.4, we argue by contradiction. Assume that they are all crossroads (if bridges, switch to the other lift of b in the double cover X). Thus, we have a (homeomorphic) lift of the union $b \cup c$ of the two loops. Hence every simple loop contained in this union lifts to a closed curve, and therefore encircles an even number of ramification points. Consider the graph Γ in the disk, defined by the union of the two loops. Let F_i be the faces (*i.e.* connected components of the complement) of Γ containing points from the complement $W_S(c) \setminus W_S(b)$. Consider the subgraph $\cup_i F_i$ of Γ, and note that

$$W_S(c) \setminus W_S(b) \subset \cup_i F_i. \tag{8.7.1}$$

Since $W_S(c) \setminus W_S(b)$ is odd, formula (8.7.1) implies that one of the faces, say F_{i_0}, must contain an odd number of points w_j from the complement $W_S(c) \setminus W_S(b)$. By construction, for each such $w_j \in F_{i_0}$, the loop c has odd winding number, while the loop b has even winding number. Any ramification point $p \in F_{i_0}$ can be connected to a w_j by a path disjoint from $b \cup c$. Hence the loops b and c have the same winding number with respect to p as with respect to w_j. Hence in fact we have $p \in W_S(c) \setminus W_S(b) \subset S$.

We conclude that F_{i_0} contains an odd number of ramification points. Hence the lift of its boundary does not close up. This contradicts the assumption that all intersection points are crossroads, and proves the proposition. □

LEMMA 8.7.5. *Let S be an odd pointset in the interior of a Riemannian disk. Consider a family of smooth simple loops $\gamma_1, \ldots, \gamma_k$ based at distinct points of the boundary. Let $L > 0$, and assume each loop is of length at most L, and that each point of S is in the interior of at least one of the loops. Then the union of the loops γ_i contains a subloop γ, also based at a point of the boundary and of length at most L, such that $W_S(\gamma)$ is odd.*

PROOF. Let S have $2n+1$ points. The proof is by induction on n. Let $b=\gamma_1$. If $W_S(b)$ is odd, the desired short loop is $\gamma=b$. Assume $W_S(b)$ is even. By the inductive hypothesis applied to the odd set

$$S_0 = S \setminus W_S(b) \tag{8.7.2}$$

in place of S, there is a short loop c such that $W_{S_0}(c)$ is odd. *A priori* the loop c might have odd winding number with respect to additional points in $W_S(b)$, but at any rate the equality

$$W_S(c) \setminus W_S(b) = W_{S_0}(c)$$

allows us to apply Lemma 8.7.2 to the pair of loops b and c, completing the proof of Lemma 8.7.5 and Step 2. □

CHAPTER 9

Hyperelliptic surfaces are Loewner

9.1. Hermite constant and Loewner surfaces

The systole, $\operatorname{sys}\pi_1(\mathcal{G})$, of a compact non simply connected Riemannian manifold $(M, \mathcal{G})$ is the least length of a noncontractible loop $\gamma \subset M$:

$$\operatorname{sys}\pi_1(\mathcal{G}) = \min_{[\gamma]\neq 0} \operatorname{length}(\gamma), \tag{9.1.1}$$

where $[\gamma] \subset \pi_1(M)$ denotes the conjugacy class containing γ. A more detailed discussion may be found in Chapter 5. We will be concerned with comparing this Riemannian invariant to the total area of the metric, as in the Loewner inequality (5.4.1) for the torus:

$$\operatorname{sys}\pi_1(\mathcal{G})^2 \leq \gamma_2 \operatorname{area}(\mathcal{G}).$$

Interesting extensions of the Loewner inequality appear in [**Amm04**]. A proof of the inequality in genus 20 or more appears below in Chapter 11. Higher dimensional optimal generalisations of the Loewner inequality are studied below in Chapters 15 through 19.

DEFINITION 9.1.1. We say that a metric $\mathcal{G}$ on a surface is *Loewner* if it satisfies the Loewner inequality (5.4.1).

QUESTION 9.1.2. It follows from Gromov's estimate (9.2.1) that every metric on an orientable surface Σ_g of genus g is Loewner whenever $g > 50$. This result is improved in Chapter 11 to $g \geq 20$. Can the genus assumption be removed altogether?

Note that a similar question for Pu's inequality (5.2.2) has an affirmative answer, *cf.* Theorem 5.2.1. We will prove the following result toward answering Question 9.1.2, *cf.* Theorem 9.3.1. These results were originally obtained in [**KatzS06a**].

THEOREM 9.1.3. *Every metric on an orientable surface is Loewner if it lies in a hyperelliptic conformal class. In particular, every metric on the genus 2 surface is Loewner.*

While the precise value of the systolic ratio in genus 2 is unknown in the class of all metrics, an optimal systolic inequality for CAT(0) metrics does exist (Chapter 10). The relevant literature and basic estimates are reviewed in Section 9.2. We will state the main theorem of this chapter in more detail in Section 9.3 and prove it for genus 3 or more. We will complete the proof in the genus 2 case in Section 9.4.

9.2. Basic estimates

DEFINITION 9.2.1. The *systolic ratio* of a metric $\mathcal{G}$ on a closed n-manifold M is defined as

$$\mathrm{SR}(M, \mathcal{G}) = \frac{\mathrm{sys}\pi_1(\mathcal{G})^n}{\mathrm{vol}_n(\mathcal{G})}.$$

The systolic ratio of a conformal class C of metrics on a closed n-manifold M, denoted $\mathrm{SR}(M, C)$, is defined as

$$\mathrm{SR}(M, C) = \sup \{\mathrm{SR}(\mathcal{G}) \mid \mathcal{G} \in C \text{ a metric on } M\}$$

(not to be confused with the conformal systole, *cf.* Definition 15.9.1). The supremum of systolic ratio over all the conformal structures of M is called the *optimal systolic ratio* of M, denoted $\mathrm{SR}(M)$.

M. Gromov [**Gro83**, p. 50] (*cf.* [**Kod87**, Theorem 4, part (1)]) proved a general estimate which implies that if Σ_g is a compact orientable surface of genus g with a Riemannian metric, then

$$\mathrm{SR}(\Sigma_g) < \frac{64}{4\sqrt{g} + 27}. \tag{9.2.1}$$

Thus the optimal systolic ratio tends to 0 as the genus increases without bound.

REMARK 9.2.2. It was shown in [**Gro83**] (see also [**Bab93**], as well as Chapter 11 and [**KatzSV07**]) that asymptotically the optimal systolic ratio of a surface of genus g behaves as $C\frac{(\log g)^2}{g}$, as $g \to \infty$.

9.3. Hyperelliptic surfaces and ε-regularity

Recall that a Riemann surface Σ is called hyperelliptic if it admits a degree 2 meromorphic function, *cf.* Section 7.1, as well as [**Mir95**, p. 60-61] and [**Mir95**, Proposition 4.11, p. 92]. The associated ramified double cover

$$Q : \Sigma \to S^2 \tag{9.3.1}$$

over the sphere S^2 is conformal, and has $2g + 2$ ramification points, where g is the genus of Σ. Its deck transformation $J : \Sigma \to \Sigma$ is called the *hyperelliptic involution*. Such a holomorphic involution, if it exists, is uniquely characterized by the property of having precisely $2g + 2$ fixed points. The fixed points of J are called Weierstrass points. Their images under the map Q of (9.3.1) will be referred to as *ramification points*.

THEOREM 9.3.1 (Hyperelliptic Loewner bound). *Let $(\Sigma, \mathcal{G})$ be an orientable surface, where the metric $\mathcal{G}$ belongs to a hyperelliptic conformal class. Then $(\Sigma, \mathcal{G})$ is Loewner.*

Recall that every genus 2 surface is hyperelliptic [**FarK92**, Proposition III.7.2, page 100]. We therefore obtain the following corollary.

COROLLARY 9.3.2. *Every metric on the genus 2 surface is Loewner.*

Note that this is the first improvement on Gromov's 3/4 bound of inequality (5.2.3) in over 20 years, for surfaces of genus below 50, *cf.* Question 9.1.2. No extremal metric has as yet been conjectured in this genus, but it cannot be flat with conical singularities [**Sab04**]. The best available lower bound for the optimal systolic ratio in genus 2 is described in Chapter 10, *cf.* [**CrK03**, section 2.2]. For genus $g \geq 3$, our Theorem 9.3.1 follows from Proposition 9.3.7, *cf.* Remark 9.2.2 and [**Kon03**]. Let us go over the definitions of [**Gro83**, 5.1].

DEFINITION 9.3.3. Given a Riemannian metric $\mathcal{G}$ on Σ, the *tension* of a noncontractible loop γ based at $x \in \Sigma$ is the upper bound of all $\delta > 0$ such that there exists a free homotopy of γ which diminishes the length of γ by δ. The tension is denoted $\mathrm{tens}_{\mathcal{G}}(\gamma)$.

DEFINITION 9.3.4. The height $h_{\mathcal{G}}(x)$ of $x \in \Sigma$ is defined as the lower bound of the tensions of the noncontractible loops based at x.

Note that the height function is 2-Lipschitz.

DEFINITION 9.3.5. A metric $\mathcal{G}$ is said to be ε-regular with $\varepsilon < \mathrm{sys}\pi_1(\mathcal{G})$ if its height function $h_{\mathcal{G}}$ is bounded from above by ε.

LEMMA 9.3.6. *Given any metric $\mathcal{G}$ in the conformal class of (Σ_g, J) and a real number $\delta > 0$, there exists an ε-regular and J-invariant metric $\bar{\mathcal{G}}$ in the same conformal class, whose systolic ratio is at least* $\mathrm{SR}(\mathcal{G}) - \delta$.

PROOF. The proof appears in [**Gro83**, 5.6.C″] in the general case, and proceeds by a (finite) sequence of modifications of the metric in suitable small disks, while staying in the same conformal class. Note that averaging the metric by J improves the systolic ratio and does not change the conformal class. Thus we can assume that $\mathcal{G}$ is J-invariant. To adapt the proof to our situation, we perform the modifications in a J-invariant way. □

PROPOSITION 9.3.7. *Every metric $\mathcal{G}$ in the conformal class of a hyperelliptic surface (Σ_g, J) of genus g satisfies the estimate*

$$\mathrm{SR}(\Sigma_g, \mathcal{G}) < \frac{4}{g+1}.$$

PROOF. By Lemma 9.3.6, we can assume that the metric $\mathcal{G}$ is J-invariant and ε-regular. Since the metric is J-invariant, the distance between any pair of the $2g+2$ Weierstrass points is at least $\frac{1}{2}\mathrm{sys}\pi_1(\mathcal{G})$. Thus, the disks of radius $R = \frac{1}{4}\mathrm{sys}\pi_1(\mathcal{G})$ centered at the Weierstrass points are disjoint. From [**Gro83**, 5.1.B] (see also [**Heb81**]), the area of such a disk on an ε-regular surface is at least

$$2R^2 - \varepsilon = \tfrac{1}{8}\ \mathrm{sys}\pi_1(\mathcal{G})^2 - \varepsilon,$$

cf. (5.2.4). Therefore, we have

$$\mathrm{area}(\mathcal{G}) \geq (2g+2)\left(\tfrac{1}{8}\ \mathrm{sys}\pi_1(\mathcal{G})^2 - \varepsilon\right).$$

Since this is true for all $\varepsilon > 0$, the proposition is proved. □

9.4. Proof of the genus two Loewner bound

Let Σ be a genus 2 surface. Recall that Σ has a hyperelliptic involution J with 6 Weierstrass points.

The idea of the proof of Theorem 9.3.1 in genus 2 is to apply the Loewner inequality (5.4.1) to certain *companion tori* of Σ, and to surger the resulting loops so as to obtain a Loewner loop on Σ. There is apparently no consensus regarding whether elliptic curves should be called hyperelliptic. At any rate, a torus admits a unique holomorphic involution, denoted $J_{\mathbb{T}^2}$, with 4 ramification points, and quotient surface S^2. We may need the following lemma.

LEMMA 9.4.1. *Let $\mathbb{T}^2$ be a torus endowed with a metric invariant under its involution $J_{\mathbb{T}^2}$, with conical singularities with total angle less than 2π around each. Then the image of a systolic loop of $\mathbb{T}^2$ in S^2 is a simple loop.*

PROOF. Let $\gamma \subset \mathbb{T}^2$ be a systolic, and therefore simple, loop. Since $J_{\mathbb{T}^2}$ induces minus the identity homomorphism on $\pi_1(\mathbb{T}^2)$, the loops γ and $-J_{\mathbb{T}^2}(\gamma)$ are homotopic. Under the hypotheses of our lemma, two homotopic systolic loops are necessarily disjoint. Hence the image of γ on S^2 is simple. □

DEFINITION 9.4.2. A *companion torus* $\mathbb{T}(a,b,c,d)$ of Σ is a torus whose ramification locus $\{a,b,c,d\} \subset S^2$ is a subset of the ramification locus of $Q : \Sigma \to S^2$.

PROOF OF THEOREM 9.3.1 FOR $g = 2$. As in the proof of Proposition 9.3.7, we can assume that the metric on Σ is invariant under J, *cf.* Chapter 8. Therefore $\mathcal{G}$ descends to a metric $\mathcal{G}_0$, of half the area, on S^2. Let's choose four of the 6 ramification points, say $a,b,c,d \in S^2$. Choose a double cover with ramification locus $\{a,b,c,d\}$, denoted

$$\mathbb{T}^2(a,b,c,d) \to S^2.$$

Pulling back the metric $\mathcal{G}_0$ to the torus $\mathbb{T}^2(a,b,c,d)$, we obtain a metric of the same area as the surface Σ itself. This metric on the torus is smooth away from the lifts to the torus of the two remaining ramification points. At the two lifts, the metric on the torus has a conical singularity with total angle π around each. Consider a Loewner loop

$$\bar{L} \subset \mathbb{T}^2(a,b,c,d)$$

on this torus, *e.g.* a systolic loop realizing (5.4.1). Let L be the projection of $\bar{L}$ to S^2.

DEFINITION 9.4.3. The simple loop $L \subset S^2$ separates the four points a,b,c,d into two pairs, say a,b on one side and c,d, on the other.

If the lift of L to Σ closes up, we obtain a Loewner loop on Σ and the theorem is proved. Thus, we may assume that the following three equivalent conditions are satisfied:

(1) the lift of L to Σ does not close up;
(2) the inverse image $Q^{-1}(L) \subset \Sigma$ under Q of (9.3.1) is connected;
(3) the loop L surrounds precisely 3 ramification points of $Q : \Sigma \to S^2$.

The last condition is equivalent to the first two since every based loop is homotopic, in the complement of the ramification points, to a composition of some loops from a finite collection of "standard" simple based loops, circling each of the ramification points. Meanwhile, going once around such a standard loop clearly switches the two sheets of the cover.

DEFINITION 9.4.4. The simple loop L partitions the sphere into two hemispheres, H_L^+ and H_L^-, with $a,b,e \in H_L^+$ and $c,d,f \in H_L^-$ where a,b,c,d,e,f are the 6 ramification points of $Q : \Sigma \to S^2$.

Using a pair of companion tori, we will construct two loops on the sphere, defining two distinct partitions of the ramification locus into a pair of triples.

EXAMPLE 9.4.5. Consider the case of a centrally symmetric 6-tuple of points, corresponding for instance to the completion of the complex curve $y^2 = x^5 - x$, and a pair of generic great circles, which cut the sphere into 4 digons, such that each of the four digons contains at least one ramification point.

We now construct a second companion torus $\mathbb{T}(a, b, e, f)$. Consider a Loewner loop

$$\bar{L}' \subset \mathbb{T}^2(a, b, e, f),$$

and its projection $L' \subset S^2$. If its lift to Σ closes up, the theorem is proved. Therefore assume that the lift of L' to Σ does not close up, *i.e.* L' surrounds exactly 3 ramification points. Now L' separates the four points a, b, e, f into two pairs. Hence L and L' define distinct splittings of the sixtuple of ramification points into two triples. The connected components of $L' \cap H_L^+$ form a nonempty finite collection of disjoint nonselfintersecting arcs $\alpha_i \subset H_L^+$.

Each arc α_i divides H_L^+ into a pair of regions homeomorphic to disks. Such regions are partially ordered by inclusion. A minimal element for the partial order is necessarily a digon. Such a digon must contain at least one ramification point of Q (otherwise exchange the two sides of the digon between the loops L and L', so as to decrease the total number of intersections, or else argue as in Lemma 9.4.1). It is clear that there are at least two such digons in H^+.

Hence one of them, denoted $D \subset H^+$, must contain precisely one of the 3 ramification points of H^+. We now exchange the two sides of D between the loops L, L', obtaining a pair of new loops M, M'. Each of the new loops surrounds a nonzero, *even* number of ramification points. Since

$$\text{length}(M) + \text{length}(M') = \text{length}(L) + \text{length}(L'),$$

the shorter one of the loops M or M' lifts to a Loewner loop on Σ. □

CHAPTER 10

An optimal inequality for CAT(0) metrics

10.1. Hyperelliptic surfaces of nonpositive curvature

Given a metric $\mathcal{G}$ on a surface, let $\mathrm{SR}(\mathcal{G})$ denote its systolic ratio

$$\mathrm{SR}(\mathcal{G}) = \frac{\mathrm{sys}\pi_1(\mathcal{G})^2}{\mathrm{area}(\mathcal{G})}.$$

Given a compact Riemann surface Σ, its *optimal systolic ratio* is defined by setting $\mathrm{SR}(\Sigma) = \sup_{\mathcal{G}} \mathrm{SR}(\mathcal{G})$, where the supremum is over all metrics in the conformal type of Σ. Finally, given a smooth compact surface M, its optimal systolic ratio is defined by setting $\mathrm{SR}(M) = \sup_{\Sigma} \mathrm{SR}(\Sigma)$, where the supremum is over all conformal structures Σ on M. The latter ratio has been computed for the Klein bottle, *cf.* (10.5.2), in addition to the torus and the real projective plane, *cf.* Chapter 5.

In the class of all metrics without any curvature restrictions, no singular flat metric on a surface of genus 2 can give the optimal systolic ratio in this genus [**Sab04**]. The best available upper bound for the systolic ratio of an arbitrary genus two surface is $\gamma_2 \simeq 1.1547$, *cf.* Chapter 9. The precise value of SR for the genus 2 surface has so far eluded researchers, *cf.* [**Cal96, Bry96**]. We propose an answer in the framework of negatively curved, or more generally, CAT(0) metrics. These results originally appeared in [**KatzS06b**].

The term "CAT(0) space" evokes an extension of the notion of a manifold of nonpositive curvature, to encompass singular spaces. Strictly speaking, it is the universal covering space that is usually referred to as a CAT(0) space, *cf.* [**BrH99**].

We will use the term to refer to surfaces with metrics with only mild quotient singularities, defined below. Here the condition of nonpositive curvature translates into a lower bound of 2π for the total angle at the singularity. We need such an extension so as to encompass the metric which attains our optimal bound (10.1.1).

A mild quotient singularity is defined as follows. Consider a smooth metric on $\mathbb{R}^2$. Let $q \geq 1$ be an integer. Consider the q-fold cover X_q of $\mathbb{R}^2 \setminus \{0\}$ with the induced metric. We compactify X_q in the neighborhood of the origin to obtain a complete metric space $X_q^c = X_q \cup \{0\}$.

DEFINITION 10.1.1. Suppose X_q^c admits an isometric action of the finite cyclic group $\mathbb{Z}_p$ fixing the origin. Then we can form the orbit space $Y_{p,q} = X_q^c/\mathbb{Z}_p$. The space $Y_{p,q}$ is then called mildly singular at the origin.

The total angle at the singularity is then $\frac{2\pi q}{p}$, and the CAT(0) condition translates into the inequality $\frac{q}{p} \geq 1$.

REMARK 10.1.2. Alternatively, a point is singular of total angle $2\pi(1+\beta)$ if the metric is of the form $e^{h(z)}|z|^{2\beta}|dz|^2$ in its neighborhood, where $|dz|^2 = dx^2 + dy^2$, *cf.* M. Troyanov [**Tro90**, p. 915].

THEOREM 10.1.3. *The systolic ratio of every* CAT(0) *metric* $\mathcal{G}$ *on a surface* Σ_2 *of genus two, satisfies the bound*

$$\mathrm{SR}(\Sigma_2, \mathcal{G}) \leq \tfrac{1}{3}\cot\tfrac{\pi}{8} = \tfrac{1}{3}\left(\sqrt{2}+1\right) = 0.8047\ldots \tag{10.1.1}$$

The optimal bound (10.1.1) *is attained by a singular flat metric, with 16 conical singularities, in the conformal class of the Bolza surface.*

The Bolza surface is described in Section 10.2. The optimal metric is described in more detail in Section 10.3. Theorem 10.1.3 is proved in Section 10.4 based on the octahedral triangulation of S^2.

REMARK 10.1.4. A similar optimal inequality can be proved for hyperelliptic surfaces of genus 5 based on the icosahedral triangulation, *cf.* [**Bav86**].

10.2. Distinguishing 16 points on the Bolza surface

The Bolza surface $\mathcal{B}$ is the smooth completion of the affine complex algebraic curve

$$y^2 = x^5 - x. \tag{10.2.1}$$

It is the unique Riemann surface of genus two with a group of holomorphic automorphisms of order 48. (A way of passing from an affine hyperelliptic surface to its smooth completion is described in [**Mir95**, p. 60-61].)

DEFINITION 10.2.1. A conformal involution J of a compact Riemann surface Σ of genus g is called *hyperelliptic* if J has precisely $2g+2$ fixed points. The fixed points of J are called the Weierstrass points of Σ.

See Chapter 7 for details. The quotient Riemann surface Σ/J is then necessarily the Riemann sphere, denoted henceforth S^2. Let $Q : \Sigma \to S^2$ be the conformal ramified double cover, with $2g+2$ branch points. Thus, J acts on Σ by sheet interchange. Recall that every surface of genus 2 is hyperelliptic, *i.e.* admits a hyperelliptic involution [**FarK92**, Proposition III.7.2].

Consider the regular octahedral triangulation of $S^2 = \mathbb{C} \cup \infty$. Its set of vertices is conformal to the set of roots of the polynomial $x^5 - x$ of formula (10.2.1), together with the unique point at infinity. Thus the six points in question can be thought of as the ramification points of the ramified conformal double cover $Q : \mathcal{B} \to S^2$. In other words, the x-coordinates of the ramification points are

$$\{0, \infty, 1, -1, i, -i\},$$

which stereographically correspond to the vertices of a regular inscribed octahedron. The conformal type therefore admits the symmetries of the cube. If one includes both the hyperelliptic involution, and the real (antiholomorphic) involution of $\mathcal{B}$ induced by the complex conjugation $(x,y) \to (\bar{x}, \bar{y})$ of $\mathbb{C}^2$, one obtains the full symmetry group $\mathrm{Aut}(\mathcal{B})$ of the Bolza surface, of order

$$|\,\mathrm{Aut}(\mathcal{B})| = 96, \tag{10.2.2}$$

see [**KuN95**, p. 404] for more details.

We make note of 16 special points on $\mathcal{B}$. We refer to a point as being special if it is a fixed point of an order 3 automorphism of $\mathcal{B}$. The 16 special points of $\mathcal{B}$ project to the 8 vertices of the cubical subdivision dual to the octahedral triangulation.

	Berger	Jenni	metric $\mathcal{G}(16, \frac{9\pi}{4})$ on Bolza
$\mathrm{SR}(\mathcal{G}) = \dfrac{\mathrm{sys}\pi_1(\mathcal{G})^2}{\mathrm{area}(\mathcal{G})}$	0.6666	0.7437	0.8047

FIGURE 10.3.1. Three systolic ratios in genus 2, see Remark 10.3.1

LEMMA 10.2.2. *The hyperbolic metric of $\mathcal{B}$ admits* 12 *systolic loops. The* 12 *loops are in one to one correspondence with the edges of the octahedral decomposition of S^2. The correspondence is given by taking the inverse image under $Q : \mathcal{B} \to S^2$ of an edge. The* 12 *systolic loops cut the surface up into* 16 *equilateral hyperbolic triangles. The centers of the triangles are the* 16 *special points.*

See P. Schmutz [**Sc93**, § 5] for further details. The Bolza surface is extremal for two distinct problems:

- systole of hyperbolic surfaces [**Bav92**], [**Sc93**, Theorem 5.2];
- conformal systole of Riemann surfaces [**BusS94**].

Note that the square of the conformal systole of a Riemann surface is also known as its Seshadri constant [**Kon03**]. A suitable singular metric on $\mathcal{B}$ is conjectured to be optimal for the first eigenvalue of the Laplacian [**JaP04, JaP05**]. The above evidence suggests that the systolically extremal surface may lie in the conformal class of $\mathcal{B}$, as well. Meanwhile, we have the following result, proved in Section 10.5.

THEOREM 10.2.3. *The Bolza surface $\mathcal{B}$ satisfies* $\mathrm{SR}(\mathcal{B}) \le \frac{\pi}{3}$.

Note that Theorems 10.2.3 and 10.1.3 imply that $\mathrm{SR}(\mathcal{B}) \in [0.8, 1.05]$.

10.3. A flat singular metric in genus two

The optimal systolic ratio of a genus 2 surface $(\Sigma_2, \mathcal{G})$ is unknown, but it satisfies the Loewner inequality, *cf.* Chapter 9. Here we discuss a *lower* bound for the optimal systolic ratio in genus 2, briefly described in [**CrK03**].

REMARK 10.3.1. The example of M. Berger (see [**Gro83**, Example 5.6.B′]) in genus 2 is a singular flat metric with conical singularities. It has a systolic ratio of 2/3. This ratio was improved by F. Jenni [**Je84**], who identified the hyperbolic genus 2 surface with the optimal systolic ratio among all hyperbolic genus 2 surfaces (see also C. Bavard [**Bav92**] and P. Schmutz [**Sc93**, Theorem 5.2]). The surface in question is a triangle surface of type (2,3,8). Its conformal class is that of the Bolza surface, *cf.* Section 10.2. It admits a regular hyperbolic octagon as a fundamental domain, and has 12 systolic loops of length $2x$, where $x = \cosh^{-1}(1+\sqrt{2})$. It has

$$\mathrm{sys}\pi_1 = 2\log\left(1+\sqrt{2}+\sqrt{2+2\sqrt{2}}\right),$$

area 4π, and systolic ratio $\mathrm{SR} \simeq 0.7437$. This ratio can be improved as follows (see the table of Figure 10.3.1).

PROPOSITION 10.3.2. *The conformal class of the Bolza surface $\mathcal{B}$ admits a metric, denoted $\mathcal{G}(16, \frac{9\pi}{4})$, with the following properties:*

(1) *the metric is singular flat, with conical singularities precisely at the* 16 *special points of Section* 10.2*;*
(2) *each singularity is of total angle $\frac{9\pi}{4}$, so that the metric $\mathcal{G}(16, \frac{9\pi}{4})$ is of type* CAT(0)*;*
(3) *the metric is glued from six flat regular octagons, centered on the Weierstrass points, while the 1-skeleton projects under $Q : \mathcal{B} \to S^2$ to that of the dual cube in S^2;*
(4) *the systolic ratio equals* $\mathrm{SR}\left(\mathcal{G}(16, \frac{9\pi}{4})\right) = \frac{1}{3}\left(\sqrt{2}+1\right) > \frac{4}{5}$.

PROOF. The octahedral triangulation of the sphere, *cf.* Section 10.2, lifts to a triangulation of $\mathcal{B}$ consisting of 16 triangles, which we think of as being "equilateral". Here 8 equilateral triangles are connected cyclically around each of the 6 Weierstrass vertices of the triangulation of $\mathcal{B}$.

We further subdivide each equilateral triangle into 3 isosceles triangles, with a common vertex at the center of the equilateral triangle. We equip each of the 48 isosceles triangles with a flat metric with obtuse angle $\frac{3\pi}{4}$.

Each of the 6 Weierstrass vertices of the original triangulation is a smooth point, since the total angle is $8\left(\frac{\pi}{4}\right) = 2\pi$. Each equilateral triangle possesses a singularity at the center with total angle $\frac{9\pi}{4} = 2\pi\left(1+\frac{1}{8}\right) > 2\pi$. Alternatively, we can apply the Gauss-Bonnet formula $\sum_\sigma \alpha(\sigma) = 2g-2$, in genus 2 with 16 isometric singularities. Here the sum is over all singularities σ of a singular flat metric on a surface of genus g, where the cone angle at singularity σ is $2\pi(1+\alpha(\sigma))$. Since the metric $\mathcal{G}(16, \frac{9\pi}{4})$ is smooth at a Weierstrass point of $\mathcal{B}$, the metric has only 16 singularities, precisely at the special points of Section 10.2, proving items 1 and 2 of the proposition.

Let x denote the sidelength of the equilateral triangle. The barycentric subdivision of each equilateral triangle consists of 6 copies of a flat right angle triangle, denoted $\measuredangle\left(\frac{x}{2}, \frac{\pi}{8}\right)$, with side $\frac{x}{2}$ and adjacent angle $\frac{\pi}{8}$. We thus obtain a decomposition of the metric $\mathcal{G}(16, \frac{9\pi}{4})$ into 96 copies of the triangle $\measuredangle\left(\frac{x}{2}, \frac{\pi}{8}\right)$, which can be thought of as a fundamental domain for the action of $\mathrm{Aut}(\mathcal{B})$, *cf.* (10.2.2).

We have $\mathrm{sys}\pi_1\left(\mathcal{G}(16, \frac{9\pi}{4})\right) = 2x$ by Proposition 10.3.3 below, proving item 4 of the proposition. To prove item 3, note that the union of the 16 triangles $\measuredangle\left(\frac{x}{2}, \frac{\pi}{8}\right)$ with a common Weierstrass vertex is a flat regular octagon. The latter is represented in Figure 10.3.2, together with the systolic loops passing through it. □

PROPOSITION 10.3.3. *The systole of the singular flat* CAT(0) *metric on the Bolza surface equals twice the distance between a pair of adjacent Weierstrass points.*

PROOF. Consider the smooth closed geodesic $\gamma \subset \mathcal{B}$ which is the inverse image $Q^{-1}(e)$, under the map $Q : \mathcal{B} \to S^2$, of an edge e of the octahedron, *cf.* Lemma 10.2.2. Let x be the distance between a pair of opposite sides of the regular flat octagon in $\mathcal{B}$, or equivalently, the distance between a pair of adjacent Weierstrass points. Thus, $\mathrm{length}(\gamma) = 2x$. Consider a loop $\alpha \subset \mathcal{B}$ whose length satisfies $\mathrm{length}(\alpha) < 2x$. We show below that if α is noncontractible, then it is freely homotopic to a closed geodesic γ. On the other hand, the geodesic γ is necessarily length minimizing in its free homotopy class, by the CAT(0) property of the metric [**BrH99**, Theorem 6.8]. □

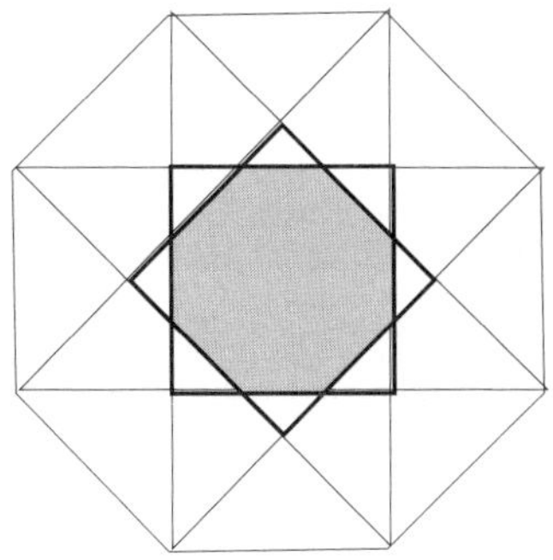

FIGURE 10.3.2. Flat regular octagon obtained as the union of 16 triangles $\triangle\left(\frac{x}{2}, \frac{\pi}{8}\right)$. The shaded interior octagon represents the region with 4 geodesic loops through every point

LEMMA 10.3.4. *A loop α of length*

$$\text{length}(\alpha) < 2x \tag{10.3.1}$$

is either contractible, or freely homotopic to one of the 12 geodesics $\gamma = Q^{-1}(e)$.

PROOF. Denote by $\mathcal{B}^{(1)}$ the graph on the surface $\mathcal{B}$ given by the inverse image under Q of the 1-skeleton of the cubical subdivision of S^2. The graph $\mathcal{B}^{(1)}$ partitions the surface into six regular octagons, denoted $\mathcal{O}_k$:

$$\mathcal{B} = \bigcup_{k=1}^{6} \mathcal{O}_k. \tag{10.3.2}$$

The boundary $\partial\mathcal{O}_{k_i}$ of each $\mathcal{O}_{k_i}$ is a polygonal loop consisting of 8 edges.

We will deform α to a loop $\beta \subset \mathcal{B}^{(1)}$, as follows. The partition (10.3.2) induces a partition of the loop α into arcs α_i, each lying in its respective octagon $\mathcal{O}_{k_i}$. The boundary points p_i, q_i of α_i split the eight-edge loop $\partial\mathcal{O}_{k_i} \subset \mathcal{B}^{(1)}$ into a pair of paths. Let

$$\beta_i \subset \partial\mathcal{O}_{k_i}$$

be a shorter of the two paths. Denote by y the distance between a pair of adjacent vertices of the octagon. Then clearly

$$\text{length}(\beta_i) \leq \frac{4y}{x}\text{length}(\alpha_i). \tag{10.3.3}$$

We first deform the loop α into the graph $\mathcal{B}^{(1)}$. The deformation fixes the intersection points $\alpha \cap \mathcal{B}^{(1)}$. Inside each $\mathcal{O}_{k_i}$, we deform the arc α_i to the path β_i. The length of the resulting loop is at most

$$\frac{4y}{x}\text{length}(\alpha) < 8y$$

by (10.3.1) and (10.3.3). Therefore, its homotopy class in $\mathcal{B}^{(1)}$ can be represented by an imbedded loop $\beta \subset \mathcal{B}^{(1)}$ of length at most $8y$. Thus, β contains fewer than 8 edges of $\mathcal{B}^{(1)}$. Since the number of edges must be even, its image under Q must retract to a circuit with at most six edges in the 1-skeleton $Q\left(\mathcal{B}^{(1)}\right)$ of the cubical subdivision of S^2.

If the number is 4, then the circuit lies in the boundary of a square face of the cube in S^2. But the boundary of a face does not lift to $\mathcal{B}$, since it surrounds a single ramification point, namely the center of the square face.

Hence there must be 6 edges in the circuit. There are two types of circuits with six edges in the 1-skeleton of the cubical subdivision of S^2:

(a) the boundary of the union $\square_A \cup \square_B$ of a pair of adjacent squares;
(b) a path consisting of the edges meeting a suitable great circle.

However, a path of type (b) surrounds an odd number, namely 3, of ramification points, and hence does not lift to the genus 2 surface. Meanwhile, a path of type (a) surrounds two ramification points, and hence does lift to the surface. Such a path is freely homotopic in $\mathcal{B}$ to the closed geodesic $\gamma = Q^{-1}(AB)$, where AB is the edge of the octahedron joining the centers of the squares $\square_A$ and $\square_B$. $\square$

10.4. Voronoi cells and Euler characteristic

The following proposition provides a preliminary lower bound on the area of hyperelliptic surfaces with nonpositive curvature.

PROPOSITION 10.4.1. *Every J-invariant* CAT(0) *metric $\mathcal{G}$ on a closed hyperelliptic surface Σ_g of genus g, satisfies the bound*

$$\mathrm{SR}(\Sigma_g, \mathcal{G}) \leq \frac{8}{(g+1)\pi}$$

PROOF. To prove this scale-invariant inequality, we normalize the metric on the surface $\Sigma = \Sigma_g$ to unit systole, *i.e.* $\mathrm{sys}\pi_1(\Sigma, \mathcal{G}) = 1$. The preimage by $Q : \Sigma \to S^2$ of an arc of S^2 joining two distinct branch points forms a noncontractible loop on Σ. Therefore, the distance between two Weierstrass points is at least $\frac{1}{2}\,\mathrm{sys}\pi_1(\Sigma, \mathcal{G}) = \frac{1}{2}$. Thus, we obtain $2g+2$ disjoint disks of radius $\frac{1}{4}$, centered at the Weierstrass points. Since the metric is CAT(0), the area of each disk is at least $\frac{\pi}{16}$. Thus, $\mathrm{area}(\Sigma, \mathcal{G}) \geq \frac{g+1}{8}\pi$. $\square$

An *optimal* lower bound requires a more precise estimate on the area of the Voronoi cells. The idea is to replace area of balls by area of polygons, where control over the number of sides is provided by the Euler characteristic, *cf.* [**Bav92**].

Denote by $u : \tilde{\Sigma} \to \Sigma$ its universal cover. Let $\{x_i \mid i \in \mathbb{N}\}$ be an enumeration of the lifts of Weierstrass points on $\tilde{\Sigma}$. The Voronoi cell $V_i \subset \tilde{\Sigma}$ centered at x_i is defined as the set of points closer to x_i than to any other lift of a Weierstrass point. In formulas,

$$V_i = \left\{ x \in \tilde{\Sigma} \,\middle|\, d(x, x_i) \leq d(x, x_j) \text{ for every } j \neq i \right\}. \tag{10.4.1}$$

The Voronoi cells on $\tilde{\Sigma}$ are polygons whose edges are arcs of the equidistant curves between a pair of lifts of Weierstrass points. Note that these edges are not necessarily geodesics. The Voronoi cells on $\tilde{\Sigma}$ are topological disks, while their projections $u(V_i) \subset \Sigma$ may have more complicated topology. Thus, the surface Σ decomposes into a union of $2g+2$ images of Voronoi cells, centered at the $2g+2$ Weierstrass points. By the number of sides of $u(V)$ we will mean the number of sides of the polygon V.

LEMMA 10.4.2. *Averaging the metric by the hyperelliptic involution does not take one out of the class of* CAT(0)*-metrics.*

PROOF. We check that CAT(0) metrics form a convex cone in the space of all metrics in a given conformal class. Thus, consider a metric $\mathcal{G}_0$ which is flat in a neighborhood, as well as another metric of type CAT(0), expressed by means of a conformal factor e^h with respect to the flat metric. Then the sum of the metrics, $(1+e^h)\mathcal{G}_0$, is also CAT(0). To see this, recall that the second metric is CAT(0) if and only if the function $h = h(x, y)$ is subharmonic. The sign of the curvature coincides with the sign of the expression $-\Delta \log(1+e^h)$, *cf.* (3.4.2).

Let $f(t) = \log(1+e^t)$. Then $f' > 0$ and $f'' > 0$. Let $g = f \circ h$. We obtain

$$\Delta g = f''(h)\|\nabla h\|^2 + f'(h)\Delta h. \tag{10.4.2}$$

If $\Delta h \geq 0$, then $\Delta g \geq 0$. It follows that the averaged metric is CAT(0), as well, if both points are regular. For singular points with positive angle excess, the CAT(0) property for the averaged metric is immediate from Remark 10.1.2. □

PROOF OF THEOREM 10.1.3. Since averaging by J can only improve the systolic ratio, we may assume without loss of generality that our metric is already J-invariant.

Consider the exponential map $\exp_A$ at a Weierstrass point $A \in \Sigma$. If A is a singular point, we lift to the appropriate smooth finite cover, and use its exponential map, *cf.* Definition 10.1.1.

Let $B \in \Sigma$ be another Weierstrass point. Consider its lift $B_0 = \exp_A^{-1}(B)$ to the tangent plane T_A. Consider the equidistant line

$$L_{(O,B_0)} \subset T_A$$

between the origin $O \in T_A$ and the point B_0.

By the Toponogov theorem applied to the hinge at O defined by the point B_0 and an arbitrary point $X \in L_{(O,B_0)}$ on the equidistant line, we have

$$\mathrm{dist}_\Sigma(A, \exp_A(X)) \leq \mathrm{dist}_\Sigma(B, \exp_A(X)). \tag{10.4.3}$$

Now consider the polygon in the tangent plane, obtained as the intersection of the halfplanes containing the origin, defined by the lines $L_{(O,B_0)}$, as B runs over all Weierstrass points. It follows from (10.4.3) that the exponential image of this flat polygon is contained in the Voronoi cell of A. Since the exponential map does not decrease distances, the area of the polygon is a lower bound for the area of the Voronoi cell. If k is the number of sides of V, then V is partitioned into k triangles with angle θ_i at O, whose area is bounded below by

$$\left(\tfrac{\mathrm{sys}\pi_1}{4}\right)^2 \tan\tfrac{\theta_i}{2}. \tag{10.4.4}$$

Consider the graph on S^2 defined by the projections of the Voronoi cells to the sphere. Thus we have $f = 6$ faces. Applying the formula $v - e + f = 2$ and the well-known fact that $3v \leq 2e$, we obtain

$$e \leq 3f - 6 = 12.$$

Hence the spherical graph has at most 12 edges. Note that the maximum is attained by the 1-skeleton of the cubical subdivision.

The area of a flat isosceles triangle with third angle θ, and with unit altitude from the third vertex, is $\tan\frac{\theta}{2}$. This formula provides a lower bound for the area of the Voronoi cells as in (10.4.4). The proof is completed by Jensen's inequality applied to the convex function $\tan(\frac{x}{2})$ when $0 < x < \pi$. In the boundary case of equality, we have $e = 12$, all angles θ_i as in (10.4.4) must be equal, curvature must be zero because of equality in Toponogov, and we easily deduce that each Voronoi cell is a regular octagon. To minimize the area of the octagon, we must choose θ as small as possible. The CAT(0) hypothesis at the center of the octagon imposes a lower bound $\theta \geq \frac{\pi}{4}$. Hence the optimal systolic ratio is achieved for the regular flat octagon with a smooth point at the center. □

10.5. Arbitrary metrics on the Bolza surface

The conformal class of the Bolza surface $\mathcal{B}$, *cf.* Section 10.2, is likely to contain a systolically optimal surface in genus 2, as discussed in Section 10.1. It therefore may be of interest to obtain bounds for the optimal systolic ratio in this particular conformal class.

THEOREM 10.5.1. *Every metric $\mathcal{G}$ in the conformal class of the Bolza surface satisfies the bound*

$$\mathrm{SR}(\mathcal{G}) \leq \tfrac{\pi}{3} = 1.0471\ldots \tag{10.5.1}$$

REMARK 10.5.2. In particular, every metric $\mathcal{G}$ in the conformal class of the Bolza surface satisfies the bound of Bavard's inequality

$$\mathrm{SR}(\mathcal{G}) \leq \frac{\pi}{2^{3/2}} \simeq 1.1107 \tag{10.5.2}$$

for the Klein bottle [**Bav86**]. This suggests a possible monotonicity of $\chi(\Sigma)$ as a function of $\mathrm{SR}(\Sigma)$.

LEMMA 10.5.3. *Let $\mathcal{G}$ be an* $\mathrm{Aut}(\mathcal{B})$*-invariant metric on $\mathcal{B}$. Consider the displacement $\delta(J)$ of J on the 1-skeleton $\mathcal{B}^{(1)}$ of the Voronoi subdivision of $\mathcal{B}$, namely*

$$\delta(J) = \min_{x \in \mathcal{B}^{(1)}} \mathrm{dist}(x, J(x)).$$

Then $\mathrm{area}(\mathcal{B}, \mathcal{G}) \geq 6\left(\frac{2}{\pi}\right)\delta(J)^2$.

PROOF. Consider the Voronoi subdivision with respect to the set of six Weierstrass points on $\mathcal{B}$. Since each Voronoi cell $\mathcal{O} \subset \mathcal{B}$ is J-invariant, we can identify all pairs of opposite points of the boundary $\partial\mathcal{O}$, to obtain a projective plane

$$\mathbb{RP}^2 = \mathcal{O}/\sim. \tag{10.5.3}$$

We now apply Pu's inequality (5.2.2) to each of the six Voronoi cells, to obtain

$$\mathrm{area}(\mathbb{RP}^2) \geq \tfrac{2}{\pi}\,\mathrm{sys}\pi_1(\mathbb{RP}^2)^2. \tag{10.5.4}$$

The lemma now follows from the bound $\mathrm{sys}\pi_1(\mathbb{RP}^2) \geq \delta(J)$. □

LEMMA 10.5.4. *Every* $\mathrm{Aut}(\mathcal{B})$*-invariant metric $\mathcal{G}$ on $\mathcal{B}$ satisfies the following bound:* $2\delta(J) \geq \mathrm{sys}\pi_1(\mathcal{B}, \mathcal{G})$.

PROOF. Let $p, J(p) \in \partial\mathcal{O}$ satisfy $\mathrm{dist}(p, J(p)) = \delta(J)$. Let $\alpha \subset \mathcal{O}$ be a minimizing path joining p to $J(p)$. Let $\mathcal{O}' \subset \mathcal{B}$ be the adjacent Voronoi cell containing this pair of boundary points, and $r : \mathcal{B} \to \mathcal{B}$ the anticonformal involution which switches $\mathcal{O}$ and $\mathcal{O}'$, and fixes their common boundary. The loop $\alpha \cup r(\alpha)$ belongs to the free homotopy class of the noncontractible loop $\gamma \subset \mathcal{B}$ obtained as the inverse image under $Q : \mathcal{B} \to S^2$ of the edge of the octahedral decomposition of S^2 joining the images of the centers of $\mathcal{O}$ and $\mathcal{O}'$, *cf.* Lemma 10.2.2. Since $\mathrm{length}(\alpha \cup r(\alpha)) = 2\delta(J)$, the lemma follows. □

PROOF OF THEOREM 10.5.1. We may assume that the metric on $\mathcal{B}$ is $\mathrm{Aut}(\mathcal{B})$-invariant, since averaging the metric by a finite group of holomorphic and antiholomorphic diffeomorphisms can only improve the systolic ratio. We combine the inequalities of Lemmas 10.5.4 and 10.5.3 to complete the proof of the theorem. □

CHAPTER 11

Volume entropy and asymptotic upper bounds

11.1. Entropy and systole

We show that the volume entropy h (together with A. Katok's optimal inequality for h) is the "right" intermediary in a transparent proof of M. Gromov's asymptotic bound for the systolic ratio of surfaces of large genus. These results originally appeared in [**KatzS05**].

Minimal entropy was defined in Section 7.4. We define the systolic ratio SR of a Riemannian n-manifold $(M, \mathcal{G})$ as

$$\mathrm{SR}(M, \mathcal{G}) = \frac{\mathrm{sys}\pi_1(M, \mathcal{G})^n}{\mathrm{vol}(M, \mathcal{G})}, \tag{11.1.1}$$

and the optimal systolic ratio of M as

$$\mathrm{SR}(M) = \sup_{\mathcal{G}} \mathrm{SR}(M, \mathcal{G}), \tag{11.1.2}$$

where $\mathcal{G}$ runs over the space of all metrics on M. C. Loewner proved the first systolic inequality (5.4.1). Thus, we have $\mathrm{SR}(\mathbb{T}^2) = \frac{2}{\sqrt{3}}$.

We will relate the minimal entropy of a closed surface to its optimal systolic ratio. Namely, we find an upper bound for the entropy of a systolically extremal surface, in terms of its systole.

PROPOSITION 11.1.1. *Every extremal metric $\mathcal{G}$ on a surface Σ satisfies*

$$h(\mathcal{G}) \leq \frac{1}{\beta\, \mathrm{sys}\pi_1(\mathcal{G})} \log\left(2\alpha^2\, \mathrm{SR}(\Sigma)\right), \tag{11.1.3}$$

whenever $\alpha, \beta > 0$ and $4\alpha + \beta < \frac{1}{2}$.

We combine this upper bound with A. Katok's optimal lower bound in terms of the volume, see (7.4.3), to obtain a simpler alternative proof of M. Gromov's asymptotic estimate for the optimal systolic ratio of surfaces of large genus. Furthermore, we improve the multiplicative constant in Gromov's theorem (see Section 11.3 for further details). Note that there are errors in exponents in [**Gro83**, Theorem 6.4.D$'$, p. 74] and [**Gro83**, Corollary 6.4.D$''$, p. 74].

THEOREM 11.1.2. *A surface Σ_g of genus g satisfies the bound*

$$\mathrm{SR}(\Sigma_g) \leq \frac{\log^2 g}{\pi g}(1 + o(1)) \text{ when } g \to \infty. \tag{11.1.4}$$

An alternative approach is taken by F. Balacheff [**Bal04**], but his constant is not as good as Gromov's. Furthermore, the approach taken in the present chapter lends itself to higher-dimensional generalisations [**Sab06**].

As an application, we show that every metric on a surface of genus at least 20 satisfies the Loewner inequality (5.4.1) for the torus. This improves the best earlier estimate, which stood at 50.

In Section 11.2, we describe the basic estimate, based on a maximal packing argument combined with area lower bounds for systolically optimal surfaces. In Section 11.3, we combine the basic estimate with Katok's inequality to prove one of our main results, Theorem 11.1.2. In Section 11.4, we prove that every surface of genus at least 20 is Loewner.

11.2. Basic estimate

In Lemma 7.4.3 we recalled the following well-known fact: the volume entropy agrees with the exponential growth rate of orbits under the action of the fundamental group in the universal cover. The volume entropy is sometimes referred to as the *critical exponent*, *cf.* [**BesCG03**].

We will also need the following estimate, *cf.* (11.2.5). M. Gromov showed in [**Gro83**] that every aspherical closed surface Σ satisfies the inequality

$$\mathrm{SR}(\Sigma) \leq \frac{4}{3}. \tag{11.2.1}$$

He also showed that every nonsimply connected closed surface admits an extremal metric in a suitable generalized sense, namely a metric $\mathcal{G}_{\mathrm{ex}}$ with optimal systolic ratio $\mathrm{SR}(\Sigma, \mathcal{G}_{\mathrm{ex}}) = \mathrm{SR}(\Sigma)$. Furthermore, the disks $D(x,r)$ of radius $r \leq \frac{1}{2}\operatorname{sys}\pi_1(\mathcal{G}_{\mathrm{ex}})$ of extremal surfaces satisfy the bound

$$\operatorname{area} D(x,r) \geq 2r^2, \tag{11.2.2}$$

cf. (5.2.4).

We will use arguments developed in [**Kat83**, p. 357] to prove Proposition 11.2.1 below. Related arguments have been exploited in the works [**BurgM02, Gela04**].

PROPOSITION 11.2.1. *Every extremal metric $\mathcal{G}$ on a surface Σ satisfies*

$$h(\mathcal{G}) \leq -\frac{1}{\beta \operatorname{sys}\pi_1(\mathcal{G})} \log\left(2\alpha^2\,\mathrm{SR}(\Sigma)\right), \tag{11.2.3}$$

whenever $\alpha, \beta > 0$ and $4\alpha + \beta < \frac{1}{2}$.

PROOF OF PROPOSITION 11.2.1. The idea is to bound from above the number of homotopy classes of based loops in Σ, by deforming such loops into a fixed subgraph in Σ. The subgraph can be allowed to be as "coarse" as the size of the systole of Σ.

More precisely, choose a basepoint $x_0 \in \Sigma$. Consider a maximal system of disjoint disks

$$D_i = D(x_i, R) \subset \Sigma \tag{11.2.4}$$

of radius $R = \alpha \operatorname{sys}\pi_1(\mathcal{G})$ and centers x_i with $i \in I$, including x_0. Since the metric $\mathcal{G}$ is assumed extremal, inequality (11.2.2) implies

$$\operatorname{area} D_i \geq 2\alpha^2 \operatorname{sys}\pi_1(\mathcal{G})^2 \quad \forall i \in I. \tag{11.2.5}$$

Therefore, this system admits at most $\frac{\operatorname{area}(\mathcal{G})}{2\alpha^2 \operatorname{sys}\pi_1(\mathcal{G})^2}$ disks. Thus,

$$|I| \leq \left(2\alpha^2 \mathrm{SR}(\Sigma)\right)^{-1}. \tag{11.2.6}$$

Let $c : [0, T] \to \Sigma$ be a geodesic loop of length T based at x_0. Let

$$m = \left[\frac{T}{\beta \operatorname{sys}\pi_1(\mathcal{G})} \right] \tag{11.2.7}$$

be the integer part. The point $p_0 = x_0$, together with the points

$$p_k = c(k\beta \operatorname{sys}\pi_1(\mathcal{G})), \quad k = 1, \ldots, m$$

and the point $p_{m+1} = x_0$, partition the loop c into $m + 1$ segments of length at most $\beta \operatorname{sys}\pi_1(\mathcal{G})$. Since the system of disks D_i is maximal, the disks of radius $2R = 2\alpha \operatorname{sys}\pi_1(\mathcal{G})$ centered at x_i cover Σ. Therefore, for every p_k, a nearest point q_k among the centers x_i, is at distance at most $2R$ from p_k. Consider the loop

$$\alpha_k = c_k \cup [p_{k+1}, q_{k+1}] \cup [q_{k+1}, q_k] \cup [q_k, p_k],$$

where c_k is the arc of c joining p_k to p_{k+1}, while $[x, y]$ denotes a minimizing path joining x to y. Then

$$\operatorname{length}(\alpha_k) \leq 2(4\alpha + \beta) \operatorname{sys}\pi_1(\mathcal{G}) < \operatorname{sys}\pi_1(\mathcal{G}),$$

by our hypothesis on α, β. Thus the loop α_k is contractible. The same is true for the loops $c_0 \cup [p_1, q_1] \cup [q_1, x_0]$ and $c_m \cup [x_m, q_m] \cup [q_m, p_m]$.

Therefore, the geodesic loop c is homotopic to a piecewise geodesic loop

$$c' = (x_0, q_1, \ldots, q_m, x_0) = [x_0, q_1] \cup [q_1, q_2] \cup \ldots \cup [q_{m-1}, q_m] \cup [q_m, x_0]. \tag{11.2.8}$$

Note that the minimizing path from p_k to q_k may not be unique, but we choose one, being careful that the same choice is used on "both sides", *i.e.* for both loops α_{k-1} and α_k.

Thus, two nonhomotopic closed geodesic loops c_1 and c_2 based at x_0, give rise to two distinct loops c_1' and c_2' as in (11.2.8). In constructing the loops c', we always choose the same minimizing path between a given pair of points q_k. Thus, the number $P'(T)$ of homotopy classes which can be represented by loops of length T based at x_0 satisfies

$$\begin{aligned} P'(T) &\leq |I|^m \\ &\leq |I|^{\frac{T}{\beta \operatorname{sys}\pi_1(\mathcal{G})}} \\ &\leq \left(2\alpha^2 \operatorname{SR}(\Sigma)\right)^{-\frac{T}{\beta \operatorname{sys}\pi_1(\mathcal{G})}}, \end{aligned} \tag{11.2.9}$$

and the proposition now follows from Lemma 7.4.3. □

REMARK 11.2.2. Instead of relying on the existence of systolically extremal surfaces, we could have exploited instead ϵ-regular surfaces satisfying (11.2.2) for $r \geq \epsilon$, whose existence is considerably easier to establish [**Gro83**, 5.6.C″]. Since the choice of α in (11.3.5) below entails exploiting packings by arbitrarily small disks in (11.2.4), we have to be careful to choose $\epsilon < \alpha \operatorname{sys}\pi_1(\Sigma)$.

REMARK 11.2.3. A similar technique can be used to bound the number of free homotopy classes of loops in Σ. In higher dimensions, the latter bound seems to be less useful, therefore we chose to bound the number of based loops.

Combining inequalities (7.4.3) and (11.2.3), we obtain the following corollary.

COROLLARY 11.2.4. *Let Σ_g be a closed orientable surface of genus g. Whenever $4\alpha+\beta<\frac{1}{2}$, we have*

$$\frac{\log^2\left(2\alpha^2\ \mathrm{SR}(\Sigma_g)\right)}{\mathrm{SR}(\Sigma_g)} \geq 4\pi\beta^2(g-1). \tag{11.2.10}$$

11.3. Asymptotic behavior of systolic ratio for large genus

We now consider the asymptotic behavior of the optimal systolic ratio of surfaces. M. Gromov [**Gro83**, p. 74] established a bound for the ratio $\mathrm{SR}(\Sigma_g)$ by using a technique known as "diffusion of chains". The multiplicative constant $\frac{1}{\pi}$ in (11.3.2) below improves the constant 4 which could be obtained from the techniques in [**Gro83**].

THEOREM 11.3.1. *Given a real number $\lambda<\pi$, every surface Σ_g of genus g satisfies*

$$\mathrm{SR}(\Sigma_g) \leq \frac{\log^2 g}{\lambda g} \tag{11.3.1}$$

for g large enough. Thus,

$$\mathrm{SR}(\Sigma_g) \lesssim \frac{\log^2 g}{\pi g} \quad \text{as } g\to\infty. \tag{11.3.2}$$

REMARK 11.3.2. A near-optimal lower bound for $\mathrm{SR}(\Sigma_g)$ was found by P. Buser and P. Sarnak [**BusS94**], *cf.* [**KatzSV07**]. Namely, they construct hyperbolic surfaces $(\Sigma_g,\mathcal{G}_g)$ of arbitrarily large genus g, obtained as congruence coverings of a fixed arithmetic Riemann surface, such that

$$\mathrm{SR}(\Sigma_g,\mathcal{G}_g) \gtrsim \frac{4}{9\pi}\frac{\log^2 g}{g} \quad \text{as } g\to\infty. \tag{11.3.3}$$

Therefore, we have

$$\frac{4}{9\pi} \leq \limsup_{g\to\infty}\left(\mathrm{SR}(\Sigma_g)\frac{g}{\log^2 g}\right) \leq \frac{1}{\pi}. \tag{11.3.4}$$

PROOF OF THEOREM 11.3.1. Let $\mathcal{G}$ be an extremal metric on Σ_g. We now apply Corollary 11.2.4. Note that $\mathrm{SR}(\Sigma_g,\mathcal{G})$ tends to zero as the genus g becomes unbounded, *cf.* (11.4.4). For a right choice of α and β, inequality (11.2.10) leads to the asymptotic implicit upper bound (11.3.6) on $\mathrm{SR}(\Sigma_g)$ below. Indeed, given $\lambda<\lambda_+<\pi$, we set

$$\beta=\sqrt{\tfrac{\lambda_+}{4\pi}}<\tfrac{1}{2},$$

and choose

$$\alpha<\tfrac{1}{4}\left(\tfrac{1}{2}-\beta\right). \tag{11.3.5}$$

Inequality (11.2.10) implies

$$\frac{\log^2(\mathrm{SR}(\Sigma_g))}{\mathrm{SR}(\Sigma_g)} \geq \lambda g \tag{11.3.6}$$

if g is large enough. Now we would like to invert this relation in order to obtain an asymptotic upper bound on $\mathrm{SR}(\Sigma_g)$. Let

$$\rho=\mathrm{SR}(\Sigma_g)^{-\frac{1}{2}}$$

and $\delta = \frac{1}{2}\sqrt{\lambda g}$. Then inequality (11.3.6) yields the following estimate:

$$\begin{aligned}\rho \log \rho &\geq \delta \\ &\geq \delta - \frac{\delta \log \log \delta}{\log \delta} \\ &= \frac{\delta}{\log \delta} \log \left(\frac{\delta}{\log \delta} \right).\end{aligned}$$

Since the function $x \log x$ is increasing for x large enough, we deduce that $\rho \geq \frac{\delta}{\log \delta}$, and the latter inequality translates back into (11.3.1). □

11.4. When is a surface Loewner?

We now extend the classical Loewner inequality (5.4.1) on the torus to surfaces of higher genus. We will say that a surface Σ is *Loewner* if $\mathrm{SR}(\Sigma) \leq \frac{2}{\sqrt{3}}$.

THEOREM 11.4.1. *Every surface of genus at least* 20 *is Loewner.*

PROOF. Let $\mathcal{G}$ be an extremal metric on Σ. If $\beta = \frac{1}{2} - 4\alpha$, inequality (11.2.10) yields

$$\frac{\log^2\left(2\alpha^2 \,\mathrm{SR}(\Sigma)\right)}{\mathrm{SR}(\Sigma)} \geq 4\pi \left(\tfrac{1}{2} - 4\alpha\right)^2 (g-1) \tag{11.4.1}$$

for every $\alpha \leq \frac{1}{8}$.

Suppose now that $\mathrm{SR}(\Sigma) > \frac{2}{\sqrt{3}}$. Since $\mathrm{SR}(\Sigma) \leq \frac{4}{3}$ by (11.2.1), we have the estimate $2\alpha^2 \,\mathrm{SR}(\Sigma) \leq 1$ for every $\alpha \leq \frac{1}{8}$. Therefore,

$$\frac{\sqrt{3}}{2} \log^2\left(\tfrac{\sqrt{3}}{4\alpha^2}\right) \geq 4\pi \left(\tfrac{1}{2} - 4\alpha\right)^2 (g-1) \tag{11.4.2}$$

for every $\alpha \leq \frac{1}{8}$. Hence,

$$\min_{0<\alpha\leq\frac{1}{8}} \frac{\sqrt{3}}{8\pi} \left(\frac{\log\left(\frac{\sqrt{3}}{4\alpha^2}\right)}{\frac{1}{2} - 4\alpha} \right)^2 \geq g-1 \tag{11.4.3}$$

For $\alpha = .031$, the expression to minimize is about 18.201. Thus, $g \leq 19$. Therefore, every surface of genus greater or equal to 20 is Loewner. □

Note that M. Gromov [**Gro83**, p. 50] (*cf.* [**Kod87**, Theorem 4, part (1)]) proved a general estimate which implies that

$$\mathrm{SR}(\Sigma_g) < \frac{64}{4\sqrt{g} + 27}. \tag{11.4.4}$$

It follows from Gromov's estimate (11.4.4) that orientable surfaces satisfy the Loewner inequality (5.4.1) if the genus g is greater than 50. Our theorem brings this bound down to 20. Furthermore, the genus 2 surface is Loewner (Chapter 9). The remaining open cases are therefore the genera $g = 3, \ldots, 19$.

Part 2

Systolic geometry and topology in n dimensions

CHAPTER 12

Systoles and their category

The Loewner inequality, as well as Pu's inequality and other inequalities for surfaces, discussed in Part 1, all involve 1-dimensional systolic invariants, measured in terms of lengths of loops in a given surface. Interesting geometric and topological phenomena arise also in the study of higher systoles, namely invariants expressed in terms of areas of cycles which are not null-homologous. See the introduction (Chapter 1) for a description of such results in their mutual relation.

12.1. Systoles

Let M be a compact smooth manifold of dimension n. Let k be an integer satisfying $1 \le k \le n$. We will now provide detailed definitions of the k-systolic invariants of a metric on M.

DEFINITION 12.1.1. The graded module of Lipschitz chains on M will be denoted $C_*(M)$, *cf.* Section 14.2.

In more detail, a choice of a Riemannian metric $\mathcal{G}$ on M allows us to define the total volume $\mathrm{vol}_n(M, \mathcal{G})$, as well as the k-volumes of k-dimensional submanifolds of M. More generally, given an integer Lipschitz k-chain

$$c = \sum_i n_i \sigma_i \in C_k(M),$$

one defines its volume as follows:

$$\mathrm{vol}_k(c) = \Sigma_i |n_i| \,\mathrm{vol}_k(\sigma_i). \tag{12.1.1}$$

Here the volume $\mathrm{vol}_k(\sigma)$ of a Lipschitz k-simplex $\sigma : \Delta^k \to M$ is the integral over the k-simplex Δ^k of the "volume form" of the pullback $\sigma^*(\mathcal{G})$.

DEFINITION 12.1.2. The (minimal) volume,

$$\mathrm{vol}_k(h),$$

of an integer homology class $h \in H_k(M, \mathbb{Z})$ is the infimum of $\mathrm{vol}_k(c)$ over all integer Lipschitz cycles c representing h.

We define the homology k-systole $\mathrm{sysh}_k(\mathcal{G})$ of $(M, \mathcal{G})$ as the minimum of the volumes of nonzero integer k-homology classes:

$$\mathrm{sysh}_k(\mathcal{G}) = \inf \left\{ \mathrm{vol}_k(h) \,\middle|\, h \in H_k(M, \mathbb{Z}) \setminus \{0\} \right\}. \tag{12.1.2}$$

In particular, we have $\mathrm{sysh}_n(\mathcal{G}) = \mathrm{vol}_n(\mathcal{G})$ if M is a closed orientable manifold. Formally, if $H_k(M, \mathbb{Z}) = 0$, we set $\mathrm{sysh}_k(\mathcal{G}) = \infty$ by definition.

REMARK 12.1.3. If M is orientable, the quantity $\mathrm{sysh}_{n-1}(\mathcal{G})$ coincides with the infimum of the $(n-1)$-volumes of compact, orientable, non-separating hypersurfaces in M.

Replacing integer cycles by real cycles, we define the stable norm as follows.

DEFINITION 12.1.4. The stable norm $\|h\|$ of a homology class $h \in H_k(M,\mathbb{R})$ is the infimum of the volumes

$$\mathrm{vol}_k(c) = \Sigma_i |r_i| \,\mathrm{vol}_k(\sigma_i) \tag{12.1.3}$$

over all real Lipschitz cycles $c = \Sigma_i r_i \sigma_i \in C_*(M)$ representing h.

Note that $\| \;\|$ is indeed a norm, *cf.* [**Fed74**] and [**Gro99**, 4.C].

We denote by $H_k(M,\mathbb{Z})_{\mathbb{R}}$ the image of $H_k(M,\mathbb{Z})$ in $H_k(M,\mathbb{R})$ and by $h_{\mathbb{R}}$ the image of $h \in H_k(M,\mathbb{Z})$ in $H_k(M,\mathbb{R})$. Recall that $H_k(M,\mathbb{Z})_{\mathbb{R}}$ is a lattice in $H_k(M,\mathbb{R})$. Obviously

$$\|h_{\mathbb{R}}\| \le \mathrm{vol}_k(h) \tag{12.1.4}$$

for all $h \in H_k(M,\mathbb{Z})$. Moreover, one has $\|h_{\mathbb{R}}\| = \mathrm{vol}_n(h)$ if $h \in H_n(M,\mathbb{Z})$. H. Federer [**Fed74**, 4.10, 5.8, 5.10] (see also [**Gro99**, 4.18 and 4.35]) investigated the relations between $\|h_{\mathbb{R}}\|$ and $\mathrm{vol}_k(h)$ and proved the following (see also [**Wh83**]).

PROPOSITION 12.1.5. *If $h \in H_k(M,\mathbb{Z})$, $1 \le k < n$, then*

$$\|h_{\mathbb{R}}\| = \lim_{i\to\infty} \frac{1}{i}\,\mathrm{vol}_k(ih). \tag{12.1.5}$$

If M is orientable and $h \in H_{n-1}(M,\mathbb{Z})$, then

$$\|h_{\mathbb{R}}\| = \mathrm{vol}_{n-1}(h). \tag{12.1.6}$$

Equation (12.1.5) is the origin of the term *stable norm* for $\| \;\|$.

DEFINITION 12.1.6. The stable k-systole $\mathrm{stsys}_k(\mathcal{G})$ of $(M,\mathcal{G})$ is the quantity

$$\begin{aligned}\mathrm{stsys}_k(\mathcal{G}) &= \min\left\{\|h\| \;\middle|\; h \in H_k(M,\mathbb{Z})_{\mathbb{R}} \setminus \{0\}\right\}\\ &= \lambda_1\left(H_k\left(M,\mathbb{Z}\right)_{\mathbb{R}}, \| \;\|\right),\end{aligned} \tag{12.1.7}$$

where λ_1 is the first successive minimum, *cf.* (4.4.1).

If $H_k(M,\mathbb{Z})$ is free abelian, then (12.1.4) implies

$$\mathrm{stsys}_k(\mathcal{G}) \le \mathrm{sys}_k(\mathcal{G}).$$

If M is orientable, then

$$\mathrm{stsys}_n(\mathcal{G}) = \mathrm{sys}_n(\mathcal{G}) = \mathrm{vol}_n(\mathcal{G}),$$

and (12.1.6) implies the equality

$$\mathrm{stsys}_{n-1}(\mathcal{G}) = \mathrm{sys}_{n-1}(\mathcal{G}).$$

We now specialize to the case $k = 1$.

DEFINITION 12.1.7. The homotopy 1-systole, denoted $\mathrm{sys}\pi_1(M,\mathcal{G})$, is the least length of a non-contractible loop for the metric $\mathcal{G}$.

The homology 1-systole $\mathrm{sysh}_1(M,\mathcal{G})$, already defined in (12.1.2), can be defined similarly to the homotopy systole, as the least length of a loop which is not zero-homologous. Clearly,

$$\mathrm{sys}\pi_1(\mathcal{G}) \le \mathrm{sysh}_1(\mathcal{G}).$$

Now consider a ring $A = \mathbb{Z}$ or $\mathbb{Z}_2$. The homology k-systoles

$$\mathrm{sysh}_k(M,\mathcal{G};A),$$

with coefficients in $A = \mathbb{Z}_2$, can be defined similarly to the integer case (12.1.2). Namely, $\operatorname{sysh}_k = \operatorname{sysh}_k(M, \mathcal{G}; A)$ is the infimum of k-volumes of k-cycles, with coefficients in A, where the infimum is over cycles which are not zero-homologous. When $A = \mathbb{Z}_2$, the absolute value $|n_i|$ in (12.1.1) is replaced by 1 whenever the coefficient n_i is nonzero. Note that we adopt the usual convention, convenient for our purposes, that the infimum over an empty set is infinity.

More generally, consider a regular covering map $M' \to M$ and let B be its group of deck transformations. Consider the homology group $H_k(M'; A)$ of M'. Then $H_k(M'; A)$ can be identified with the homology group, denoted $H_k(M; A[B])$, of M with twisted coefficients in the group ring $A[B]$. Allowing more general coefficients in such a group ring, we can define the corresponding systole

$$\operatorname{sysh}_k(M, \mathcal{G}; A[B]). \tag{12.1.8}$$

All systolic notions can be defined similarly on polyhedra, as well (see M. Gromov [**Gro96**] and I. Babenko [**Bab02a**]).

In Section 12.2, we state Gromov's spectacular inequality for the 1-systole of essential manifolds [**Gro83**], and present an outline of a proof.

In Section 12.3, we define a numerical invariant called *systolic category*, first introduced in [**KatzS06a**].

In Section 12.4, we discuss examples pointing toward a relation between systolic category and the classical Lusternik-Schnirelmann category.

In Section 12.5, we prove that for manifolds of any dimension, the two categories attain the maximal value simultaneously. The latter result relies on a compression theorem of Section 12.6.

Section 12.7 contains a proof of the agreement of the two categories in dimension 3.

In Section 12.8, we compare the two categories in the case of a simply connected manifold of low dimension.

12.2. Gromov's spectacular inequality for the 1-systole

The deepest result in systolic geometry is Gromov's inequality for the homotopy 1-systole of essential manifolds. Gromov's original definition of an essential manifold M depends on the choice of the coefficient ring A, taken to be $\mathbb{Z}$ if M is orientable, or $\mathbb{Z}_2$ otherwise.

DEFINITION 12.2.1. A closed n-dimensional manifold M is called *essential* if there exists a map from M to a suitable Eilenberg-MacLane space $K(\pi, 1)$ such that the induced homomorphism

$$h : H_n(M, A) \to H_n(K(\pi, 1), A)$$

maps the fundamental homology class $[M]$ of M to a nonzero class in the homology group $H_n(K(\pi, 1), A)$, *i.e.* $h([M]) \neq 0$.

A more general definition of an n-essential space X, in the context of an arbitrary polyhedron X, appears in Section 12.5. The following theorem was proved in [**Gro83**, Section 0] and [**Gro83**, Appendix 2, p. 139, item B_1'].

THEOREM 12.2.2 (M. Gromov). *Every n-essential, compact, n-dimensional polyhedron X satisfies the inequality*

$$\operatorname{sys}\pi_1(X)^n \leq C_n \operatorname{vol}_n(X). \tag{12.2.1}$$

where the constant C_n depends only on n. If X is a manifold, the constant C_n can be chosen to be on the order of n^{2n^2}.

In other words, the quotient

$$\frac{\mathrm{vol}_n}{(\mathrm{sys}\pi_1)^n} > 0 \tag{12.2.2}$$

is bounded away from zero for such polyhedra.

OUTLINE OF PROOF OF THEOREM 12.2.2. We sketch a proof in the manifold case, relying on the results of S. Wenger [**Wen05**], who proved the following items:

- an isoperimetric inequality in Banach space;
- the existence of a minimizing current under suitable hypotheses;
- a monotonicity-type lower bound (with the Euclidean exponent) for masses of balls in a minimizing current.

Immediately following Lemma 3.4 in [**Wen05**], and preceding its proof, there is an allusion to the filling radius, without mentioning the invariant itself. Wenger's work builds upon earlier work in geometric measure theory by L. Ambrosio and B. Kirchheim [**AmK00**, Theorem 10.6].

As an application, we can now prove Gromov's 1983 inequality in a rather transparent way as follows. Let M^n be a compact Riemannian manifold. Denote by $L^\infty(M)$ the space of bounded Borel functions on M, equipped with the sup-norm $\|\ \|$. Let $p \in M$. Let $\vee_p \in L^\infty(M)$ be the function defined by

$$\vee_p(x) = \mathrm{dist}(p, x).$$

The notation is inspired by the fact that if M is 1-dimensional, the distance function from a point looks like the absolute value function. The map

$$M \to L^\infty(M), \quad p \mapsto \vee_p$$

is a strongly isometric imbedding, in the sense that

$$\mathrm{dist}(p, q) = \| \vee_p - \vee_q \|.$$

Its universal properties are discussed in detail in [**Gro83**].

DEFINITION 12.2.3. The *filling radius*, FillRad(M), of M is the infimum of all $\epsilon > 0$ such that the inclusion of M in the ϵ-neighborhood of its image in $L^\infty(M)$, sends the fundamental homology class $[M]$ to the zero class (the coefficients in homology have to be chosen as a function of the orientability of M).

Relying upon Wenger's work, we now proceed as follows:

(1) imbed M^n in $L = L^\infty(M)$;
(2) take a minimizing current $T \subset L$ spanning M;
(3) choose a point $\tau \in T$ at maximal distance $r_T > 0$ from $M \subset L$;
(4) apply the monotonicity estimate at τ (replace τ by a nearby point of spt(T) if necessary);
(5) deduce an upper bound for the distance r_T in terms of the mass $\mathbf{M}(T)$;
(6) apply the isoperimetric inequality (with Euclidean exponent), so as to deduce an upper bound for the mass $\mathbf{M}(T)$ in terms of vol(M);
(7) conclude that

$$\mathrm{FillRad}(M) < r_T < C_n \mathbf{M}(T)^{1/(n+1)} < C'_n \,\mathrm{vol}(M)^{1/n}$$

for all M, where the constant depends only on n;

(8) apply the usual argument for an essential manifold M to bound the systole by 6 times the filling radius, *cf.* [**Gro83, CrK03**].

Note that $L^\infty(M)$ can easily be replaced by a finite dimensional Banach space, so that functional analytic difficulties are not relevant here. Namely, let $s = \mathrm{sys}\pi_1(M)$, and consider a maximal $\frac{s}{10}$-separated set $S \subset M$. If $\gamma \subset M$ is a systolic loop, decompose it as into a pair of minimizing arcs $\gamma = \gamma_1 \cup \gamma_2$, joining a pair of "opposite" points $a, b \in \gamma$. If $a_0, b_0 \in S$ are the corresponding "nearest" points of S, then by the triangle inequality, we have

$$\mathrm{dist}(a_0, b_0) \geq \tfrac{3s}{10}.$$

Similarly, the distance between the images of a and b in $L^\infty(S)$ is at least $\frac{s}{10}$. Hence each γ_i has length at least $\frac{s}{10}$ in $L^\infty(S)$, and therefore the image of γ in the finite dimensional Banach space $L^\infty(S)$ has length at least $\frac{s}{5}$. The estimate can obviously be improved to be as close as one wishes to the value s, at the cost of raising the dimension of the Banach space. Thus, to prove Gromov's inequality, it suffices to prove the necessary estimates in finite-dimensional Banach space.

Note that the results in [**Wen05**] are stated for integral currents. Thus as stated, our argument works only for orientable M. The experts feel confident that the results in [**Wen05**] are valid for currents with coefficients modulo 2, as well. □

In terms of systolic category defined in the next section, the result can be restated as follows.

COROLLARY 12.2.4. *Every n-essential polyhedron X of dimension n satisfies* $\mathrm{cat}_{\mathrm{sys}}(X) = n$.

Note that better bounds are available as the topological complexity of the space increases [**Gro99, Sab06**].

REMARK 12.2.5. Gromov's influential theorem was exploited in [**BesCG03**], where the authors use (a generalisation of) Gromov's volume lower bound (12.2.1) in the proof of their Lemma 1.3 at the bottom of page 259, imbedded in the proof of their Theorem 1.2, without making any further comment elsewhere in their article, showing conclusively that the lower bound has become part of mathematical folklore.

The converse was proved in [**Bab93**]. Namely, systolic category is strictly smaller than n for n-inessential n-manifolds, *cf.* Section 12.6. Combined with Theorem 12.5.2, this yields the following theorem, in harmony with equality (12.4.2).

THEOREM 12.2.6. *Let M be a n-manifold. Then* $\mathrm{cat}_{\mathrm{sys}}(M) = n$ *if and only if* $\mathrm{cat}_{\mathrm{LS}}(M) = n$.

12.3. Systolic category

Let $(M^n, \mathcal{G})$ be a Riemannian manifold. Recall that, in our convention, the systolic invariants are infinite when defined over an empty set of loops or cycles.

DEFINITION 12.3.1. Given $k \in \mathbb{N}$, $k > 1$ we set

$$\mathrm{sys}_k(M, \mathcal{G}) = \inf\left\{\mathrm{sysh}_k(M, \mathcal{G}; \mathbb{Z}[B]),\ \mathrm{sysh}_k(M, \mathcal{G}; \mathbb{Z}_2[B]),\ \mathrm{stsys}_k(M, \mathcal{G})\right\},$$

where the infimum is over all regular covers with deck group B, as in (12.1.8). Furthermore, we define

$$\operatorname{sys}_1(M,\mathcal{G}) = \min\left\{\operatorname{sys}\pi_1(M,\mathcal{G}),\ \operatorname{stsys}_1(M,\mathcal{G})\right\}.$$

In the sequel, we abbreviate $\operatorname{sys}_k(M,\mathcal{G})$ as $\operatorname{sys}_k(M)$, $\operatorname{sys}_k(\mathcal{G})$, or even sys_k, depending on the context.

It can be proved that the systolic invariants thus defined are strictly positive (or infinite), *cf.* Section 18.2, and [**KatzR07**].

Let $d \geq 2$ be an integer. Consider a partition

$$n = k_1 + \ldots + k_d, \tag{12.3.1}$$

where $k_i \geq 1$ for all $i = 1, \ldots, d$. We seek scale-invariant inequalities "of length d" of the following type:

$$\operatorname{sys}_{k_1}(\mathcal{G})\operatorname{sys}_{k_2}(\mathcal{G})\ldots\operatorname{sys}_{k_d}(\mathcal{G}) \leq C(M)\operatorname{vol}_n(\mathcal{G}). \tag{12.3.2}$$

Such inequalities are expected to be satisfied by all metrics $\mathcal{G}$ on M, where the (finite!) constant $C(M)$ is expected to depend only on the topological type of M, but not on the metric $\mathcal{G}$. Here the quantity sys_k denotes the minimum of all non-vanishing systolic invariants in dimension k, as in Definition 12.3.1.

DEFINITION 12.3.2. The Lusternik-Schnirelmann category $\operatorname{cat}_{\mathrm{LS}}(X)$ of a topological space X is the least integer n such that there exists a covering $U_0, \ldots, U_n$ of X by open sets, with each U_i contractible to a point in the space X, *cf.* [**CorT03**, Definition 1.1].

Note that this commonly accepted definition differs by a unit from the original definition given by L. Lusternik and L. Schnirelmann [**LuS34**]. In analogy with $\operatorname{cat}_{\mathrm{LS}}$, we define an invariant called *systolic category* as follows.

DEFINITION 12.3.3 ([**KatzR06**]). The *systolic category* of a manifold M, denoted

$$\operatorname{cat}_{\mathrm{sys}}(M),$$

is the largest integer d such that there exists a partition of type (12.3.1) with

$$\prod_{i=1}^{d}\operatorname{sys}_{k_i}(M,\mathcal{G}) \leq C(M)\operatorname{vol}_n(M,\mathcal{G})$$

for all metrics $\mathcal{G}$ on M.

In paricular, $\operatorname{cat}_{\mathrm{sys}} M \leq \dim M$.

EXAMPLE 12.3.4. Every orientable n-dimensional manifold M with positive Betti number $b_k(M) \geq 1$ satisfies Gromov's stable systolic inequality

$$\operatorname{stsys}_k(M)\operatorname{stsys}_{n-k}(M) < C(b_k(M))\operatorname{vol}_n(M). \tag{12.3.3}$$

SKETCH OF PROOF. By Poincaré duality, the fundamental cohomology class of M decomposes as a product $\alpha \cup \beta$, where $\alpha \in H^k(M)$ while $\beta \in H^{n-k}(M)$. Recall that the cohomology groups are equipped with the comass norm. We now choose quasiorthogonal bases

$$\{\alpha_i\} \subset H^k(M,\mathbb{Z})_{\mathbb{R}}$$

and $\{\beta_j\} \subset H^k(M, \mathbb{Z})_{\mathbb{R}}$. We decompose α and β with respect to these bases, and apply the distributive law, to obtain a nonzero product $\alpha_{i_0} \cup \beta_{j_0} \neq 0$. Inequality (12.3.3) now results from the duality of the comass norm in cohomology and the stable norm in homology, *cf.* Theorem 15.2.5. □

In our notation, the existence of inequality (12.3.3) implies the following lower bound for the systolic category of an orientable manifold:

$$\mathrm{cat}_{\mathrm{sys}}(M) \geq 2 \ \text{ if } b_k(M) \geq 1 \text{ for some } 1 \leq k \leq n-1, \tag{12.3.4}$$

see Corollary 15.2.3 for a generalisation in terms of real cup-length.

We note the following similarities between systolic category and Lusternik-Schnirelmann category:

- the two categories coincide in dimensions ≤ 3;
- the two categories attain the maximal value (*i.e.* dimension) simultaneously;
- both categories admit a lower bound in terms of the real cup-length;
- both categories turn out to be sensitive to Massey products, see Section 14.3 and Appendix B, item B.1.2.

Remark 12.3.5. There exist lower bounds for systolic category which are stronger than real cup-length, *cf.* Corollary 17.2.3.

12.4. Some examples and questions

In general, systolic inequalities involving any higher (unstable) k-systole are notoriously difficult to obtain, in view of the widespread phenomenon of *systolic freedom*, see [**Katz95a**], [**BabK98**], [**Gro99**, Systolic reminiscences, p. 271], as well as [**Gro99**, Appendix D], [**Bab02a, Katz02**], summarized in [**CrK03**]. These results tend to indicate that the "right" definition of systole in dimension $k \geq 2$ is the stable one.

Example 12.4.1. It is still unknown whether $\mathbb{RP}^3$ admits a (1,2)-systolic inequality, or whether on the contrary it is (1,2)-systolically free. Here, of course, the systoles are defined in terms of cycles with $\mathbb{Z}_2$ coefficients. The analogous question for the 3-manifold $S^1 \times S^2$ was resolved, in favor of freedom, by M. Freedman [**Fre99**]. The possible (1,2)-freedom of $\mathbb{RP}^3$ does not affect the value of its categories, since $\mathbb{RP}^3$ is essential and therefore

$$\mathrm{cat}_{\mathrm{LS}}(\mathbb{RP}^3) = \mathrm{cat}_{\mathrm{sys}}(\mathbb{RP}^3) = 3,$$

cf. Theorem 12.2.2. Thus, the question of the existence of a (1,2)-systolic inequality for $\mathbb{RP}^3$ does not affect the value of its systolic category. On the other hand, the case of $\mathbb{RP}^2 \times S^2$ is a simple case that is still open. Smale's spin rational homology 5-sphere M_k (see [**Sm62**]) of Example 12.8.4 satisfies $\mathrm{cat}_{\mathrm{LS}}(M_k) = 2$ by Theorem 12.8.1, but its systolic category is inaccessible with the techniques currently available.

Question 12.4.2. It is clear that

$$\mathrm{cat}_{\mathrm{sys}}(M \times N) \leq \mathrm{cat}_{\mathrm{sys}}(M) + \mathrm{cat}_{\mathrm{sys}}(N). \tag{12.4.1}$$

Ganea conjectured that taking the Cartesian product of a space with a sphere increases the Lusternik-Schnirelmann category of the space by a unit. In analogy with the Ganea conjecture, we can ask the following. How far can the inequality in (12.4.1) be from equality? Are there examples with $\operatorname{cat}_{\text{sys}}(M \times S^k) = \operatorname{cat}_{\text{sys}} M$?

EXAMPLE 12.4.3 (Categories agree in dimension 1). Every finite metric 1-dimensional polyhedron X satisfies the equality

$$\operatorname{cat}_{\text{sys}} X = \operatorname{cat}_{\text{LS}} X. \tag{12.4.2}$$

Namely, the common value is 0 if X is contractible, and 1 if X is homotopy equivalent to a wedge of circles. In other words, a tree does not satisfy a systolic inequality, while a graph (containing cycles) does. Namely, the 1-systole of a tree is infinite since all loops are contractible and therefore the systole is defined by an infimum over an empty set. Meanwhile, for (non-contractible) graphs Γ, we have $\operatorname{sys}_1(\Gamma) \leq \operatorname{length}(\Gamma)$ since the systole is realized by an imbedded loop.

EXAMPLE 12.4.4 (Categories agree in dimension 2). Every compact surface Σ (orientable or not) with infinite fundamental group satisfies Gromov's inequality (5.2.3). For the projective plane $\mathbb{RP}^2$, we have Pu's optimal inequality (5.2.2). Similarly, we have $\operatorname{cat}_{\text{LS}}\Sigma = 1$ for $\Sigma = S^2$ and $\operatorname{cat}_{\text{LS}}\Sigma = 2$ for all $\Sigma \neq S^2$. Thus, we have $\operatorname{cat}_{\text{LS}}\Sigma = \operatorname{cat}_{\text{sys}} \Sigma$ for all closed surfaces Σ. It can be shown that the two categories agree for arbitrary connected 2-complexes X, as well. Namely, we have an inequality $\operatorname{SR}(X) \leq 12$ if and only if $\pi_1(X)$ is not free and non-trivial [**KatzRS06**]. However, one cannot expect equality (12.4.2) to hold in all dimensions, *cf.* Example 12.8.5.

12.5. Essentialness and Lusternik–Schnirelmann category

Given a group π, recall that the Eilenberg-MacLane space $K(\pi, 1)$ is a connected CW-space with $\pi_1(K(\pi,1)) = \pi$ and $\pi_i(K(\pi,1)) = 0$ for all $i \geq 2$. Given a connected topological space X with $\pi_1(X) = \pi$, we define a *classifying map* of X to be any map $f : X \to K(\pi, 1)$ such that the induced map

$$f_* : \pi_1(X, x) \to \pi_1(K(\pi,1), f(x))$$

is an isomorphism for all $x \in X$. Every CW space X admits a classifying map, and for every pair of classifying maps $f, g : X \to K(\pi, 1)$, there exists a homotopy self-equivalence $h : K(\pi,1) \to K(\pi,1)$ such that hf and g are homotopic.

DEFINITION 12.5.1 ([**Gro83**]). Following M. Gromov [**Gro99**, item 4.40], we say that a manifold or polyhedron M is *n-essential* if no classifying map $f : M \to K(\pi_1(M), 1)$ can be deformed into the $(n-1)$-skeleton of $K(\pi_1(M), 1)$.

In the language of [**OR01, Rum79**], a manifold or polyhedron M is n-essential if there exists a cohomology class in $H^n(M; A)$ which is a detecting element for M of category weight n.

THEOREM 12.5.2. *Let M be a manifold of dimension n, and let π denote its fundamental group $\pi_1(M)$. The following three conditions are equivalent:*

(1) *The manifold M is n-essential;*
(2) *there exists a (twisted) coefficient system A on $K(\pi, 1)$ such that, for any classifying map $f : M \to K(\pi,1)$, the induced homomorphism $f^* : H^n(K(\pi,1); A) \to H^n(M; f^*A)$ is non-zero;*
(3) *we have $\operatorname{cat}_{\text{LS}} M = n = \dim M$.*

PROOF. The cases $n = 1$ and $n = 2$ are trivial, so we assume that $n \geq 3$.

The equivalence of (2) and (3) for $n \geq 3$ is proved in [**Bers76**], see also [**CorT03**, Theorem 2.51]. In fact, I. Berstein [**Bers76**] proved the following: given a connected CW-space Y of dimension n, consider the augmentation ideal I of the group ring $\mathbb{Z}[\pi_1(Y)]$. Then one has a twisted coefficient system I on Y and an element $a \in H^1(Y; I)$, the characteristic class of the universal cover, and, for $n > 2$, $\mathrm{cat}_{\mathrm{LS}} Y = \dim Y = n$ if and only if $0 \neq a^n \in H^n(Y; I^{\otimes n})$.

It is clear that (2) $\Rightarrow$ (1), while the converse follows by an obstruction theoretic argument, *cf.* [**Bab93**, Lemma 8.4]. □

REMARK 12.5.3. Theorem 12.5.2 remains valid for polyhedra, as well. Namely, for $n > 2$ the same proof is applicable. For $n = 2$ the equivalence of (1) and (3) follows from results of C. Wall [**Wal65**], see [**KatzR07**] for details. Clearly, (2) implies (1), and we can deduce (2) from (1) via obstruction theoretic arguments if we use 2-dimensional cohomology with non-abelian coefficients, *cf.* [**BrowHK84, Ell88, Whi49**].

12.6. Inessential manifolds and pullback metrics

We prove a compression result valid for any of our systolic notions, including the homotopy 1-systole, relying essentially on I. Babenko's proof of [**Bab93**, Lemma 8.4], *cf.* [**Bab93**, Proposition 2.7].

The main idea is the following. If M can be deformed, in the classifying space, into a skeleton of positive codimension, then it admits metrics of zero volume and positive systole obtained by pullback from a metric on the skeleton, thus violating the systolic inequality.

THEOREM 12.6.1 (Compression theorem). *Let $n = k_1 + \ldots + k_d$. Let $f : M^n \to K$ be a map of polyhedra with the following three properties:*

(1) *the map f induces an isomorphism of fundamental groups;*
(2) *we have $f(M) \subset K^{(n-1)}$;*
(3) *the induced homomorphism f_* is injective in k_i-dimensional homology for each $k_i < n$, $i = 1, \ldots, d$.*

Then the inequality

$$\prod_i \mathrm{sys}_{k_i}(M) \leq C \,\mathrm{vol}_n(M) \tag{12.6.1}$$

is violated, for any $C < \infty$, by a suitable metric on M.

PROOF. Without loss of generality we can assume that f is linear on each simplex, and on each top dimensional simplex f is a projection to a face of positive codimension. Choose a fixed piecewise linear metric $\mathcal{G}_{\mathrm{PL}}$ on $K^{(n-1)}$. This amounts to a choice of a positive quadratic form on each simplex, such that the forms of neighboring simplices agree on their common face. Then the pullback $f^*(\mathcal{G}_{\mathrm{PL}})$ is a non-negative quadratic form on M which is of rank at most $(n-1)$ at every point of M. Nonetheless, the volume of a Lipschitz simplex can be defined as usual with respect to this form. Thus the k-systole of $(M, f^*(\mathcal{G}_{\mathrm{PL}}))$ can be defined as in (12.1.2), in terms of Lipschitz chains $c \in C_k(M)$. Since f induces a monomorphism f_* by hypothesis, each systole $\mathrm{sys}_{k_i}(M, f^*(\mathcal{G}_{\mathrm{PL}}))$ is finite, and

$$\mathrm{sys}_k\left(M, f^*(\mathcal{G}_{\mathrm{PL}})\right) \geq \mathrm{sys}_k\left(K^{(n-1)}, \mathcal{G}_{\mathrm{PL}}\right) > 0.$$

Then the inequality (12.6.1) is certainly violated, to the extent that the left hand side is positive (*cf.* [**KatzR07**]), whereas the right hand side vanishes. The "metric" $f^*(\mathcal{G}_{\rm PL})$ has two shortcomings. First, it is only defined on the full tangent space at $x \in M$ when $x \notin M^{(n-1)}$, *i.e.* outside the codimension 1 skeleton of M. Furthermore, the non-negative quadratic form is not definite, even in the interior of a top dimensional cell. Both shortcomings are overcome by the metric

$$\phi_{M^{(n-1)}} f^*(\mathcal{G}_{\rm PL}) + \left(1 - \phi_{M^{(n-1)}}\right) \mathcal{G}_0, \tag{12.6.2}$$

where $\mathcal{G}_0$ is a fixed smooth "background" metric on M dominating the metric $\mathcal{G}_{\rm PL}$, while $\phi_{M^{(n-1)}}$ is a function of "bump" type, which equals 1 outside of an ε-neighborhood of the codimension 1 skeleton, and vanishes in an $\varepsilon/2$-neighborhood. The volume becomes arbitrarily small when ε tends to zero, thus violating (12.6.1), *cf.* [**Bab93**], [**BabK98**, section 6]. □

The following corollary, due to I. Babenko [**Bab93**], is a converse of Theorem 12.2.2.

COROLLARY 12.6.2. *If M^n is not n-essential, then it fails to satisfy Gromov's inequality* (12.2.1) *for the homotopy* 1*-systole.*

PROOF. Indeed, we can find a classifying map

$$f : M \to K(\pi_1(M), 1) = K$$

such that $f(M) \subset K^{(n-1)}$, and we apply the compression theorem to the homotopy 1-systole. □

12.7. Manifolds of dimension 3

The main result of the article [**GG92**], *cf.* also [**OR01**], is the following theorem.

THEOREM 12.7.1. *Let M be a 3-manifold. Then $\mathrm{cat}_{\rm LS} M = 1$ if M is simply connected, while $\mathrm{cat}_{\rm LS} M = 2$ if $\pi_1(M)$ is a free group, and $\mathrm{cat}_{\rm LS} M = 3$ otherwise.* □

The proof, in the orientable case, relies on the following algebraic alternative for 3-manifolds, which persists to a certain extent for PD_3 spaces, *cf.* J. Hillman's work [**Hi03**].

PROPOSITION 12.7.2. *Let π be the fundamental group of a closed orientable 3-manifold. Then either π is free, or the image of the fundamental homology class in $H_3(K(\pi, 1), \mathbb{Z})$ under a classifying map is non-zero.*

PROOF. Decompose our 3-manifold M as a connected sum of prime manifolds, see [**Hem76**, Theorems 3.15 and 3.21]. If $\pi_1(M)$ is not free, then $M = X \# Y$ where X has finite fundamental group or is aspherical. Let $\tau = \pi_1(X)$. If X is aspherical, then $H_3(\tau; \mathbb{Z}) = H_3(X; \mathbb{Z}) \neq 0$. If τ is finite, we have $\pi_2(X) = 0$, since X is covered by the homotopy 3-sphere. Hence we have the Hopf exact sequence

$$\pi_3(X) \to H_3(X; \mathbb{Z}) \to H_3(\tau; \mathbb{Z}) \to 0$$

of the form

$$\mathbb{Z} \to \mathbb{Z} \to \mathbb{Z}_d \to 0$$

and the degree d of the first (Hurewicz) map is equal to the order of fundamental group τ. In either case, if the map $g : X \to K(\tau, 1)$ induces an isomorphism of fundamental groups, then

$$g_* : H_3(X; \mathbb{Z}) \to H_3(\pi; \mathbb{Z}) \tag{12.7.1}$$

is surjective.

Now, the map $c : M \to X$ (contracting Y) induces an isomorphism $H_3(M; \mathbb{Z}) \to H_3(X; \mathbb{Z})$. We have $\pi = \tau * G$ where $G = \pi_1(Y)$, and the imbedding $\tau \to \tau * G = \pi$ induces a map

$$h : K(\tau, 1) \to K(\tau, 1) \vee K(G, 1) = K(\pi, 1)$$

which, in turn, induces a monomorphism

$$h_* : H_*(\tau) \to H_*(\tau) \oplus H_*(G) = H_*(\pi).$$

Consider the composition

$$H_3(M; \mathbb{Z}) \xrightarrow{\ c\ } H_3(X; \mathbb{Z}) \xrightarrow{\ g_*\ } H_3(\tau; \mathbb{Z}) \xrightarrow{\ h_*\ } H_3(\pi; \mathbb{Z})$$

with g_* as in (12.7.1). This composition is non-zero. Since the map

$$M \xrightarrow{\ c_*\ } X \xrightarrow{\ g\ } K(\tau, 1)) \xrightarrow{\ h\ } K(\pi, 1)$$

factors through a classifying map $M \to K(\pi, 1)$, the proposition is proved. □

In the case of free fundamental group, the theorem follows from the fact that the manifold is homotopy equivalent to a connected sum of 2-sphere bundles over the circle.

COROLLARY 12.7.3. *Let M be a 3-manifold. Then:*
(i) *if M is simply connected then* $\operatorname{cat}_{\mathrm{LS}} M = \operatorname{cat}_{\mathrm{sys}} M = 1$*;*
(ii) *if $\pi_1(M)$ is not a free group then* $\operatorname{cat}_{\mathrm{LS}} M = \operatorname{cat}_{\mathrm{sys}} M = 3$*;*
(iii) *if M is orientable and $\pi_1(M)$ is a free group, then* $\operatorname{cat}_{\mathrm{LS}} M = \operatorname{cat}_{\mathrm{sys}} M = 2$.

REMARK 12.7.4. For non-orientable 3-manifolds with free fundamental group, we only prove that $1 \le \operatorname{cat}_{\mathrm{sys}} M \le \operatorname{cat}_{\mathrm{LS}} M = 2$. This follows from Theorem 12.2.6. In fact, one has $\operatorname{cat}_{\mathrm{sys}} M = 2$ in this case, as well [**KatzR07**].

PROOF OF COROLLARY 12.7.3. Item (i) is obvious. To prove item (ii), note that the equality $\operatorname{cat}_{\mathrm{LS}} M = 3$ follows from Theorem 12.7.1, and we apply Theorem 12.2.6.

To prove item (iii), note that if the fundamental group of M is free, then we have $\operatorname{cat}_{\mathrm{LS}} M = 2$ by Theorem 12.7.1. Since $b_1(M) \ge 1$, we have $\operatorname{cat}_{\mathrm{sys}} M \ge 2$ by (12.3.3), and $\operatorname{cat}_{\mathrm{sys}} M \le 2$ by Theorem 12.2.6. □

Thus, a 3-manifold with systolic category 1 is diffeomorphic to a 3-sphere by the work of G. Perelman [**Pe02**].

REMARK 12.7.5. Let $f : M \to K(\pi_1(M), 1)$ be a classifying map. An interesting role in dimension 3 is played by the lowest dimension $d_{\min}$ of the skeleton of $K(\pi_1(M), 1)$ that f can be deformed into, *i.e.* the maximal d such that M is d-essential. For 3-manifolds, this dimension determines both categories. Namely, if $d_{\min} = 1$ then both categories are 2, whereas if $d_{\min} = 3$, both categories are also 3. Here $d_{\min}$ cannot be equal to 2. (Indeed, if f can be deformed into the 2-skeleton then M is not 3-essential, hence $\operatorname{cat}_{\mathrm{LS}} M \le 2$, and so $\pi_1(M)$ is free, hence $K(\pi_1(M),1)$ is homotopy equivalent to a wedge of circles.) The case of maximal $d_{\min}=n=\dim(M)$ also characterizes the case when both categories equal n, by Theorem 12.2.6. It would be interesting to explore the general case $d_{\min}<n$, *cf.* Appendix B, item B.1.2.

12.8. Category of simply connected manifolds

We first recall the following general result [**CorT03**, Theorem 1.50].

THEOREM 12.8.1. *Every $(k-1)$-connected CW-space X satisfies the inequality* $\mathrm{cat}_{\mathrm{LS}}X \le k^{-1}\dim X$. □

COROLLARY 12.8.2. *If $M^n, n=4,5$ is a simply connected manifold which is not homotopy equivalent to the sphere S^n, then* $\mathrm{cat}_{\mathrm{LS}}M=2$.

PROOF. Since M is not a homotopy sphere, we have $H^2(M;\mathbb{Z}_p)\neq 0$ for some prime p. Indeed, otherwise, by Poincaré duality, we have $H^{n-2}(M;\mathbb{Z}_p)=0$ for all prime p. In other words, $H^i(M;\mathbb{Z}_p)=0$ for all primes p and all $i\neq 0,n$. By the Universal Coefficient Theorem, we have $H_i(M)=0$ for $i\neq 0,n$, and so M is a homotopy sphere because it is simply connected.

Let $x\in H^2(M;\mathbb{Z}_p)\setminus\{0\}$. By Poincaré duality, there exists an element $y\in H^{n-2}(M;\mathbb{Z}_p)$ such that $xy\neq 0$. Now the cup-length estimate implies $\mathrm{cat}_{\mathrm{LS}}M\ge 2$, and the claim follows from Theorem 12.8.1. □

EXAMPLE 12.8.3 (Four-manifolds with category 2). If M^4 is simply connected and is not homotopy equivalent to S^4, then the second Betti number is positive, and M satisfies the stable systolic inequality in middle dimension:

$$\mathrm{stsys}_2(M)^2 \le C(b_2(M))\,\mathrm{vol}_4(M), \tag{12.8.1}$$

cf. inequalities (12.3.3) and (13.2.4), and hence $\mathrm{cat}_{\mathrm{sys}}(M)=2$. On the other hand, one has $\mathrm{cat}_{\mathrm{LS}}M=2$ by Corollary 12.8.2, in harmony with equally (12.4.2).

EXAMPLE 12.8.4. By Smale's work [**Sm62**], a simply connected spin 5-manifold which is a rational homology sphere is a connected sum of manifolds M_k, $k\ge 2$, where

$$H_2(M_k;\mathbb{Z})=\mathbb{Z}_k+\mathbb{Z}_k.$$

By Corollary 12.8.2, all these manifolds have Lusternik-Schnirelmann category equal to 2. It is unknown whether any systolic inequalities are satisfied by M_k. Based on the value of its $\mathbb{Z}_k$-cup-length, we could conjecture the existence of a systolic inequality

$$\mathrm{sys}_2\,\mathrm{sys}_3 \le C(M_k)\,\mathrm{vol}_5, \tag{12.8.2}$$

where the systoles are calculated over the cycles with $\mathbb{Z}_k$ coefficients which are not zero-homologous, *cf.* (15.2.2). For a discussion of the notion of volume for cycles with torsion coefficents, see B. White's work [**Wh99**].

An interesting family of metrics on M_k was studied in [**PatP03**, Theorem 8.2]. Does it satisfy inequality (12.8.2)?

EXAMPLE 12.8.5. There are examples of S^2-bundles M over spheres satisfying the equality $\mathrm{cat}_{\mathrm{LS}}(M)=3$ [**Sin78**, §3]. Such a bundle $M^{16}\to S^{14}$ can be induced by an element of $\pi_{14}(S^4)$ from the bundle $\mathbb{CP}^3\to S^4$ obtained as a circle quotient of the quaternionic Hopf bundle $S^7 \to S^4$, *cf.* Iwase's work [**Iw02**] in the context of the Ganea conjecture. Now let $u \in H^2(M^{16};\mathbb{Z}) = \mathbb{Z}$ be a generator, and let

$$f : M \to K(\mathbb{Z},2) = \mathbb{CP}^\infty$$

classify u. By the construction of M^{16}, the map f factors through a skeleton $\mathbb{CP}^3 \subset K(\mathbb{Z},2)$.

Hence M^{16} admits a positive quadratic form of zero "volume" (as well as zero 14-systole), but positive 2-systole, *cf.* Section 12.6. By compression theorem 12.6.1, there could be no inequality of type $\operatorname{sys}_2^8 \leq C \operatorname{vol}_{16}$. By Poincaré duality and (12.3.4), the manifold M^{16} has stable systolic category 2. Since there is no torsion in homology, no higher value could be expected for systolic category. Thus, we have $\operatorname{cat}_{\rm sys} M = 2 < 3 = \operatorname{cat}_{\rm LS} M$, in contrast with (12.4.2). It is unknown whether the strict inequality can occur in dimension lower than 16.

CHAPTER 13

Gromov's optimal stable systolic inequality for $\mathbb{CP}^n$

In Section 13.2, we present Gromov's proof of the optimal stable systolic inequality (13.2.4) for all metrics on complex projective space $\mathbb{CP}^n$, *cf.* [**Gro99**, Theorem 4.36], based on the cup product decomposition of its fundamental class, *cf.* (15.2.2). The proof relies upon the Wirtinger inequality, proved in Section 13.1. For background on differential forms and the de Rham theorem, see [**BoT82**], *cf.* Section 14.2.

Generalisations exist for the other two-point homogeneous spaces, but they are not yet optimal [**Gro96**, Section 4], [**Gro99**, Remark 4.37, p. 262], see Section 13.3 below.

In Section 14.3 of the next chapter, we will adapt Gromov's proof of the optimal stable systolic inequality for $\mathbb{CP}^n$, to a situation where the cup product is trivial, but a nontrivial Massey product is available. The result is a stable systolic inequality mildly dependent on the metric via an isoperimetric quotient. An example of such an inequality in the case of a homogeneous manifold is described in Section 14.4.

13.1. Federer's proof of the Wirtinger inequality

Following H. Federer [**Fed69**, p. 40], we prove an optimal upper bound for the comass norm $\| \ \|$, *cf.* (15.5.2) below, of the exterior powers of an alternating 2-form.

Let V be a vector space over $\mathbb{C}$. Let $H = H(v, w)$ be a Hermitian product on V, with real part $v \cdot w$, and imaginary part $A = A(v, w)$, where $A \in \bigwedge^2 V$, the second exterior power of V. Here we adopt Federer's convention that H is complex linear in the *second* variable.

EXAMPLE 13.1.1. Let $Z_1, \ldots, Z_\nu \in \bigwedge^1(\mathbb{C}^\nu, \mathbb{C})$ be the coordinate functions in $\mathbb{C}^\nu$. We then have the standard (symplectic) 2-form $A \in \bigwedge^2(\mathbb{C}^\nu, \mathbb{C})$, given by

$$A = \tfrac{i}{2} \sum_{j=1}^{\nu} Z_j \wedge \bar{Z}_j. \tag{13.1.1}$$

Recall that an alternating form is called *simple* if it can be expressed as a wedge product of 1-forms.

DEFINITION 13.1.2. The *comass norm* of a linear p-form ω on $\mathbb{R}^n$ is

$$\|\omega\| = \max \left\{ \omega(e_1, \ldots, e_p) \mid e_i \in \mathbb{R}^n \right\}, \tag{13.1.2}$$

where $|e_i| = 1$ for $1 \le i \le p$.

LEMMA 13.1.3. *The comass of the form A of* (13.1.1) *satisfies* $\|A\| = 1$.

PROOF. We can set $\xi = v \wedge w$, where v and w are orthonormal. We then have $H(v, w) = iA(v, w)$, hence

$$\langle \xi, A \rangle = A(v, w) = H(iv, w) = (iv) \cdot w \le 1,$$

where equality is attained if and only if $iv = w$. □

PROPOSITION 13.1.4 (Wirtinger inequality). *Let $\mu \geq 1$. If $\xi \in \bigwedge_{2\mu} V$ and ξ is simple, then*

$$\langle \xi, A^\mu \rangle \leq \mu! \, |\xi|;$$

equality holds if and only if there exist elements $v_1, \ldots, v_\mu \in V$ such that

$$\xi = v_1 \wedge (iv_1) \wedge \cdots \wedge v_\mu \wedge (iv_\mu).$$

Consequently, we have $\|A^\mu\| = \mu!$

PROOF. We assume that $|\xi| = 1$, where $| \ |$ is the natural Euclidean norm in $\bigwedge_{2\mu} V$. The case $\mu = 1$ was treated in Lemma 13.1.3.

In the general case $\mu \geq 1$, we consider the 2μ dimensional subspace T associated with ξ. Let $f : T \to V$ be the inclusion map, and consider the pullback 2-form $(\wedge^2 f)A \in \bigwedge^2 T$. Next, we orthogonally diagonalize the skew-symmetric 2-form, *i.e.* decompose it into 2×2 diagonal blocks. Thus, we can choose dual orthonormal bases $e_1, \ldots, e_{2\mu}$ of T and $\omega_1, \ldots, \omega_{2\mu}$ of $\bigwedge^1 T$, and nonnegative numbers $\lambda_1, \ldots, \lambda_\mu$, so that

$$(\wedge^2 f)A = \sum_{j=1}^{\mu} \lambda_j \left(\omega_{2j-1} \wedge \omega_{2j}\right). \tag{13.1.3}$$

By Lemma 13.1.3, we have

$$\lambda_j = A(e_{2j-1}, e_{2j}) \leq \|A\| = 1 \tag{13.1.4}$$

for each j. Noting that $\xi = \epsilon e_1 \wedge \cdots \wedge e_{2\mu}$ with $\epsilon = \pm 1$, we compute

$$\left(\wedge^{2\mu} f\right) A^\mu = \mu! \, \lambda_1 \ldots \lambda_\mu \omega_1 \wedge \cdots \wedge \omega_{2\mu},$$

and therefore

$$\langle \xi, A^\mu \rangle = \epsilon \mu! \, \lambda_1 \ldots \lambda_\mu \leq \mu! \tag{13.1.5}$$

Note that equality occurs in (13.1.5) if and only if $\epsilon = 1$ and $\lambda_j = 1$. Applying the proof of Lemma 13.1.3, we conclude that $e_{2j} = ie_{2j-1}$, for each j. □

COROLLARY 13.1.5. *Every 2-form ω satisfies the comass bound*

$$\|\omega^\mu\| \leq \mu! \|\omega\|^\mu. \tag{13.1.6}$$

PROOF. An inspection of the proof Proposition 13.1.4 reveals that the orthogonal diagonalisation argument, *cf.* (13.1.4), applies to an arbitrary 2-form ω with comass $\|\omega\| = 1$. □

13.2. Optimal inequality for complex projective space

THEOREM 13.2.1 (M. Gromov). *Every Riemannian metric $\mathcal{G}$ on complex projective space $\mathbb{CP}^n$ satisfies the optimal inequality*

$$\operatorname{stsys}_2(\mathcal{G})^n \leq n! \operatorname{vol}_{2n}(\mathcal{G});$$

equality holds for the Fubini-Study metric on $\mathbb{CP}^n$.

PROOF. We follow M. Gromov's notation in [**Gro99**, Theorem 4.36]. We choose a homology generator $\alpha \in H_2(\mathbb{CP}^n;\mathbb{Z}) = \mathbb{Z}$. Let $\omega \in H^2(\mathbb{CP}^n;\mathbb{Z}) = \mathbb{Z}$ be the dual generator in cohomology. Then the cup power ω^n is a generator of $H^{2n}(\mathbb{CP}^n;\mathbb{Z}) = \mathbb{Z}$, which may be assumed to be positively oriented. Let $\eta \in \Omega^2(\mathbb{CP}^n)$ be a closed differential 2-form representing the class ω, viewed as a de Rham cohomology class, *cf.* Section 14.2. The wedge product of the de Rham differential graded associative algebra $\Omega^*(X)$ descends to cup product in $H^*(X,\mathbb{R})$, *cf.* Section 14.2. Therefore we have

$$1 = \int_{\mathbb{CP}^n} \eta^{\wedge n}. \tag{13.2.1}$$

Now let $\mathcal{G}$ be a metric on $\mathbb{CP}^n$. Recall that the pointwise comass norm for a simple k-form coincides with the natural Euclidean norm on k-forms associated with $\mathcal{G}$. In general, it is defined by evaluating on k-tuples of unit vectors and taking the maximal value, *cf.* (13.1.2) and (15.5.2). The comass norm $\| \ \|_\infty$ of a differential k-form is, by definition, the supremum of the pointwise comass norms. Then by the Wirtinger inequality and Corollary 13.1.5, we obtain

$$\begin{aligned} 1 &\le \int_{\mathbb{CP}^n} \|\eta^{\wedge n}\| \, d\mathrm{vol} \\ &\le n! \left(\|\eta\|_\infty\right)^n \mathrm{vol}_{2n}(\mathbb{CP}^n, \mathcal{G}) \end{aligned} \tag{13.2.2}$$

(see [**Gro99**, Remark 4.37] for a discussion of the constant in the context of the Wirtinger inequality). The infimum of (13.2.2) over all 2-forms $\eta \in \omega$ gives

$$1 \le n! \left(\|\omega\|^*\right)^n \mathrm{vol}_{2n}\left(\mathbb{CP}^n, \mathcal{G}\right), \tag{13.2.3}$$

where $\| \ \|^*$ is the comass norm in cohomology. Denote by $\| \ \|$ the stable norm in homology. Recall that the normed lattices $(H_2(M;\mathbb{Z})_\mathbb{R}, \| \ \|)$ and $(H^2(M;\mathbb{Z})_\mathbb{R}, \| \ \|^*)$ are dual to each other [**Fed69**], *cf.* Chapter 15. Therefore

$$\|\alpha\| = \frac{1}{\|\omega\|^*},$$

and hence

$$\mathrm{stsys}_2(\mathcal{G})^n = \|\alpha\|^n \le n! \, \mathrm{vol}_{2n}(\mathcal{G}). \tag{13.2.4}$$

Equality is attained by the two-point homogeneous Fubini-Study metric, since the standard $\mathbb{CP}^1 \subset \mathbb{CP}^n$ is calibrated by the Fubini-Study Kahler 2-form, which satisfies equality in the Wirtinger inequality at every point. □

EXAMPLE 13.2.2. For the class of all metrics $\mathcal{G}$ on the complex projective plane, we obtain the optimal inequality

$$\mathrm{stsys}_2(\mathbb{CP}^2, \mathcal{G})^2 \le 2\,\mathrm{vol}_4(\mathbb{CP}^2, \mathcal{G}),$$

which in fact generalizes to the manifold obtained as the connected sum of a finite number of copies of $\mathbb{CP}^2$, *cf.* (20.5.2). The situation in the quaternionic projective plane is discussed in Section 13.3.

In harmony with equality (12.4.2), we obtain the following corollary.

COROLLARY 13.2.3. *Systolic category of complex projective space $\mathbb{CP}^n$ equals n, and thus agrees with $\mathrm{cat}_{\mathrm{LS}}(\mathbb{CP}^n)$.*

13.3. Quaternionic projective plane

The quaternionic projective plane $\mathbb{HP}^2$ has volume $\mathrm{vol}_8(\mathbb{HP}^2) = \frac{\pi^4}{5!}$ for the symmetric metric with sectional curvature $1 \le K \le 4$, while for the projective line with $K \equiv 4$ we have $\mathrm{vol}_4(\mathbb{HP}^1) = \frac{\pi^2}{3!}$, *cf.* [**Berg72a**, formula (3.10)]. This gives a systolic ratio

$$\frac{\mathrm{stsys}_4(\mathbb{HP}^2)^2}{\mathrm{vol}_8(\mathbb{HP}^2)} = \frac{10}{3} \tag{13.3.1}$$

for the symmetric metric.

In more detail, we endow $\mathbb{HP}^n$ with the natural metric as the base space of the Riemannian submersion from the unit sphere

$$S^{4n+3} \subset \mathbb{H}^{n+1}.$$

A projective line $\mathbb{HP}^1 \subset \mathbb{HP}^n$ is a round 4-sphere of (Riemannian) diameter $\frac{\pi}{2}$ and sectional curvature $+4$, attaining the maximum of sectional curvatures of $\mathbb{HP}^n$. The extension of scalars from $\mathbb{R}$ to $\mathbb{H}$ gives rise to an inclusion $\mathbb{R}^3 \hookrightarrow \mathbb{H}^3$, and thus an inclusion $\mathbb{RP}^2 \hookrightarrow \mathbb{HP}^2$. Then $\mathbb{RP}^2 \subset \mathbb{HP}^2$ is a totally geodesic submanifold of diameter $\frac{\pi}{2}$ and curvature $+1$, attaining the minimum of the sectional curvatures of $\mathbb{HP}^2$, *cf.* [**ChE75**, p. 73].

The following proposition was essentially proved by V. Kraines [**Kr66**] and M. Berger [**Berg72a**]. The invariant 4-form was briefly discussed in [**HarL82**, p. 152]. A discussion of the dga algebra $\Omega^*(M)$ appears in Section 14.2.

PROPOSITION 13.3.1. *There is a differential 4-form $\kappa_{\mathbb{HP}} \in \Omega^4(\mathbb{HP}^2)$, representing a generator of $H^4(\mathbb{HP}^2, \mathbb{Z}) = \mathbb{Z}$, with*

$$|\kappa_{\mathbb{HP}}{}^2| = \tfrac{10}{3}\|\kappa_{\mathbb{HP}}\|^2 \tag{13.3.2}$$

and

$$|\kappa_{\mathbb{HP}}|^2 = \tfrac{10}{3}\|\kappa_{\mathbb{HP}}\|^2, \tag{13.3.3}$$

where $|\ |$ and $\|\ \|$ are, respectively, the Euclidean norm and the comass of the unit volume symmetric metric on $\mathbb{HP}^2$.

PROOF. Consider the quaternionic vector space $\mathbb{H}^n = \mathbb{R}^{4n}$. Each of the three quaternions i, j, and k defines a complex structure on $\mathbb{H}^n$, *i.e.* an identification of $\mathbb{H}^n$ with $\mathbb{C}^{2n}$. The imaginary part of the associated Hermitian inner product on $\mathbb{C}^{2n}$ is the standard symplectic exterior 2-form, *cf.* Example 13.1.1. Let ω_i, ω_j, and ω_k be the triple of 2-forms on $\mathbb{H}^n$ defined by the three complex structures. We take their wedge squares ω_i^2, ω_j^2, and ω_k^2. Note that ω_i is the sum of 4 monomial terms, while ω_i^2 is twice the sum of 6 monomial terms. We define an exterior 4-form κ_n, first written down explicitly by V. Kraines [**Kr66**], by setting

$$\kappa_n = \tfrac{1}{6}\left(\omega_i^2 + \omega_j^2 + \omega_k^2\right). \tag{13.3.4}$$

The coefficient $\frac{1}{6}$ normalizes the form to unit comass, *cf.* Lemma 13.1.3. The form κ_n is invariant under transformations in $Sp(n) \times Sp(1)$ [**Kr66**, Theorem 1.9]. The corresponding parallel differential 4-form $\kappa_{\mathbb{HP}} \in \Omega^4\mathbb{HP}^n$ is closed. The form $\kappa_{\mathbb{HP}}$ represents a nonzero element of the line $H^4(\mathbb{HP}^n, \mathbb{R}) = \mathbb{R}$.

The 4-form $3\kappa_2$ on $\mathbb{H}^2$ decomposes into a sum of 18 simple 4-forms m, *i.e.* "monomials" in the 8 coordinate 1-forms. The 18 monomials are not all distinct.

Two of them, denoted m_0 and its Hodge star $*m_0$, occur with multiplicity 3. Thus, we obtain a decomposition as a linear combination of seven pairs

$$3\kappa_2 = 3(m_0 + *m_0) + \sum_{\ell=1}^{6}(m_\ell + *m_\ell), \tag{13.3.5}$$

where $*$ is the Hodge star operator, *cf.* (15.3.1).

LEMMA 13.3.2. *The Kraines form has unit comass:* $\|\kappa_2\| = 1$.

This was proved in [**Berg72a, DaHM88**]. Meanwhile, from (13.3.5) we have

$$(3\kappa_2)^2 = 2\,(9\,\mathrm{vol} + 6\,\mathrm{vol})\,,$$

where $\mathrm{vol} = e_1 \wedge e_2 \wedge \cdots \wedge e_8$ is the volume form of $\mathbb{H}^2 = \mathbb{R}^8$. Hence

$$\left|(3\kappa_2)^2\right| = 2 \cdot 15 = 30,$$

proving identity (13.3.2). Meanwhile, $|3\kappa_2|^2 = 9 + 9 + 12 = 30$, proving identity (13.3.3). □

REMARK 13.3.3. There is a misprint in the calculation of the systolic constants in [**Berg72a**, Theorem 6.3], as is evident from [**Berg72a**, formula (6.14)]. Namely, in the last line on page [**Berg72a**, p. 12], the formula for the coefficient $s_{4,b}$ lacks the exponent b over the constant 6 appearing in the numerator. The formula should appear as follows:

$$s_{4,b} = \frac{6^b}{(2b+1)!}.$$

DEFINITION 13.3.4. The *Wirtinger constant* Wirt_n of $\mathbb{R}^{2n}$ is the maximal ratio $\frac{\|\omega^2\|}{\|\omega\|^2}$ over all n-forms $\omega \in \Lambda^n\mathbb{R}^{2n}$. The *modified Wirtinger constant* Wirt'_n is the maximal ratio $\frac{|\omega|^2}{\|\omega\|^2}$ over all n-forms ω on $\mathbb{R}^{2n}$.

If ω is a middle-dimensional form, then

$$\|\omega^2\| = |\omega^2| = \langle \omega, *\omega \rangle \le |\omega|\,|*\,\omega| = |\omega|^2, \tag{13.3.6}$$

proving that $\mathrm{Wirt}_n \le \mathrm{Wirt}'_n$.

By the Wirtinger inequality and Corollary 13.1.5, we have $\mathrm{Wirt}_2 = \mathrm{Wirt}'_2 = 2$.

DEFINITION 13.3.5. Let X be an orientable, closed manifold of dimension $2n$. We define the systolic ratio $\mathrm{SR}_{2n,n}$ by setting

$$\mathrm{SR}_{2n,n}(X) = \sup_{\mathcal{G}} \frac{\mathrm{stsys}_n(\mathcal{G})^2}{\mathrm{vol}_{2n}(\mathcal{G})}$$

where the supremum is over all metrics $\mathcal{G}$ on X.

Thus, $\mathrm{SR}_{4,2}(\mathbb{CP}^2) = 2$.

PROPOSITION 13.3.6. *Let X be an orientable, closed manifold of dimension $2n$, with $b_n(X) = 1$. Then*

$$\mathrm{SR}_{2n,n}(X) \le \mathrm{Wirt}_n.$$

PROOF. The fundamental cohomology class of $H^{2n}(X,\mathbb{Z}) = \mathbb{Z}$ is the cup square of a generator of $H^n(X,\mathbb{Z})_\mathbb{R} = \mathbb{Z}$, by Poincaré duality. The inequality is now immediate by applying the method of proof of (13.2.2). If a form realizing Wirt'_n happens to be self-dual, so that we have $\mathrm{Wirt}_n = \mathrm{Wirt}'_n$, *cf.* (13.3.6), then we can also apply the method of proof of (20.5.2). □

COROLLARY 13.3.7. *Let $n \in \mathbb{N}$. We have the following bounds for the systolic ratio:*

$$\mathrm{SR}_{8n,4n}(\mathbb{HP}^{2n}) \leq \mathrm{Wirt}_{4n}$$
$$\mathrm{SR}_{4n,2n}(\mathbb{CP}^{2n}) \leq \mathrm{Wirt}_{2n}$$
$$\mathrm{SR}_{16,8}(M^{16}) \leq \mathrm{Wirt}_8$$

where M^{16} is the Cayley projective plane.

REMARK 13.3.8. The n-th power of the Kahler form on $\mathbb{CP}^{2n}$ is the middle-dimensional generator in $H^{2n}(\mathbb{CP}^{2n},\mathbb{Z})$. Therefore one easily calculates the value of the systolic ratio of the symmetric metric, namely $(2n)!/(n!)^2$. For $2n = 2$, we obtain the value 6, so that $\mathrm{Wirt}_4 \geq 6$. Since

$$6 > \tfrac{10}{3},$$

the symmetric metric of the space $\mathbb{CP}^4$ has higher systolic ratio than the symmetric metric of $\mathbb{HP}^2$. In contrast, it is shown in [**BaKSS06**] that

$$\mathrm{SR}_{8,4}(\mathbb{HP}^2) = \mathrm{SR}_{8,4}(\mathbb{CP}^4).$$

Furthermore, it is shown in [**BaKSS06**] that $\mathrm{Wirt}_4 = 14$. Thus, it is in principle impossible to prove an optimal stable systolic inequality for an arbitrary metric on $\mathbb{CP}^4$, by any direct generalisation of the method of proof of (20.5.2) or (13.2.2).

CHAPTER 14

Systolic inequalities dependent on Massey products

In Section 13.2, we described Gromov's optimal stable systolic inequality for complex projective space, which is the prototype both for a number of other systolic inequalities in a manifold M, and for lower bounds for the systolic category of M. The latter depend on nontrivial cup product relations in the cohomology algebra of M.

In the present chapter, we will describe a related phenomenon in the context of Massey products. Namely, the existence of certain systolic inequalities for M is conditional upon the existence of nontrivial Massey products in the cohomology algebra of M.

By Stokes' theorem, the natural integration map $\oint$ from the de Rham complex Ω to the (smooth) singular cochain complex C^*, commutes with the *differentials* of these complexes, *i.e.* it is a cochain map. On the other hand, the integration map does not respect the *multiplicative structures* of the two dga algebras whose underlying cochain complexes are Ω and C^*. In other words, $\oint$ is not a morphism of dga algebras. The effect of $\oint$ on Massey products is, therefore, not obvious.

To handle this difficulty, and following a suggestion of R. Hain, we will describe a succession of five dga quasi-isomorphisms linking the two dga algebras, constructed in **[FelHT01]**, *cf.* Remark 14.2.3.

14.1. Massey Products via Differential Graded Associative Algebras

DEFINITION 14.1.1. A *differential graded associative algebra* (dga algebra for short) is a pair (A, d), where A is an associative algebra with a grading, while d is a differential of degree 1. The *homology algebra* of (A, d) is the algebra $H = \operatorname{Ker} d / \operatorname{Im} d$.

DEFINITION 14.1.2. Given (homogeneous) elements $u, v, w \in H$ with $uv = 0 = vw$, the *Massey triple product*

$$\langle u, v, w\rangle = \langle u, v, w\rangle_A \subset H$$

(or simply *Massey product* for short) is defined as follows. Choose elements $a, b, c \in A$ representing the homology classes u, v, w, respectively. Then $dx = ab$ and $dy = bc$ for suitable $x, y \in A$, and the element $xc - (-1)^{|u|}ay$ is a cycle of A. We now define $\langle u, v, w\rangle$ to be the set of all elements of the form

$$\left[xc - (-1)^{|u|}ay\right] \in H,$$

with all possible choices of a, b, c, x, y as above.

The set $\langle u, v, w\rangle$ is a coset with respect to the so-called *indeterminacy subgroup*, denoted $\mathrm{Indet} \subset H^{|u|+|v|+|w|-1}$, and defined as follows:

$$\mathrm{Indet} = uH^{|v|+|w|-1} + H^{|u|+|v|-1}w.$$

A Massey product is said to be *non-trivial* if $0 \notin \langle u, v, w\rangle$, *i.e.* if we have

$$\langle u, v, w\rangle \cap \mathrm{Indet} = \emptyset.$$

See [**May69**], [**RudT00**] for more details.

DEFINITION 14.1.3. A *morphism* $f : A \to A'$ *of dga algebras* is a homomorphism of algebras that is also a chain map. A morphism $f : A \to A'$ of dga algebras is called a *quasi-isomorphism* if the induced homomorphism $f_* : H \to H'$ is an isomorphism.

PROPOSITION 14.1.4. *Each morphism $f : A \to A'$ of dga algebras induces a map of Massey products. In other words, if the Massey product $\langle u, v, w\rangle$ with $u, v, w \in H(A)$ is defined, then the corresponding Massey product $\langle f_*u, f_*v, f_*w\rangle$ is defined, and we have an inclusion*

$$f_*\langle u, v, w\rangle \subset \langle f_*u, f_*v, f_*w\rangle. \tag{14.1.1}$$

Moreover, if f is a quasi-isomorphism, then $f_* : H(A) \to H(A')$ induces a bijection of Massey products, i.e. the inclusion (14.1.1) *is bijective.*

PROOF. Let a, b, c be elements of A whose homology classes are u, v, w, respectively, while $x, y \in A$ satisfy $dx = ab$ and $dy = bc$, so that we have an element

$$\left[xc - (-1)^{|u|}ay\right] \in \langle u, v, w\rangle.$$

Since f is, in particular, a homomorphism of cochain complexes, we have $df(x) = f(dx)$ and $df(y) = f(dy)$. Furthermore, being a dga homomorphism, f respects the multiplication. Therefore

$$df(x) = f(dx) = f(ab) = f(a)f(b),$$

and similarly for y. It follows that we have an inclusion

$$\left[f(x)f(c) - (-1)^{|u|}f(a)f(y)\right] \in \langle f(u), f(v), f(w)\rangle,$$

proving the first part of the proposition.

Next, if f is a quasi-isomorphism then

$$\begin{aligned} f_*(uH(A) + H(A)w) &= f_*(u)f_*(H(A)) + f_*(H(A))f_*(w) \\ &= f_*(u)H(A') + H(A')f_*(w). \end{aligned}$$

Thus, f_* maps indeterminacies isomorphically, proving the lemma. □

DEFINITION 14.1.5. We say that two dga algebras B and B' are *quasi-isomorphic* if there exists a succession of homomorphisms

$$B = B_0 \longrightarrow B_1 \longleftarrow B_2 \longrightarrow \cdots \longleftarrow B_n = B' \tag{14.1.2}$$

where each of the homomorphisms is a quasi-isomorphism.

By Proposition 14.1.4, every diagram of the form (14.1.2) specifies a bijection between Massey products in $H(B)$ and $H(B')$.

14.2. Integrality of de Rham Massey products

Let X be a finite CW-space. Let $C^*(X;\mathbb{Z})$ and $C^*(X;\mathbb{R})$ be the singular cochain complexes with coefficients, respectively, in $\mathbb{Z}$ and $\mathbb{R}$. We equip these cochain complexes with the Alexander-Whitney products compatible with the ring homomorphism $\mathbb{Z} \to \mathbb{R}$. Note that the Alexander-Whitney product is associative but *not* (anti)commutative.

LEMMA 14.2.1. *Consider a Massey triple product $\langle u, v, w\rangle \subset H^*(X;\mathbb{Z})$ of elements $u, v, w \in H^*(X;\mathbb{Z})$. Let* Indet *be its indeterminacy subgroup. Then the Massey product*

$$\langle u_{\mathbb{R}}, v_{\mathbb{R}}, w_{\mathbb{R}}\rangle \subset H^*(X;\mathbb{R})$$

is defined, and we have

$$\langle u_{\mathbb{R}}, v_{\mathbb{R}}, w_{\mathbb{R}}\rangle = \langle u, v, w\rangle_{\mathbb{R}} + \mathrm{Indet} \otimes \mathbb{R}.$$

PROOF. If $x \in \langle u, v, w\rangle$, then $x_{\mathbb{R}} \in \langle u_{\mathbb{R}}, v_{\mathbb{R}}, w_{\mathbb{R}}\rangle$ by Proposition 14.1.4. Furthermore, the indeterminacy subgroup for $\langle u_{\mathbb{R}}, v_{\mathbb{R}}, w_{\mathbb{R}}\rangle$ is the subgroup

$$u_{\mathbb{R}} H^{|v|+|w|-1}(X;\mathbb{R}) + H^{|u|+|v|-1|}(X;\mathbb{R}) w_{\mathbb{R}},$$

and the result follows. □

Let M be a manifold. Let $\Omega^*(M)$ be the de Rham algebra of differential forms on M. Let $C^*(M;\mathbb{R})$ be the singular cochain complex equiped with the Alexander–Whitney product.

THEOREM 14.2.2. *The dga algebras $C^*(M;\mathbb{R})$ and $\Omega^*(M)$ are quasi-isomorphic.*

PROOF. We refer the reader to the book [**FelHT01**, Corollary 10.10 and Theorem 11.4], where the authors construct a suitable succession of quasi-isomorphisms linking our pair of dga algebras (we retain the notation of [**FelHT01**]). Our dga algebras are linked via Sullivan's A_{PL}, as follows. We have a pair of quasi-isomorphisms

$$C^*(M;\mathbb{R}) \longrightarrow (C_{\mathrm{PL}} \otimes A_{\mathrm{PL}}) \otimes \mathbb{R} \longleftarrow A_{\mathrm{PL}}(M;\mathbb{R}) \tag{14.2.1}$$

as well as the triple of quasi isomorphisms

$$A_{\mathrm{PL}}(M;\mathbb{R}) \to A_{\mathrm{PL}}(S_* M;\mathbb{R}) \to A_{\mathrm{dR}}(S_*^{\infty}(M)) \leftarrow A_{\mathrm{dR}}(M), \tag{14.2.2}$$

where $A_{\mathrm{dR}}(M) = \Omega^*(M)$, see also [**Hal83**]. □

Let $\rho : H^*(M;\mathbb{R}) \to H^*_{\mathrm{dR}}(M)$ be the isomorphism specified by the successions (14.2.1) and (14.2.2).

REMARK 14.2.3. It is proved in [**FelHT01**] that the isomorphism ρ of singular cohomology and de Rham cohomology, specified by the succession of five quasi-isomorphisms, coincides with the integration map $\oint$. The relevant commutative diagram appears as part of [**FelHT01**, Remark, p. 130].

Consider the map

$$r : H^*(M;\mathbb{Z}) \longrightarrow H^*(M;\mathbb{R}) \xrightarrow{\ \rho\ } H^*_{\mathrm{dR}}(M) \tag{14.2.3}$$

where the first map has the form

$$H^*(M;\mathbb{Z}) \to H^*(M;\mathbb{Z}) \otimes \mathbb{R} = H^*(M;\mathbb{R}), \quad a \mapsto a \otimes 1.$$

Recall that a de Rham cohomology class is called *integral* if its values on all integral cycles are integral. By the universal coefficient theorem, a class is integral if and only if it belongs to the image of the map r as in (14.2.3).

LEMMA 14.2.4. *Let $u, v, w \in H^*_{\mathrm{dR}}(M)$ be integral classes and assume that the groups $H^{|u|+|v|}(M;\mathbb{Z})$ and $H^{|v|+|w|}(M;\mathbb{Z})$ are torsion free. If the Massey product $\langle u, v, w\rangle \subset H^*_{\mathrm{dR}}(M)$ is defined, then it contains an integral class.*

PROOF. Let $\overline{u}, \overline{v}, \overline{w} \in H^*(M;\mathbb{Z})$ be elements such that $r(\overline{x}) = x$ for $x = u, v, w$. By our torsion hypotheses, we conclude that $\overline{u}\,\overline{v} = 0 = \overline{v}\,\overline{w}$, and therefore the Massey product $\langle \overline{u}, \overline{v}, \overline{w}\rangle \subset H^*(M;\mathbb{Z})$ is defined. By Theorem 14.2.2, ρ yields an isomorphism of Massey products. But, by Lemma 14.2.1, we have the inclusion

$$r(\langle \overline{u}, \overline{v}, \overline{w}\rangle) \subset \langle u, v, w\rangle,$$

and the lemma follows from the fact that the set $r(\langle \overline{u}, \overline{v}, \overline{w}\rangle)$ consists of integral classes. □

14.3. Gromov's calculation in the presence of a Massey

In this section, we deal with an extension of the notion of systolic category defined in Chapter 12. Namely, we will consider related systolic inequalities, but of a slightly different type, insofar as they involve another invariant of the manifold, in addition to the systoles. One could modify the definition of systolic category to incorporate such inequalities, as in (14.4.3) below.

We will follow Gromov's suggestion [**Gro83**, 7.4.C', p. 96] of exploiting non-trivial Massey products, in combination with suitable isoperimetric quotients, so as to obtain geometric inequalities. Some examples appear in [**DR03**]. We will use in an essential way the identification of two distinct Massey product theories, *cf.* (14.3.3).

Given a metric $\mathcal{G}$ on M, let $\mathrm{IQ}(M,\mathcal{G})$ be its isoperimetric quotient, defined as the maximin of quotients of the comass norm of a $(k-1)$-dimensional differential form which is a primitive, by the comass norm of an exact k-form on M, over all k. Namely,

$$\mathrm{IQ}(\mathcal{G}) = \max_k \sup_{\alpha\in\Omega^k} \inf_\beta \left\{ \tfrac{\|\beta\|^*}{\|\alpha\|^*} \,\middle|\, d\beta = \alpha \right\}, \tag{14.3.1}$$

cf. Section 14.2. The relation of such quotients to filling inequalities is studied in [**Sik05**]. Note that a metric with high IQ must cleverly impede a cycle from being efficiently filled by a chain.

THEOREM 14.3.1. *Let M be an orientable n-manifold and p_1, p_2 positive integers. Let $p_3 = n - (2p_1 + p_2 - 1)$. We make the following assumptions:*

1. *the Betti numbers satisfy $b_{p_1} = b_{p_2} = b_{p_3} = 1$;*
2. *the Massey triple product $\langle \omega_1, \omega_1, \omega_2\rangle$ is defined and nontrivial, where $\omega_i \in H^{p_i}_{\mathrm{dR}}(M)$, $i = 1, 2$ is the image of an integral generator ;*
3. *the group $H^{p_1+p_j}(M;\mathbb{Z})$, $j = 1, 2$ is torsion free.*

Set

$$A_j = \frac{n!}{p_j!\,(p_1+p_j)!\,p_3!}\binom{p_1+p_j}{p_1}, \quad j = 1, 2.$$

Then every metric $\mathcal{G}$ on M satisfies the inequality

$$\mathrm{stsys}_{p_1}(\mathcal{G})^2\,\mathrm{stsys}_{p_2}(\mathcal{G})\,\mathrm{stsys}_{p_3}(\mathcal{G}) \le (A_1 + A_2)\mathrm{IQ}(\mathcal{G})\,\mathrm{vol}_n(\mathcal{G}).$$

REMARK 14.3.2. A similar inequality holds in the presence of a more general nontrivial Massey product $\langle\omega_1, \omega_2, \omega_3\rangle$, modulo suitable torsion free hypotheses. We do not consider such a generalisation so as to avoid long formulas. A generalisation to arbitrary Betti numbers $b_{p_i} \geq 1$ appears in [**Katz07**].

PROOF of Theorem 14.3.1. Throughout the proof, we denote the de Rham cohomology group $H^*_{\mathrm{dR}}(M)$ by $H^*(M;\mathbb{R})$. We have $b_{n-p_3} = b_{p_3} = 1$ by Poincaré duality. Recall that the indeterminacy of the Massey product $\langle\omega_1, \omega_1, \omega_2\rangle$ is an $\mathbb{R}$-vector subspace of $H^{n-p_3}(M;\mathbb{R}) = \mathbb{R}$. Since the Massey product is assumed nontrivial, it must have zero indeterminacy.

Choose a representative $\eta_i \in \omega_i$. By hypothesis, the (p_1+p_j)-form $\eta_1 \wedge \eta_j$ is exact for $j = 1, 2$. Choose a primitive η_{1j} satisfying $d\eta_{1j} = \eta_1 \wedge \eta_j$ of minimal comass norm, so that

$$\frac{\|\eta_{1j}\|_\infty}{\|\eta_1 \wedge \eta_j\|_\infty} \leq \mathrm{IQ}(M, \mathcal{G}). \tag{14.3.2}$$

Notice that the cohomology class ω of the closed form

$$\eta_{11} \wedge \eta_2 - (-1)^{p_1}\eta_1 \wedge \eta_{12}$$

is equal to the Massey product $\langle\omega_1, \omega_1, \omega_2\rangle_{\mathrm{dR}} \in H^{n-p_3}(M;\mathbb{R})$. By Poincaré duality, there is a class $\omega_3 \in H^{p_3}(M;\mathbb{R}) = \mathbb{R}$ which is the image of an integral generator, such that

$$(\omega_3 \cup \omega)[M] > 0.$$

The compatibility of the de Rham and the singular cochain theories of Massey products was clarified in Section 14.2. By Lemma 14.2.4, the class ω is integral, and hence the class $\omega_3 \cup \omega$ is integral. We therefore obtain the following lower bound:

$$(\omega_3 \cup \omega)[M] \geq 1. \tag{14.3.3}$$

Choose a p_3-form η_3 in the cohomology class ω_3. Then

$$1 \leq \int_M (\eta_{11} \wedge \eta_2 - (-1)^{p_1}\eta_1 \wedge \eta_{12}) \wedge \eta_3. \tag{14.3.4}$$

Similarly to (13.2.2) and exploiting (14.3.2), we obtain

$$\begin{aligned} 1 &\leq \sum_{j=1}^{2} \frac{n!}{p_j!\,(p_1+p_j-1)!\,p_3!} \|\eta_j\|_\infty\,\|\eta_{1j}\|_\infty\,\|\eta_3\|_\infty \operatorname{vol}_n(\mathcal{G}) \\ &\leq (A_1 + A_2)\,\mathrm{IQ}(\mathcal{G})\,\|\eta_1\|_\infty\,\|\eta_1\|_\infty\,\|\eta_2\|_\infty\|\,\eta_3\|_\infty \operatorname{vol}_n(\mathcal{G}). \end{aligned}$$

Hence, similarly to (13.2.3), we obtain

$$1 \leq (A_1 + A_2)\,\mathrm{IQ}(\mathcal{G})\,(\|\omega_1\|^*)^2\,\|\omega_2\|^*\,\|\omega_3\|^* \operatorname{vol}_n(\mathcal{G}).$$

Now let $\alpha_i \in H_{p_i}(M;\mathbb{R})$ be the generator dual to ω_i. By duality of comass and stable norm, and since $b_{p_i} = 1$, we have $\|\alpha_i\| = \frac{1}{\|\omega_i\|^*}$. Thus, similarly to (13.2.4) we obtain the inequality

$$\operatorname{stsys}_{p_1}(\mathcal{G})^2 \operatorname{stsys}_{p_2}(\mathcal{G}) \operatorname{stsys}_{p_3}(\mathcal{G}) \leq (A_1 + A_2)\,\mathrm{IQ}(\mathcal{G}) \operatorname{vol}_n(\mathcal{G}),$$

as required. □

14.4. A homogeneous example

Consider the homogeneous manifold

$$M^{n^2+2n-16} = SU(n+1)/SU(3) \times SU(3),$$

where $n \geq 5$ [**Tra93**], [**GreHV76**, Chapter 11]. Denote by ω_i the generator of $H^i(M)$, for $i = 4, 6$. The Massey triple product

$$\langle \omega_4, \omega_4, \omega_6 \rangle \in H^{13}(M; \mathbb{Z})$$

is nontrivial. For $n = 5$, we obtain that every metric $\mathcal{G}$ on the 19-dimensional manifold

$$SU(6)/SU(3) \times SU(3) \tag{14.4.1}$$

satisfies

$$\mathrm{stsys}_4(\mathcal{G})^2 \, \mathrm{stsys}_6(\mathcal{G})^2 \leq 19! \; \mathrm{IQ}(\mathcal{G}) \, \mathrm{vol}_{19}(\mathcal{G}). \tag{14.4.2}$$

REMARK 14.4.1. This inequality is not exactly of the form envisioned above in (12.3.2), since the constant is metric-dependent via the isoperimetric quotient IQ factor. The IQ factors appear as the result of integrations involved in defining a Massey product.

Applying this technique to the standard 3-dimensional nilmanifold, we get a lower bound given by a product of *four* systoles, for the quantity $\mathrm{IQ}(\mathcal{G}) \, \mathrm{vol}(\mathcal{G})$. Since we expect the category of a 3-manifold to be at most 3, it is natural to define an analogue of systolic category in the presence of a Massey product, by *subtracting* the number of IQ factors, as in (14.4.3) below. This would be consistent with the lower bound for $\mathrm{cat}_{\mathrm{LS}}(M)$ in terms of weights of Massey products. Along these lines, inequality (14.4.2) can be restated as a lower bound for an IQ-modified systolic category:

$$\mathrm{cat}_{\mathrm{sys}}^{\mathrm{IQ}}(M^{19}) \geq 4 - 1 = 3. \tag{14.4.3}$$

On the other hand, it can be shown that $3 \leq \mathrm{cat}_{\mathrm{LS}} M^{19} \leq 4$, and we expect that that $\mathrm{cat}_{\mathrm{LS}} = 4$. Meanwhile, J. Oprea proved Proposition 14.4.2 below. The definition and properties of rational Lusternik–Schnirelmann category can be found in [**CorT03**].

PROPOSITION 14.4.2 (J. Oprea). *The rational Lusternik–Schnirelmann category of the manifold M^{19} is equal to* 3.

PROOF. Let M be a simply connected manifold. Then the rational Lusternik–Schnirelmann category of M is equal to its rational Toomer invariant $e_0(M)$, see [**FelHL98**]. To compute the invariant $e_0(M)$, we notice that the Sullivan model of the space (14.4.1) has the form

$$\lambda(x_4, x_6, y_7, y_9, y_{11})$$

with differential

$$dx_i = 0, \; dy_7 = x_4^2, \; dy_9 = x_4 x_6, \; dy_{11} = x_6^2.$$

Therefore a top class is given by

$$x_4^2 y_{11} - x_4 x_6 y_9, \tag{14.4.4}$$

and so $e_0(M^{19}) = 3$ by [**CorT03**, Proposition 5.23]. □

CHAPTER 15

Cup products and stable systoles

15.1. Introduction

Let $(M, \mathcal{G})$ be a Riemannian manifold. Let $n = \dim(M)$ and $b = b_1(M)$. The optimal inequality (5.4.1) of C. Loewner can be generalized by optimal inequalities in two different ways as follows:

$$\frac{\deg(\mathcal{A}_M)\, \mathrm{stsys}_1(\mathcal{G})^b}{\mathrm{vol}_n(\mathcal{G})} \le (\gamma_b)^{\frac{b}{2}} \tag{15.1.1}$$

Chapter 17 ↗

$$\text{Loewner}: \ \frac{\mathrm{sys}_1(\mathcal{G})^2}{\mathrm{area}(\mathcal{G})} \le \gamma_2$$

Chapter 16 ↘

$$\frac{\mathrm{stsys}_1(\mathcal{G})\, \mathrm{sys}_{n-1}(\mathcal{G})}{\mathrm{vol}_n(\mathcal{G})} \le \gamma_b'$$

Here $\mathcal{A}_M : M \to \mathbb{T}^b$ is the Abel-Jacobi map (16.5.3), while $\deg(\mathcal{A}_M)$ denotes the infimum of $(n-b)$-volumes of typical fibers of the map $\mathcal{A}_M$. The case $n = b$ corresponds to the algebraic degree.

The two directions are explored, respectively, in Chapter 16 and Chapter 17. More generally, one can study systolic inequalities associated with any nontrivial real cup product.

Multiplicative relations in the cohomology ring of a manifold impose constraints upon its stable systoles. Given a compact Riemannian manifold $(M, \mathcal{G})$, its real homology $H_*(M, \mathbb{R})$ is naturally endowed with the stable norm, *cf.* Definition 12.1.4. Briefly, if $h \in H_k(M, \mathbb{R})$ then the stable norm of h is the infimum of the Riemannian k-volumes of real cycles representing h. The stable k-systole is the minimum of the stable norm over nonzero elements in the lattice of integral classes in $H_k(M, \mathbb{R})$, *cf.* (12.1.7). Relying on results from the geometry of numbers due to W. Banaszczyk, and extending work by M. Gromov and J. Hebda, we prove metric-independent inequalities for products of stable systoles, where the product can be as long as the real cup-length of M. These results first appeared in [**BaK03**].

Our goal is to explore the hidden power of the seminal calculation contained in lines 7-12, p. 262 of M. Gromov's book [**Gro99**]. The reader should familiarize himself with Gromov's seminal calculation for complex projective space, already presented in Section 13.2 of the previous chapter, which serves as a model for the calculations in the present chapter, *cf.* (15.5.5) below.

We rely on results of W. Banaszczyk [**Bana93, Bana96**] from the geometry of numbers, which provide a bound for certain products of the successive minima of a pair of dual lattices, see (15.5.1) and (15.8.1). We thus extend the work of

M. Gromov [**Gro83**] and J. Hebda [**Heb86**] to prove metric-independent inequalities for products of stable systoles, where the product can be as long as the real cup-length of M. In particular, for the pair of stable systoles of dimension and codimension 1, we prove an optimal bound, which is attained, for instance, by flat tori defined as quotients of the so-called dual-critical lattices of A.-M. Bergé and J. Martinet [**BeM89**].

Geometric Measure Theory provides a framework in which the existence of minimal representatives in homology classes can be proved. Given a homology clas $h \in H_k(M, \mathbb{Z})$, the quantity $\mathrm{vol}_k(h)$ (*cf.* Definition 12.1.2) is the minimal mass of a closed, integer-multiplicity, rectifiable k-current representing h, *cf.* [**Fed69**, 5.1.6]. If $h \in H_k(M, \mathbb{R})$ then the stable norm (or mass norm) of h (*cf.* Definition 12.1.4) is the minimal mass of a closed (normal) k-current representing the class h.

In the cases $k = 1$ and $k = n - 1$, the transition from the Riemannian length (respectively, the $(n-1)$-volume functional) to the corresponding stable norm is closely related to the process of *homogenization* studied in Analysis. For $k = n - 1$ this shows up in J. Moser's work [**Mo86**, Sect. 7], [**MS92**], see [**Sen91a, Sen91b**] for more details. It is through these remarks by J. Moser that the first-named author of [**BaK03, BaK04**] originally came in contact with the study of the stable norm.

While the present chapter studies systolic ramifications of multiplicative relations in the cohomology ring with respect to the cup product, Section 14.3 of the previous chapter explores the systolic influence of Massey products, *cf.* [**Katz07**].

The main results are stated in Section 15.2 and Section 15.3. Some topological preliminaries are presented in Section 15.4. The inequality of W. Banaszczyk, a crucial ingredient in our technique, appears in Section 15.5, together with the proof of Theorem 15.2.1. Theorem 15.2.5 is proved in Section 15.6, Corollary 15.2.6 in Section 15.7, Theorem 15.3.2 in Section 15.8, and Theorem 15.3.4 in Section 15.10.

15.2. Statement of main results

Assuming that the fundamental cohomology class of a compact, oriented manifold M can be written as a cup product of classes of dimensions $k_1, \ldots, k_m$, M. Gromov proved an upper bound for the product of the stable systoles $\Pi_j \,\mathrm{stsys}_{k_j}(\mathcal{G})$ in terms of the volume of M, see [**Gro83**, 7.4.C] and also [**Gro99**, 4.38]. We extend this result, provide a simpler proof and analyze the dependence of the constant on the Betti numbers involved.

THEOREM 15.2.1 (Ring structure-dependent bound). *Let M be a compact manifold and let $k \geq 1$ be an integer such that $H^k(M, \mathbb{R})$ is not zero and spanned by cup products of classes of dimensions $k_1, \ldots, k_m$, where $\Sigma_j k_j = k$. Then every Riemannian metric $\mathcal{G}$ on M satisfies*

$$\prod_{j=1}^{m} \mathrm{stsys}_{k_j}(\mathcal{G}) \leq C(k) \left(\prod\nolimits_j b_{k_j}(M) \left(1 + \log b_{k_j}(M)\right) \right) \mathrm{stsys}_k(\mathcal{G}) \qquad (15.2.1)$$

where $C(k)$ is a constant only dependent on k.

Regardless of the nature of the dependence of the multiplicative constant in inequality (15.2.1) on the Betti numbers, we point out the following. The existence of Gromov's inequality, corresponding to the longest decomposition of the fundamental cohomology class as a product of lower dimensional classes, yields the bound of Corollary 15.2.3 below for systolic category, see Definition 12.3.3.

DEFINITION 15.2.2. We say that M has real cup-length $\geq n$ if there exist classes $\alpha_1 \in H^{k_1}(M, \mathbb{R}), \ldots, \alpha_n \in H^{k_n}(M, \mathbb{R})$ such that

$$\alpha_1 \cup \ldots \cup \alpha_n \neq 0.$$

COROLLARY 15.2.3. *The systolic category of a closed orientable manifold is bounded below by its real cup-length.*

Pu's inequality (5.2.2) for $\mathbb{RP}^2$ as well as its generalisations by Gromov lend support to the following conjectured generalisation of Corollary 15.2.3:

$$\operatorname{cat}_{\mathrm{sys}}(M) \geq \text{cup-length}(M), \tag{15.2.2}$$

which would be implied by a corresponding equality in (12.4.2). An interesting test case is (12.8.2).

REMARK 15.2.4. There exist lower bounds for systolic category which are stronger than real cup-length, *cf.* Corollary 17.2.3.

Using Poincaré duality, J. Hebda [**Heb86**, Proposition 6] bounds the product of the stable systoles in complementary dimensions by the volume. We generalize his result as follows:

THEOREM 15.2.5. *Let M be a compact manifold and let p, q be integers, $p+q \leq \dim M$. Suppose there exists $h \in H_{p+q}(M, \mathbb{Z})_{\mathbb{R}}$ such that the cap product with h induces an injective map*

$$\alpha \in H^p(M, \mathbb{Z})_{\mathbb{R}} \to h \cap \alpha \in H_q(M, \mathbb{Z})_{\mathbb{R}}.$$

Then, for every Riemannian metric $\mathcal{G}$ on M, we have

$$\operatorname{stsys}_p(\mathcal{G}) \operatorname{stsys}_q(\mathcal{G}) \leq C(p, q) b_p(M) \|h\|$$

for a finite constant $C(p, q)$ only depending on p and q. If our assumption is satisfied for all $h \subset H_{p+q}(M, \mathbb{Z})_{\mathbb{R}} \setminus \{0\}$ then

$$\operatorname{stsys}_p(\mathcal{G}) \operatorname{stsys}_q(\mathcal{G}) \leq C(p, q) b_p(M) \operatorname{stsys}_{p+q}(\mathcal{G}). \tag{15.2.3}$$

Examples of flat tori show that an inequality of type (15.2.3) has to depend on the Betti number b_p linearly. We do not know whether the dependence on the Betti numbers is optimal in Theorem 15.2.1. Explicit values for the constants $C(p, q)$ can be computed. Unless $\{p, q\} = \{1, \dim M - 1\}$, the sharp constants $C(p, q)$ are unknown to us. In the case $\{p, q\} = \{1, \dim M - 1\}$ we have the following sharp result, generalizing [**Heb86**, Theorem A] to the case $b_1(M) > 1$.

COROLLARY 15.2.6. *Let M be a compact, orientable manifold of dimension n, with positive first Betti number $b = b_1(M)$. Then every Riemannian metric $\mathcal{G}$ on M satisfies the inequality*

$$\operatorname{stsys}_1(\mathcal{G}) \operatorname{sys}_{n-1}(\mathcal{G}) \leq \gamma'_b \operatorname{vol}_n(\mathcal{G}), \tag{15.2.4}$$

where γ'_b is the Bergé-Martinet constant (5.3.3). *Equality is attained for a flat torus $\mathbb{R}^n/L$ where $L \subseteq \mathbb{R}^n$ is a lattice with $\lambda_1(L)\lambda_1(L^*) = \gamma'_n$.*

These inequalities can be seen as analogues of the optimal inequalities of C. Loewner (5.4.1), P. Pu (5.2.2), and C. Bavard [**Bav88**], and as generalisations of the results of R. Accola [**Ac60**] and C. Blatter [**Bl61a**] (*cf.* [**Berg93**, p. 290]) in dimension 2. A relative version of Corollary 15.2.6 is studied in [**Bab02b**].

Note that we have examples of manifolds M with equality in Corollary 15.2.6 only if $\dim M = n \geq b = \dim H_1(M, \mathbb{R})$. If we fix $n = \dim M$, there might exist a better estimate, perhaps even one independent of b. In particular we ask the following question, *cf.* [**Gro83**, 7.4.C]:

QUESTION 15.2.7. Do all orientable Riemannian 3-manifolds $(M, \mathcal{G})$ satisfy a systolic inequality

$$\operatorname{stsys}_1(\mathcal{G}) \operatorname{sys}_2(\mathcal{G}) \leq C \operatorname{vol}_3(\mathcal{G})$$

with a uniform ("universal") constant $C > 0$?

The manifolds $(M, \mathcal{G})$ for which equality holds in Corollary 15.2.6 are investigated in [**BaK04**] and Chapter 16. In the case $b_1(M) = 1$, J. Hebda [**Heb86**, Theorem A] proved that one has equality

$$\operatorname{stsys}_1(\mathcal{G}) \operatorname{sys}_{n-1}(\mathcal{G}) = \operatorname{vol}_n(\mathcal{G})$$

if and only if M admits a Riemannian submersion $F : M \to S^1$ onto a circle, such that all fibers $F^{-1}(s)$, $s \in S^1$, are minimal hypersurfaces in M.

15.3. Results for the conformal systole

Now let $n = 2p$, and consider the L^2-norm $|\ |_{L^2}$ in the homology group $H_p(M, \mathbb{R})$ dual to the norm on harmonic p-forms $f \in \Omega^p(M)$, *cf.* Section 14.2, defined by

$$|f|_{L^2}^2 = \int_M f \wedge *f,$$

where $*$ is the Hodge star operator of the metric $\mathcal{G}$. The operator $*$ satisfies the condition

$$f \wedge *f = |f|^2 \operatorname{vol}_{\mathcal{G}}, \tag{15.3.1}$$

where $|f|$ is the pointwise Euclidean norm, while $\operatorname{vol}_{\mathcal{G}}$ is the pointwise volume form of the (oriented) metric. A definition of $*$ and a detailed discussion may be found in [**LawM89**].

Consider also the conformally invariant norm $\|\ \|_{L^2}^*$ in $H^p_{dR}(M)$ defined by taking the infimum, over all representatives ω, of the quantity

$$\|\omega\|_{L^2} = \left(\int_M \|\omega_x\|^2 d \operatorname{vol}_n(x) \right)^{\frac{1}{2}},$$

where $\|\omega_x\|$ is the pointwise comass, defined by (13.1.2).

Let $\|\ \|_{L^2}$ be the dual norm in homology, *cf.* (15.8.4) below.

DEFINITION 15.3.1. The *Conformal systole* $\operatorname{Confsys}_p$ is defined by setting

$$\operatorname{Confsys}_p(\mathcal{G}) = \min \left\{ \|h\|_{L^2} \,\middle|\, h \in H_p(M, \mathbb{Z})_{\mathbb{R}} \setminus \{0\} \right\}.$$

A more detailed discussion of this invariant and the related *conformal systole*, denoted $\operatorname{confsys}_p$, appears below in Definition 15.9.1.

THEOREM 15.3.2. *Let $p = \frac{n}{2}$ and $b_p = b_p(M)$. Then*

$$\mathrm{Confsys}_p(\mathcal{G})^2 \le \frac{\lambda_1}{\lambda_{b_p}} \binom{n}{p} b_p$$

where $\lambda_i = \lambda_i(H_p(M,\mathbb{Z})_{\mathbb{R}}, |\ |_{L^2})$.

Note that Chapter 20 studies *lower* bounds for the quantity $\sup_{\mathcal{G}} \mathrm{confsys}_2(\mathcal{G})$, for 4-manifolds $(M,\mathcal{G})$, where M is the blowup of the complex projective plane, the bound being *polynomial* of order $n^{\frac{1}{4}}$ in the number n of blown-up points. Whether or not such lower bounds exist for the stable 2-systole is unknown.

QUESTION 15.3.3. Does there exist a constant $C < \infty$ such that

$$\mathrm{stsys}_2(\mathcal{G}) \le C\sqrt{\mathrm{vol}_4(\mathcal{G})}$$

for every four-dimensional, compact, orientable Riemannian manifold $(M,\mathcal{G})$?

Note that the answer is affirmative in the case of a definite intersection form, *cf.* (20.5.2).

For the first stable systole, one has the Burago-Ivanov-Gromov inequality, *cf.* (17.1.1):

$$\mathrm{stsys}_1(\mathcal{G}) \le \sqrt{\gamma_n}\, \mathrm{vol}_n(\mathcal{G})^{\frac{1}{n}} \tag{15.3.2}$$

if $\mathcal{G}$ is a Riemannian metric on a compact manifold M with $b_1(M) = n = \dim M$, for which there are classes $\alpha_1, \ldots, \alpha_n \in H^1(M,\mathbb{R})$ with $\alpha_1 \cup \ldots \cup \alpha_n \ne 0$. Here γ_n is the Hermite constant (5.3.1). See [**Gro96**, 1.A$_1$], [**Gro99**, pp. 259-260], [**Berg93**, p. 283] and additionally [**Bab93**, Theorem 8.8]. Combining the methods used to prove Theorem 15.2.1 and Theorem 15.2.5, we obtain a result similar to (15.3.2) for higher-dimensional stable systoles:

THEOREM 15.3.4. *Given a compact orientable manifold M of dimension satisfying $\dim(M) = n = kp$, suppose there are classes $\beta_1, \ldots, \beta_k \in H^p(M,\mathbb{R})$ such that $\beta_1 \cup \ldots \cup \beta_k \ne 0$. Then every Riemannian metric $\mathcal{G}$ on M satisfies*

$$\mathrm{stsys}_p(\mathcal{G}) \le C(n)\left(b_p(M)(1+\log b_p(M))\right)^{\frac{k-1}{k}} \mathrm{vol}_n(\mathcal{G})^{\frac{1}{k}}$$

for a finite constant $C(n)$ only depending on n.

DEFINITION 15.3.5. We say that M^n has cup-length n if for some field F, there exist classes $\alpha_1, \ldots, \alpha_n \in H^1(M,F)$ such that $\alpha_1 \cup \ldots \cup \alpha_n \ne 0$.

It is a natural question whether similar inequalities exist if one replaces the stable systoles by the ordinary systoles, *cf.* Section 12.1. The most prominent example is Gromov's systolic inequality [**Gro96**, 3.C_1]

$$\mathrm{sys}_1(\mathcal{G}) \le c_n (\mathrm{vol}_n(\mathcal{G}))^{\frac{1}{n}} \tag{15.3.3}$$

which holds for all Riemannian metrics on compact n-dimensional manifolds M of cup-length n, see Definition 15.3.5.

See (15.9.2) below for a refined inequality of this type for surfaces.

15.4. Some topological preliminaries

We recall some facts from homology theory that will be useful in the sequel. Given a compact manifold M, we consider the homology modules derived from the singular (co-)chain complexes with $\mathbb{Z}$- or $\mathbb{R}$-coefficients.

Namely, the cup product on cochains is by definition dual to the Alexander-Whitney comultiplication on chains, *cf.* Section 14.2. The cup product is associative. The cup product together with the coboundary operator equip the cochain module $C^*(M)$ with the structure of a differential graded associative (dga) algebra, *cf.* [**FelHT01**, p. 65], [**War83**, p. 212]. The cohomology of the dga algebra $C^*(M)$ is by definition the singular cohomology of M, see Section 14.1 above.

We assume that the singular simplices are Lipschitz. Since we will often use results that are actually proved for the (continuous) singular or the C^∞-singular (co-)homology theories, we note that these are naturally isomorphic to those arising from Lipschitz simplices.

By de Rham's Theorem, see *e.g.* [**War83**, 5.45], the singular cohomology algebra $H^*(M, \mathbb{R})$ is naturally isomorphic to the de Rham cohomology algebra $H^*_{\mathrm{dR}}(M)$, *cf.* [**FelHT01**, Corollary 10.10] and Section 14.2. In the de Rham theory, the cup product is defined by the wedge product of closed forms representing the cohomology classes. The (nondegenerate) bilinear pairing with real homology is given by integrating the representative forms, over representative Lipschitz cycles, *cf.* [**War83**, 4.17].

The bilinear pairing, known as Kronecker product, between homology and cohomology will be denoted by $(\ , \ )$, cup products by $\cup$, and cap products by $\cap$.

REMARK 15.4.1. The maps induced from the universal coefficient theorem are compatible both with Kronecker pairing, and with cup and cap products.

More specifically, the map

$$\iota_* : H_*(M, \mathbb{Z}) \to H_*(M, \mathbb{R}), \ h \to h_{\mathbb{R}}$$

is defined by inclusion at the level of chains. The map

$$\iota^* : H^*(M, \mathbb{Z}) \to H^*(M, \mathbb{R}) \simeq H^*_{\mathrm{dR}}(M), \ \alpha \to \alpha_{\mathbb{R}}$$

corresponds to the natural extension of a cochain $c \in \mathrm{Hom}(C(M, \mathbb{Z}), \mathbb{Z})$ to a cochain in $\mathrm{Hom}(C(M, \mathbb{R}), \mathbb{R})$. According to the universal coefficient theorem in homology, *cf.* [**Gre67**, 29.12], the image $\mathrm{Im}(\iota_*)$ is a lattice in $H_*(M, \mathbb{R})$. Let $H^*(M, \mathbb{Z})_{\mathbb{R}} := \mathrm{Im}(\iota^*)$.

LEMMA 15.4.2. *The lattice $H^*(M, \mathbb{Z})_{\mathbb{R}}$ is dual to $H_*(M, \mathbb{Z})_{\mathbb{R}}$ under the Kronecker pairing $(\ , \ )$.*

PROOF. By naturality mentioned in Remark 15.4.1, we have

$$(h_{\mathbb{R}}, \alpha_{\mathbb{R}}) = (h, \alpha) \in \mathbb{Z}$$

for all $h \in H_*(M, \mathbb{Z})$, $\alpha \in H^*(M, \mathbb{Z})$. Conversely, suppose we have a cohomology class $\tilde{\alpha} \in H^*(M, \mathbb{R})$, such that $(h_{\mathbb{R}}, \tilde{\alpha}) \in \mathbb{Z}$ for all $h \in H_*(M, \mathbb{Z})$. By the surjectivity result [**Gre67**, 23.9] for a PID coefficient ring, there exists $\alpha \in H^*(M, \mathbb{Z})$ such that $(h, \alpha) = (h_{\mathbb{R}}, \tilde{\alpha})$ for all $h \in H_*(M, \mathbb{Z})$. The nondegeneracy of $(\ , \ )$ now implies that $\tilde{\alpha} = \alpha_{\mathbb{R}} \in H^*(M, \mathbb{Z})_{\mathbb{R}}$. □

Thus the set of de Rham classes $\alpha \in H^*_{\mathrm{dR}}(M)$, that are represented by closed forms whose integrals over integer cycles are integers, coincides with $\mathrm{Im}(\iota^*)$. These classes will henceforth be called *integer classes.*

If M is a compact, oriented manifold of dimension n, then the Poincaré duality isomorphism $H^p(M,\mathbb{R}) \to H_{n-p}(M,\mathbb{R})$ is defined by

$$\mathrm{PD}_{\mathbb{R}} : \alpha \in H^p(M,\mathbb{R}) \to \xi_{\mathbb{R}} \cap \alpha \in H_{n-p}(M,\mathbb{R}),$$

where ξ denotes the fundamental class of M. Since the corresponding map

$$\mathrm{PD} : \alpha \in H^p(M,\mathbb{Z}) \to \xi \cap \alpha \in H_{n-p}(M,\mathbb{Z})$$

is an isomorphism, *cf.* [**Gre67**, 26.6], we see that $\mathrm{PD}_{\mathbb{R}}$ induces an isomorphism between $H^p(M,\mathbb{Z})_{\mathbb{R}}$ and $H_{n-p}(M,\mathbb{Z})_{\mathbb{R}}$. In particular, Lemma 15.4.2 and the preceding argument show the following.

PROPOSITION 15.4.3. *A de Rham cohomology class $\alpha \in H^p_{\mathrm{dR}}(M)$ is an integer class if* $(\mathrm{PD}_{\mathbb{R}}(\alpha),\beta) \in \mathbb{Z}$ *for all integer classes $\beta \in H^{n-p}_{\mathrm{dR}}(M)$.*

This fact was introduced as a hypothesis in [**Berg72b**, p. 253], and named a "dual lattice condition" in [**Heb86**, p. 344]. The preceding discussion shows that the condition is always satisfied.

15.5. Ring structure-dependent bound via Banaszczyk

In this section, we prove Theorem 15.2.1. The following result by W. Banaszczyk [**Bana96**] is a crucial ingredient in the proof of Theorem 15.2.1. By [**Bana96**, Corollary 2], there exists a constant $C > 0$ such that

$$\lambda_i(L, \|\ \|)\lambda_{b-i+1}(L^*, \|\ \|^*) \le C\, b(1 + \log b) \tag{15.5.1}$$

whenever $b \in \mathbb{N}$, $i \in \{1,\dots,b\}$ and L is a lattice in a b-dimensional normed space $(V, \|\ \|)$, *cf.* (4.4.1).

We apply this result with $V = H_p(M,\mathbb{R})$ and $L = H_p(M,\mathbb{Z})_{\mathbb{R}}$, where $\|\ \|$ is the stable norm on $H_p(M,\mathbb{R})$. Thus, V^* can be identified with the de Rham cohomology space $H^p_{\mathrm{dR}}(M)$, and then Lemma 15.4.2 implies that $L^* = H^p(M,\mathbb{Z})_{\mathbb{R}}$ is the lattice of integer de Rham classes.

The norm on $H^p_{\mathrm{dR}}(M)$ dual to the stable norm on $H_p(M,\mathbb{R})$ is the comass norm, *cf.* [**Fed74**, 4.10] and [**Gro00**, 4.35], generalizing the linear algebraic comass of Definition 13.1.2.

DEFINITION 15.5.1. The *comass norm* $\|\ \|_\infty$ of a p-form $\omega \in \Omega^p(M)$ on a compact Riemannian manifold M is

$$\|\omega\|_\infty = \max\{\omega_x(e_1,\dots,e_p) \mid x \in M, e_i \in \mathbb{T}M_x\}, \tag{15.5.2}$$

where $|e_i| = 1$ for $1 \le i \le p$.

Meanwhile, the comass of a cohomology class $\alpha \in H^p_{\mathrm{dR}}(M)$ is

$$\|\alpha\|^* = \inf\left\{ \|\omega\|_\infty \,\middle|\, \omega \text{ a closed } p\text{-form representing } \alpha \right\}.$$

EXAMPLE 15.5.2. For a flat torus $(\mathbb{T}^n, \mathcal{G})$ the calculation of the stable norm and the comass norm reduces to the calculation of the corresponding pointwise quantities [**Law75**, Corollary 4.4]. If we normalize $\mathcal{G}$ so that $\mathrm{vol}_n(\mathcal{G}) = 1$, and represent $(\mathbb{T}^n, \mathcal{G})$ as the quotient of Euclidean space $\mathbb{R}^n$ by a lattice L, then the standard isomorphisms

$$H_p(\mathbb{T}^n, \mathbb{R}) \simeq \Lambda^p \mathbb{R}^n$$
$$H^p(\mathbb{T}^n, \mathbb{R}) \simeq \Lambda^p \mathbb{R}^n$$

convert the stable norm on $H_p(\mathbb{T}^n, \mathbb{R})$ into the "mass norm" on $\Lambda^p\mathbb{R}^n$ (induced by the Euclidean structure on $\mathbb{R}^n$), and the comass norm on $H^p(\mathbb{T}^n, \mathbb{R})$ into the comass norm on $\Lambda^p\mathbb{R}^n$. The cases $p = 1$ and $p = n - 1$ are particularly simple. Namely, the mass and comass norms on $\Lambda^1\mathbb{R}^n \simeq \Lambda^{n-1}\mathbb{R}^n \simeq \mathbb{R}^n$ coincide with the Euclidean norm. Moreover the lattices $H_1(\mathbb{T}^n, \mathbb{Z})_\mathbb{R} \subset H_1(\mathbb{T}^n, \mathbb{R}) \simeq \mathbb{R}^n$ and

$$H_{n-1}(\mathbb{T}^n, \mathbb{Z})_\mathbb{R} \subset H_{n-1}(\mathbb{T}^n, \mathbb{R}) \simeq \mathbb{R}^n$$

correspond to L, respectively L^*, under these isomorphisms. The cases $1 < p < n-1$ are considerably more complicated [**Law75**].

We are now in a position to prove our first theorem.

PROOF OF THEOREM 15.2.1. The proof is a generalisation of Gromov's argument in the case of unit Betti number, *cf.* Section 13.2. For notational reasons, we present the proof for the case that $H^k(M, \mathbb{R})$ is non-zero, and is spanned by cup products of classes of dimensions j and l, with $j + l = k$. The generalisation to more factors is obvious. Since $H^k(M, \mathbb{R}) \neq \{0\}$, there exists a class $h \in H_k(M, \mathbb{Z})_\mathbb{R}$ such that

$$\|h\| = \operatorname{stsys}_k(\mathcal{G}). \tag{15.5.3}$$

Let $b_j = b_j(M)$ and similarly $b_\ell = b_\ell(M)$. We set $\lambda^* = \lambda_{b_j}(H^j(M, \mathbb{Z})_\mathbb{R}, \| \ \|^*)$ and $\mu^* = \lambda_{b_l}(H^l(M, \mathbb{Z})_\mathbb{R}, \| \ \|^*)$. Then we can find integral classes $\alpha_1, \ldots, \alpha_{b_j}$ in $H^j(M, \mathbb{Z})_\mathbb{R}$ spanning $H^j(M, \mathbb{R})$ and $\beta_1, \ldots, \beta_{b_l}$ in $H^l(M, \mathbb{Z})_\mathbb{R}$ spanning $H^l(M, \mathbb{R})$ such that

$$\|\alpha_s\|^* \leq \lambda^* \text{ for } 1 \leq s \leq b_j, \text{ and } \|\beta_t\|^* \leq \mu^* \text{ for } 1 \leq t \leq b_l. \tag{15.5.4}$$

By assumption, the vector space $H^k(M, \mathbb{R})$ is spanned by cup products $\alpha_s \cup \beta_t$, where $1 \leq s \leq b_j$ and $1 \leq t \leq b_l$. Hence there exist indices s and t such that

$$(h, \alpha_s \cup \beta_t) = a \neq 0.$$

By Remark 15.4.1, we have $a \in \mathbb{Z}$. Now we complete the proof by a calculation analogous to Gromov's calculation in [**Gro99**, Theorem 4.36, p. 262], *cf.* Section 13.2. Let $c \in C_*(M)$ be a real Lipschitz cycle representing the class h. If ω and π are closed forms representing α_s and β_t, then

$$1 \leq |a| = \left| \int_c \omega \wedge \pi \right| \leq \frac{k!}{j!\,l!} \operatorname{vol}_k(c) \|\omega\|_\infty \|\pi\|_\infty, \tag{15.5.5}$$

see [**Fed69**, 1.8.1] for the factor $\frac{k!}{j!\,l!} = \binom{k}{j}$. Using this and (15.5.3) and (15.5.4), we obtain

$$1 \leq |a| \leq \frac{k!}{j!\,l!} \operatorname{stsys}_k(\mathcal{G}) \lambda^* \mu^*.$$

Since $H^p(M, \mathbb{Z})_\mathbb{R}$ is the lattice dual to $H_p(M, \mathbb{Z})_\mathbb{R}$ by Lemma 15.4.2, we can use Banaszczyk's inequality (15.5.1) to conclude

$$\operatorname{stsys}_j(\mathcal{G}) \operatorname{stsys}_l(\mathcal{G}) \leq \frac{k!}{j!\,l!} C^2\, b_j(1 + \log b_j) b_l (1 + \log b_l) \operatorname{stsys}_k(\mathcal{G}).$$

This is the claim of Theorem 15.2.1 in the case $m = 2$. For arbitrary $m \geq 2$ we obtain the constant

$$\frac{k!}{\Pi_j k_j!} C^m$$

in this inequality, by a similar calculation. □

15.6. Inequalities based on cap products, Poincaré duality

In this section we prove Theorem 15.2.5. The proof of our Theorem 15.2.1 was based on Banaszczyk's inequality (15.5.1) applied to the stable norm. For Euclidean norms, Banaszczyk proved a sharper estimate which is linear in the dimension, *cf.* (15.8.1) below. Much older, linear estimates for the Hermite constant itself were reviewed in (5.3.4) and (5.3.5).

Replacing the stable norm by the L^2-norm, J. Hebda [**Heb86**] was able to apply such estimates in complementary dimensions: if $\mathrm{vol}_n(\mathcal{G})$ is normalized to one, then the L^2-norm on forms is bounded above by (a constant times) the comass norm and consequently the stable norm is bounded above by (a constant times) the dual L^2-norm. A simple estimate shows that the stable norm of the Poincaré dual of a cohomology class is bounded above by (a constant times) its L^2-norm. Thus, every inequality of type (15.5.1) leads to a systolic inequality in complementary dimensions. In Theorem 15.2.5 we generalize this procedure, by replacing the fundamental cycle used in Poincaré duality by a cycle of arbitrary dimension.

First we discuss a generalisation of the L^2-norm that is appropriate in this context. We stay in the well-known realm of Lipschitz cycles, although at some points the presentation would be more elegant if we used closed currents of finite mass.

Given a real Lipschitz $(p+q)$-cycle $z = \Sigma_i r_i \sigma_i \in C_*(M)$ in a compact Riemannian manifold $(M, \mathcal{G})$, we define a positive semidefinite, symmetric bilinear form

$$\langle\ ,\ \rangle_{L^2(z)}$$

on the space $\Omega^p(M)$ of p-forms as follows, *cf.* Section 14.2. Let $g_i = \sigma_i^* \mathcal{G}$ denote the pull-back of $\mathcal{G}$ by the Lipschitz simplex σ_i. Thus, g_i is a positive semidefinite, symmetric (0,2)-tensor field that is bounded, measurable, and defined almost everywhere on the standard $(p+q)$-simplex Δ^{p+q}. Let $\langle,\rangle_{g_i}$ denote the bilinear form induced by g_i on the bundle of alternating p-tensors over Δ^{p+q}. Let $d\,\mathrm{vol}_{g_i}$ denote the "volume element" induced by g_i on Δ^{p+q}, in particular $\mathrm{vol}_{g_i}(\Delta^{p+q}) = \mathrm{vol}_{p+q}(\sigma_i)$. If ω and $\overline{\omega}$ are p-forms on M, we set

$$\langle \omega, \overline{\omega} \rangle_{L^2(z)} = \sum_i |r_i| \int_{\Delta^{p+q}} \langle \sigma_i^* \omega, \sigma_i^* \overline{\omega} \rangle_{g_i} d\,\mathrm{vol}_{g_i} .$$

Since $\langle (\sigma_i^* \omega)_x, (\sigma_i^* \omega)_x \rangle_{g_i} \leq \binom{p+q}{p} \|\omega_{\sigma_i(x)}\|^2$ for $x \in \Delta^{p+q}$, *cf.* [**Fed69**, 1.8.1], we have

$$\langle \omega, \omega \rangle_{L^2(z)} \leq \binom{p+q}{p} \|\omega\|_\infty^2 \,\mathrm{vol}_{p+q}(z). \tag{15.6.1}$$

More generally, if π is an additional q-form, we can estimate

$$\left(\int_z \omega \wedge \pi \right)^2 \leq \binom{p+q}{p} \|\omega\|_\infty^2 \langle \pi, \pi \rangle_{L^2(z)} \,\mathrm{vol}_{p+q}(z). \tag{15.6.2}$$

Obviously, we can interchange the roles of ω and π in this inequality. If one has a strict inequality $p+q<n=\dim M$, there will be many p-forms ω such that the support of ω is disjoint from all $\sigma_i(\Delta^{p+q})$, and hence $\langle\omega,\omega\rangle_{L^2(z)}=0$. However, under the appropriate homological conditions on z, the semidefinite form $\langle\ ,\ \rangle_{L^2(z)}$ induces a scalar product on $H^p(M,\mathbb{R})$.

LEMMA 15.6.1. *Suppose $h\in H_{p+q}(M,\mathbb{R})$ satisfies $h\cap\alpha\neq 0$ for all $\alpha\in H^p(M,\mathbb{R})\setminus\{0\}$. If z is a Lipschitz cycle representing h, then there exists a scalar product on $H^p(M,\mathbb{R})$, denoted by $\langle\ ,\ \rangle_{L^2(z)}$ as well, such that for all $\alpha\in H^p(M,\mathbb{R})$, we have*

$$\langle\alpha,\alpha\rangle_{L^2(z)}=\inf\left\{\langle\omega,\omega\rangle_{L^2(z)}\,\middle|\,\omega \text{ a closed } p\text{-form representing } \alpha\right\}.$$

PROOF. The form $\langle\ ,\ \rangle_{L^2(z)}$ on the space $Z^p(M)$ of closed p-forms descends to a scalar product $\langle\ ,\ \rangle_{L^2(z)}$ on $Z^p(M)/N$, where

$$N=\left\{\omega\in Z^p(M)\mid\langle\omega,\omega\rangle_{L^2(z)}=0\right\}.$$

Now N is contained in the space $B^p(M)$ of exact forms. Indeed, if a differential form $\omega\in Z^p(M)\setminus B^p(M)$ represents a class $\alpha\in H^p(M,\mathbb{R})\setminus\{0\}$ then $h\cap\alpha\neq 0$ by assumption. Hence there exists $\beta\in H^p(M,\mathbb{R})$ such that $(h\cap\alpha,\beta)\neq 0$. It follows from [**Gre67**, 24.19] that $(h\cap\alpha,\beta)=(h,\alpha\cup\beta)$. If π is a closed q-form representing β, then by (15.6.2) we obtain

$$0<(h,\alpha\cup\beta)^2\leq\binom{p+q}{p}\|\pi\|_\infty^2\langle\omega,\omega\rangle_{L^2(z)}\operatorname{vol}_{p+q}(z)$$

and hence $\langle\omega,\omega\rangle_{L^2(z)}>0$.

Arguing in the completion of $\left(Z^p(M)/N,\langle,\rangle_{L^2(z)}\right)$, and using the fact that the quotient

$$Z^p(M)/B^p(M)\simeq(Z^p(M)/N)/(B^p(M)/N)$$

is finite-dimensional, one can easily complete the proof. □

We are now in a position to prove our second theorem.

PROOF OF THEOREM 15.2.5. Let z $C_*(M)$ be a Lipschitz cycle representing h. By the preceding lemma, we can consider the scalar product $\langle\ ,\ \rangle_{L^2(z)}$ on $H^p_{\mathrm{dR}}(M)$. From (15.6.1) we obtain

$$|\alpha|_{L^2(z)}\leq\binom{p+q}{p}^{\frac{1}{2}}\operatorname{vol}_{p+q}(z)^{\frac{1}{2}}\|\alpha\|^* \tag{15.6.3}$$

for $\alpha\in H^p_{\mathrm{dR}}(M)$. Hence the dual Euclidean norm on $H_p(M,\mathbb{R})$, also denoted by $|\ |_{L^2(z)}$, satisfies

$$\|k\|\leq\binom{p+q}{p}^{\frac{1}{2}}\operatorname{vol}_{p+q}(z)^{\frac{1}{2}}|k|_{L^2(z)} \tag{15.6.4}$$

for $k\in H_p(M,\mathbb{R})$. If $\alpha\in H^p_{\mathrm{dR}}(M)$ then

$$\|h\cap\alpha\|=\sup\left\{(h,\alpha\cup\beta)\mid\beta\in H^q(M,\mathbb{R}),\|\beta\|^*\leq 1\right\}$$

since the stable norm $\|\ \|$ and the comass norm $\|\ \|^*$ are dual to each other and one has $(h\cap\alpha,\beta)=(h,\alpha\cup\beta)$. Now (15.6.2) implies

$$\|h\cap\alpha\|\leq\binom{p+q}{p}^{\frac{1}{2}}\operatorname{vol}_{p+q}(z)^{\frac{1}{2}}\,|\alpha|_{L^2(z)}. \tag{15.6.5}$$

Set $b = b_p(M)$. By the definition of the constant γ'_b and Lemma 15.4.2, there exist $k \in H_p(M,\mathbb{Z})_{\mathbb{R}} \setminus \{0\}$ and $\alpha \in H^p(M,\mathbb{Z})_{\mathbb{R}} \setminus \{0\}$ such that

$$|k|_{L^2(z)}\, |\alpha|_{L^2(z)} \le \gamma'_b.$$

Using (15.6.4) and (15.6.5) we conclude

$$\|k\|\, \|h \cap \alpha\| \le \binom{p+q}{p} \gamma'_b \operatorname{vol}_{p+q}(z).$$

Since $h \cap \alpha \neq 0$ by assumption and $h \cap \alpha \in H_q(M,\mathbb{Z})_{\mathbb{R}}$ by Remark 15.4.1 we conclude

$$\operatorname{stsys}_p(\mathcal{G}) \operatorname{stsys}_q(\mathcal{G}) \le \binom{p+q}{p} \gamma'_b \|h\|. \tag{15.6.6}$$

Together with (5.3.4) this proves Theorem 15.2.5. □

REMARK 15.6.2. In [**Heb86**], Proposition 6, J. Hebda essentially proved Theorem 15.2.5 in the case $n = p + q$. Indeed, his arguments imply that

$$\operatorname{stsys}_p(\mathcal{G}) \operatorname{stsys}_{n-p}(\mathcal{G}) \le \binom{n}{p} \gamma_b \operatorname{vol}_n(\mathcal{G})$$

for every Riemannian metric $\mathcal{G}$ on a compact, orientable manifold M with Betti number $b = b_p(M) > 0$. Theorem 15.2.5, however, can be applied in many situations that are not covered by [**Heb86**, Proposition 6]. In general, the resulting inequality will be sharper than the inequality provided by Theorem 15.2.1, see example below.

EXAMPLE 15.6.3. We choose integers $p \ge 1$, $q \ge 1$ and m with $m > p + q$ and apply (15.6.6) to the manifold $M = T^{p+q} \times S^m$. Recalling that $\gamma'_b \le \frac{2}{3} b$ for $b \ge 2$, *cf.* (5.3.4), we obtain

$$\operatorname{stsys}_p(\mathcal{G}) \operatorname{stsys}_q(\mathcal{G}) \le \frac{2}{3} \binom{p+q}{p}^2 \operatorname{stsys}_{p+q}(\mathcal{G}) \tag{15.6.7}$$

for every metric $\mathcal{G}$ on $M = T^{p+q} \times S^m$. Here, the dependence on $b_p(M) = b_q(M) = \binom{p+q}{p}$ is quadratic, while the inequality following from the proof of Theorem 15.2.1 is

$$\operatorname{stsys}_p(\mathcal{G}) \operatorname{stsys}_q(\mathcal{G}) \le C^2 \binom{p+q}{p} b_p(M)^2 (1 + \log b_p(M))^2 \operatorname{stsys}_{p+q}(\mathcal{G})$$

where C is the universal constant in Banaszczyk's inequality (15.5.1).

15.7. A sharp inequality in codimension 1

We apply the arguments of Section 15.6 to a fundamental cycle representing the class $h = \xi_{\mathbb{R}} \in H_n(M,\mathbb{Z})_{\mathbb{R}}$ of M.

PROOF OF COROLLARY 15.2.6. Since $p = 1$, $q = n - 1$, we can replace the factors $\binom{p+q}{p}^{\frac{1}{2}}$ in (15.6.4) and (15.6.5) by one, *cf.* [**Fed69**, 1.8.1]. Thus, our final statement is

$$\operatorname{stsys}_1(\mathcal{G}) \operatorname{sys}_{n-1}(\mathcal{G}) \le \gamma'_b \operatorname{vol}_n(\mathcal{G}). \tag{15.7.1}$$

A dual-critical lattice L in b-dimensional Euclidean space $\mathbb{R}^b$ satisfies the identity $\lambda_1(L)\lambda_1(L^*) = \gamma'_b$, *cf.* [**BeM89**]. Consider the corresponding flat torus $(\mathbb{T}^b, \mathcal{G})$. Then Example 15.5.2 shows that for such tori equality holds in (15.7.1). □

The metrics which attain the optimal bound (15.7.1) are studied in [**BaK04**] and Chapter 16.

15.8. A conformally invariant inequality in middle dimension

Instead of using the constant γ'_b together with inequality (5.3.4) in the proof of Theorem 15.2.5, we can also employ the following estimate by W. Banaszczyk [**Bana93**, Theorem 2.1], in the case $i = b$.

If L is a lattice in b-dimensional Euclidean space (with the standard innner product norm) and $i \in \{1, \ldots, b\}$, then

$$1 \leq \lambda_i(L)\lambda_{b-i+1}(L^*) \leq b. \tag{15.8.1}$$

Specialized to the case $p + q = n$, $h = \xi_{\mathbb{R}}$, the proof of Theorem 15.2.5, combined with (15.8.1), leads to the estimate

$$\operatorname{stsys}_p(\mathcal{G})\ \operatorname{stsys}_{n-p}(\mathcal{G}) \leq \frac{\lambda_1}{\lambda_{b_p}} \binom{n}{p} b_p \operatorname{vol}_n(\mathcal{G}) \tag{15.8.2}$$

where $\lambda_i = \lambda_i(H_p(M, \mathbb{Z})_{\mathbb{R}}, |\ |_{L^2})$ and $b_p = b_p(M)$, *cf.* proof of Theorem 15.3.2 below.

In the case of middle dimension $p = \frac{n}{2}$ a similar argument leads to a conformally invariant estimate, *cf.* [**Gro83**, 7.4.A]. This case is special in the following sense. The L^2-norms of p-forms $\omega \in \Omega^p(M)$ (*cf.* Section 14.2),

$$|w|_{L^2} = \left(\int_M \langle w, w \rangle d \operatorname{vol}_n \right)^{\frac{1}{2}}$$

and

$$\|\omega\|_{L^2} = \left(\int_M (\|\omega_x\|)^2 d \operatorname{vol}_n(x) \right)^{\frac{1}{2}},$$

are invariant under conformal changes of the metric $\mathcal{G}$. They induce respectively the usual L^2-norm $|\ |_{L^2}$ on $H^p(M, \mathbb{R})$, for which the harmonic forms are minimizing representatives of the cohomology classes, and a norm $\|\ \|^*_{L^2}$ on $H^p(M, \mathbb{R})$ such that

$$\binom{n}{p}^{-\frac{1}{2}} |\alpha|_{L^2} \leq \|\alpha\|^*_{L^2} \leq |\alpha|_{L^2}.$$

We also consider the conformally invariant dual norms on $H_p(M, \mathbb{R})$ that satisfy

$$|h|_{L^2} \leq \|h\|_{L^2} \leq \binom{n}{p}^{\frac{1}{2}} |h|_{L^2}. \tag{15.8.3}$$

It is not difficult to prove that

$$\|h\|_{L^2} = \sup \left\{ \|h\|_{g'} \operatorname{vol}_n(g')^{-\frac{1}{2}} \mid g' \text{ conformal to } \mathcal{G} \right\}, \tag{15.8.4}$$

cf. [**Gro83**, 7.4.A].

15.9. A pair of conformal systoles

Let $(M, \mathcal{G})$ be a Riemannian manifold of even dimension $n = 2p$. We now use the norms developed in the previous section, to define a pair of conformal invariants of M.

DEFINITION 15.9.1. The middle-dimensional conformal invariant called *Conformal systole* $\mathrm{Confsys}_p$ is defined by setting

$$\mathrm{Confsys}_p(\mathcal{G}) = \min\left\{ \|h\|_{L^2} \,\middle|\, h \in H_p(M,\mathbb{Z})_{\mathbb{R}} \setminus \{0\} \right\},$$

and, similarly, the *conformal systole* $\mathrm{Confsys}_p$ is defined by setting

$$\mathrm{confsys}_p(\mathcal{G}) = \min\left\{ |h|_{L^2} \,\middle|\, h \in H_p(M,\mathbb{Z})_{\mathbb{R}} \setminus \{0\} \right\}.$$

See [**Gro83**, Warning, p. 91] for a discussion of the two L^2-norms.

If $n = 2$, we have $\mathrm{confsys}_1 = \mathrm{Confsys}_1$. Note that (15.8.4) implies that

$$\mathrm{stsys}_p(\mathcal{G}') \,\mathrm{vol}_n(g')^{-\frac{1}{2}} \leq \mathrm{Confsys}_p(\mathcal{G}) \tag{15.9.1}$$

for every pair $\mathcal{G}'$, $\mathcal{G}$ of conformal Riemannian metrics. However, a strict inequality can occur:

$$\sup\left\{ \mathrm{stsys}_p(\mathcal{G}') \,\mathrm{vol}_n(\mathcal{G}')^{-\frac{1}{2}} \,\middle|\, \mathcal{G}' \text{ conformal to } \mathcal{G} \right\} < \mathrm{Confsys}_p(\mathcal{G}).$$

EXAMPLE 15.9.2. Let us compare the following two bounds:

- Gromov's universal upper bound contained in inequality (+) of [**Gro96**, 2.*C*] for the 1-systole, and
- the lower bound for $\mathrm{confsys}_1$ of surfaces which follows from the work by P. Buser and P. Sarnak [**BusS94**].

Gromov asserts the existence of a universal constant C such that

$$\mathrm{sys}_1(\mathcal{G}) \,\mathrm{vol}_2(\mathcal{G})^{-\frac{1}{2}} \leq C \, \frac{\log g}{\sqrt{g}} \tag{15.9.2}$$

whenever $\mathcal{G}$ is a Riemannian metric on a closed, orientable surface of genus $g \geq 2$, *cf.* (15.3.3) above and Chapter 11. On the other hand, [**BusS94**, 1.13] states that the supremum of $\mathrm{confsys}_1(\mathcal{G})^2$ over all Riemannian metrics on a closed, orientable surface of genus g is bounded below by $c \log(g)$ for some universal constant $c > 0$, *cf.* [**KatzSV07**].

Thus, *a priori*, an upper bound for the conformal systole $\mathrm{confsys}_p(\mathcal{G})$ as given in Theorem 15.3.2 is stronger than the same upper bound for $\mathrm{stsys}_p(\mathcal{G}) \,\mathrm{vol}_n(\mathcal{G})^{-\frac{1}{2}}$, *cf.* Chapter 20.

PROOF OF THEOREM 15.3.2. Note that the map

$$\mathrm{PD}_{\mathbb{R}} : (H^p(M,\mathbb{R}), |\;|_{L^2}) \to (H_{n-p}(M,\mathbb{R}), |\;|_{L^2})$$

is an isometry. Indeed, if $\alpha \in H^p(M,\mathbb{R})$ and if ω is the harmonic p-form representing α, then

$$\begin{aligned} |\,\mathrm{PD}_{\mathbb{R}}(\alpha)|_{L^2} &= \sup\left\{ \int_M \omega \wedge \pi \,\middle|\, \pi \in Z^{n-p}(M), |\pi|_{L^2} = 1 \right\} \\ &= \sup\left\{ \langle \omega, *\pi \rangle_{L^2} \,\middle|\, \pi \in Z^{n-p}(M), |\pi|_{L^2} = 1 \right\} \\ &= |\alpha|_{L^2}. \end{aligned}$$

Setting $\lambda_1^* = \lambda_1(H^p(M,\mathbb{Z})_{\mathbb{R}}, |\;|_{L^2})$ and $b = b_p(M)$, we conclude $\lambda_1 = \lambda_1^*$ and $\lambda_1^2 \leq \frac{\lambda_1}{\lambda_b} b$ by Banaszczyk's inequality (15.8.1). Now (15.8.3) implies our claim. □

One can also apply inequality (15.8.2) in the opposite direction, as follows.

COROLLARY 15.9.3. *Let M be compact, orientable, $2p$-dimensional, and satisfying $b_p(M) > 0$. Given $D > 0$, let $\mathcal{G}_D$ denote the set of Riemannian metrics $\mathcal{G}$ on M satisfying the following two bounds:*

(1) $\mathrm{vol}_n(\mathcal{G}) \le D$;
(2) $\mathrm{stsys}_p(\mathcal{G}) \ge D^{-1}$.

Then the set of flat Finsler metrics defined by the stable norms of metrics $\mathcal{G} \in \mathcal{G}_D$ on the Jacobian torus $H_p(M,\mathbb{R})/H_p(M,\mathbb{Z})_\mathbb{R}$ is contained in a compact part of the set of all flat Finsler metrics on the Jacobi torus.

PROOF. First we will prove the existence of a basis of $H_p(M,\mathbb{Z})_\mathbb{R}$ whose elements have stable norm bounded by some function of the quantities D, p, and $b_p(M)$. We use Banaszczyk's inequality (15.5.1) in the case $i = b = b_p(M)$ for the stable norm $\| \ \|$ on $L = H_p(M,\mathbb{Z})_\mathbb{R}$ and the comass norm $\| \ \|^*$ on $L^* = H^p(M,\mathbb{Z})_\mathbb{R}$ to obtain

$$\lambda_b \lambda_1^* \le Cb(1 + \log b) \tag{15.9.3}$$

where $\lambda_i = \lambda_i(H_p(M,\mathbb{Z})_\mathbb{R}, \| \ \|)$, and $\lambda_i^* = \lambda_i(H^p(M,\mathbb{Z})_\mathbb{R}, \| \ \|^*)$. If $\alpha \in H^p(M,\mathbb{Z})_\mathbb{R}$ and $\|\alpha\|^* = \lambda_1^*$ then (15.6.3) and (15.6.5) imply

$$\| \mathrm{PD}_\mathbb{R}(\alpha) \| \le \binom{n}{p} \mathrm{vol}_n(\mathcal{G}) \|\alpha\|^*$$

and hence

$$\mathrm{stsys}_p(\mathcal{G}) \le \binom{n}{p} \mathrm{vol}_n(\mathcal{G}) \lambda_1^*.$$

Since $\mathrm{stsys}_p(\mathcal{G}) = \lambda_1$, we can use (15.9.3) to conclude

$$\mathrm{stsys}_p(\mathcal{G})^2 \le \frac{\lambda_1}{\lambda_b} \binom{n}{p} C\, b(1 + \log b)\, \mathrm{vol}_n(\mathcal{G}). \tag{15.9.4}$$

Now assume $\mathcal{G} \in \mathcal{G}_D$. Then (15.9.4) implies $\lambda_b \le \binom{n}{p} C\, b(1 + \log b) D^2$. Using [**Cas71**, p.135, Lemma 8], we obtain a basis $v_1, \ldots, v_b$ for the lattice $H_p(M,\mathbb{Z})_\mathbb{R}$ such that

$$\| \, v_i \, \| \le \binom{n}{p} C\, b^2 (1 + \log b) D^2 = E \tag{15.9.5}$$

for $1 \le i \le b$. Consider the isomorphism $H_p(M,\mathbb{R}) \to \mathbb{R}^b$ mapping $(v_1, \ldots, v_b)$ to the standard basis $(e_1, \ldots, e_b)$ and consider the induced norm on $\mathbb{R}^b$, denoted by $\| \ \|$ as well. Then the unit $\| \ \|$-ball $B = \{x \in \mathbb{R}^b \mid \|x\| \le 1\}$ satisfies

$$\left\{ x \in \mathbb{R}^b \,\middle|\, \Sigma_{i=1}^b |x_i| \le E^{-1} \right\} \subseteq B. \tag{15.9.6}$$

On the other hand, we have

$$\min \left\{ \|x\| \,\middle|\, x \in \mathbb{Z}^b \setminus \{0\} \right\} = \lambda_1 \ge D^{-1},$$

and hence Minkowski's Theorem, *cf.* [**Cas71**, p.71, Theorem II], implies

$$\mathrm{vol}_b(B) \le (2D)^b \tag{15.9.7}$$

where vol_b denotes the usual Lebesgue measure on $\mathbb{R}^b$. Now our claim follows from the fact that for fixed numbers D and $E > 0$, the set of convex bodies in $\mathbb{R}^b$ satisfying (15.9.6) and (15.9.7) is compact with respect to the Hausdorff metric. □

15.10. A sublinear estimate for a single systole

Given a compact, orientable manifold M of $\dim M = n = kp$, possessing classes $\alpha_1, \ldots, \alpha_k \in H^p(M, \mathbb{R})$ such that $\alpha_1 \cup \ldots \cup \alpha_k \neq 0$, we can apply Theorem 15.2.1 to conclude that

$$\operatorname{stsys}_p(\mathcal{G}) \leq \tilde{C}(n)\, b_p(M)(1 + \log b_p(M)) \operatorname{vol}_n(\mathcal{G})^{\frac{1}{k}}$$

for every Riemannian metric $\mathcal{G}$ on M. Using Poincaré duality as in Theorem 15.2.5, we can improve this estimate to sublinear dependence on the Betti number $b = b_p(M)$.

Proof of Theorem 15.3.4. We start by choosing linearly independent cohomology classes $\alpha_1, \ldots, \alpha_b$ in $H^p(M, \mathbb{Z})_{\mathbb{R}}$ such that

$$\|\alpha_i\|^* = \lambda_b^* = \lambda_b(H^p(M, \mathbb{Z})_{\mathbb{R}}, \| \ \|^*).$$

Our hypothesis implies that there exist indices $i_1, \ldots, i_k$ in $\{1, \ldots, b\}$ such that

$$\alpha_{i_1} \cup \ldots \cup \alpha_{i_k} \neq 0.$$

We may assume that M is connected and oriented. Let $h \in H_p(M, \mathbb{Z})_{\mathbb{R}}$ denote the Poincaré dual of the $(n-p)$-class $\alpha_{i_1} \cup \ldots \cup \alpha_{i_{k-1}}$. Then $h \neq 0$. Let

$$\omega_1, \ldots, \omega_b \in \Omega^p(M)$$

be p-forms representing $\alpha_1, \ldots, \alpha_b$, and let ω be an arbitrary closed p-form representing some $\alpha \in H^p(M, \mathbb{R})$. Then

$$\begin{aligned}(h, \alpha) &= \left(\xi_{\mathbb{R}}, \alpha_{i_1} \cup \ldots \cup \alpha_{i_{k-1}} \cup \alpha\right) = \textstyle\int_M \omega_{i_1} \wedge \ldots \wedge \omega_{i_{k-1}} \wedge \omega \\ &\leq \tfrac{n!}{(p!)^k} \left(\prod_{j=1}^{k-1} \|\omega_{i_j}\|_\infty \right) \|\omega\|_\infty \operatorname{vol}_n(M).\end{aligned}$$

Since $\|h\| = \sup\limits_{\|\alpha\|^* \leq 1} (h, \alpha)$, we obtain

$$\|h\| \leq \frac{n!}{(p!)^k} \left(\prod_{j=1}^{k-1} \|\alpha_{i_j}\|^* \right) \operatorname{vol}_n(M) \leq \frac{n!}{(p!)^k} \left(\lambda_b^*\right)^{k-1} \operatorname{vol}_n(M).$$

Now Banaszczyk's inequality (15.5.1) implies $\operatorname{stsys}_p(\mathcal{G})\lambda_b^* \leq Cb(1+\log b)$, and hence

$$\operatorname{stsys}_p(\mathcal{G})^k \leq \operatorname{stsys}_p(\mathcal{G})^{k-1}\|h\| \leq C^{k-1}\frac{n!}{(p!)^k}(b(1+\log b))^{k-1} \operatorname{vol}_n(M).$$

This proves Theorem 15.3.4. □

Note that for every $r > 1 - \frac{1}{k}$ the factor $(b(1+\log b))^{\frac{k-1}{k}}$ grows less fast than b^r when $b \to \infty$.

If $\dim M = n = 3p$ we can improve Theorem 15.3.4 as follows:

Theorem 15.10.1. *Let M be a compact, orientable manifold. Assume M is of dimension $n = 3p$. Suppose there exist classes $\beta_1, \beta_2, \beta_3 \in H^p(M, \mathbb{R})$ such that $\beta_1 \cup \beta_2 \cup \beta_3 \neq 0$. Then for every Riemannian metric $\mathcal{G}$ on M we have*

$$\operatorname{stsys}_p(\mathcal{G}) \leq \left(\binom{n}{p} \frac{n!}{(p!)^3} \right)^{\frac{1}{3}} b_p(M)^{\frac{2}{3}} \operatorname{vol}_n(\mathcal{G})^{\frac{1}{3}}.$$

PROOF. Set $\lambda_1 = \lambda_1(H_p(M,\mathbb{Z})_{\mathbb{R}}, |\;|_{L^2})$. Let $b = b_p(M)$, and set also

$$\lambda_b^* = \lambda_b\left(H^p(M,\mathbb{Z})_{\mathbb{R}}, |\;|_{L^2}\right).$$

We choose linearly independent classes $\alpha_1, \ldots, \alpha_b$ in $H^p(M,\mathbb{Z})_{\mathbb{R}}$ with $|\alpha_i|_{L^2} \le \lambda_b^*$. As in the proof of Theorem 15.3.4, we choose indices i_1, i_2, i_3 in $\{1, \ldots, b\}$ such that

$$\alpha_{i_1} \cup \alpha_{i_2} \cup \alpha_{i_3} \neq 0.$$

We may assume that M is connected and oriented. Let

$$h = \mathrm{PD}_{\mathbb{R}}(\alpha_{i_1} \cup \alpha_{i_2}) \in H_p(M,\mathbb{Z})_{\mathbb{R}} \setminus \{0\}.$$

Consider harmonic forms $\omega_{i_1}, \omega_{i_2}$ representing $\alpha_{i_1}, \alpha_{i_2}$, and a closed p-form ω representing an arbitrary $\alpha \in H^p(M,\mathbb{R})$. Then we have

$$(h,\alpha) = \int_M \omega_{i_1} \wedge \omega_{i_2} \wedge \omega \le \frac{n!}{(p!)^3} \|\omega\|_\infty \int_M |\omega_{i_1}||\omega_{i_2}| d\,\mathrm{vol}_n,$$

cf. [**Fed69**, 1.7.5 and 1.8.1] for the constant. By definition of comass

$$\|h\| = \sup_{\|\alpha\|^* \le 1} (h,\alpha),$$

we obtain

$$\|h\| \le \frac{n!}{(p!)^3} |\alpha_{i_1}|_{L^2} \, |\alpha_{i_2}|_{L^2} \le \frac{n!}{(p!)^3} (\lambda_b^*)^2.$$

We apply (15.6.4) to the fundamental cycle z of M to obtain

$$\mathrm{stsys}_p(\mathcal{G}) \le \binom{n}{p}^{\frac{1}{2}} \lambda_1 \, \mathrm{vol}_n(\mathcal{G})^{\frac{1}{2}}.$$

The preceding inequalities imply

$$\mathrm{stsys}_p(\mathcal{G})^3 \le \mathrm{stsys}_p(\mathcal{G})^2 \|h\| \le \binom{n}{p} \frac{n!}{(p!)^3} \left(\lambda_1 \lambda_b^*\right)^2 \mathrm{vol}_n(\mathcal{G}).$$

Now we apply Banaszczyk's inequality $\lambda_1 \lambda_b^* \le b$, *cf.* (15.8.1), to complete the proof of the theorem. □

CHAPTER 16

Dual-critical lattices and systoles

16.1. Introduction

Systolic inequalities are metric-independent inequalities involving minimal volumes of homology (or 1-dimensional homotopy) classes in a Riemannian manifold. They have a certain similarity to the classical isoperimetric inequality, *cf.* Chapter 5. Over the past few years it has become clear that systolic inequalities are systematically violated as soon as an *unstable* k-systole with $k \geq 2$ is involved; see [**CrK03**] and the references therein.

On the other hand, a number of systolic inequalities, including those involving stable and conformal systoles, may be found in the survey [**Gro96**]. More recent inequalities have been discussed in Chapter 1. This phenomenon tends to indicate that it is the stable systoles, rather than the systoles originally defined by M. Berger in [**Berg72a, Berg72b**], that are the "right" invariants to study when $k \geq 2$.

Systolic inequalities with *optimal constants* are rare. In addition to the ones by C. Loewner (5.4.1) and J. Hebda (16.4.3), one has Pu's inequality (5.2.2) for the real projective plane, as well as Bavard's inequality [**Bav86**], [**Sak88**] for the Klein bottle, *cf.* Figure 6.6.1. An optimal inequality, which could be called the Burago-Ivanov-Gromov inequality (17.1.1), or BIG inequality for short, is implied by the discussion [**Gro99**, pp. 259-260] for higher dimensional tori, *cf.* [**CrK03**, inequality (5.14)].

We present a new optimal systolic inequality, which generalizes a number of earlier inequalities. These are the inequalities of C. Loewner (5.4.1), J. Hebda [**Heb86**, Theorem A], certain results of G. Paternain [**Pat01**], and also inequality (15.2.4) of the previous chapter. These results first appeared in [**BaK04**].

A different optimal generalisation of the Loewner inequality, as well as the BIG inequality, is studied in Chapters 17–19, *cf.* diagram (15.1.1).

16.2. Statement of main theorems

Given a compact, oriented Riemannian manifold M with positive first Betti number $b > 0$, our optimal inequality (16.2.1) is a scale-invariant inequality providing a lower bound for the total volume of M, in terms of the product of its conformal 1-systole and its systole of codimension 1. The definitions of the systolic invariants involved appear in Section 16.3.

The inequality (15.2.4) of the previous chapter generalizes J. Hebda's inequality, see (16.4.3) below, to arbitrary Betti number. In the present work, we strengthen the inequality by replacing the stable 1-systole by its conformal analogue.

DEFINITION 16.2.1. Let C be a conformal class of metrics on an n-manifold M. The conformal 1-systole $\operatorname{confsys}_1(M,C)$ is defined by setting

$$\operatorname{confsys}_1(M,C) = \lambda_1\left(H_1(M,\mathbb{Z})_{\mathbb{R}}, \|\ \|_n\right),$$

where $\|\ \|_n$ is the L^n-norm on $H_1(M,\mathbb{R})$ dual to the norm on $H^1(M,\mathbb{R})$ defined as follows with respect to any metric representing the class C. Given $\alpha \in H^1(M,\mathbb{R})$, we set

$$\|\alpha\|_n = \inf\left\{ \left(\int_M |\omega_x|^n d\operatorname{vol}_M(x)\right)^{1/n} \middle| \omega \in \alpha \right\},$$

where $\omega \in \Omega^1(M)$ is a differential 1-form, *cf.* Section 14.2.

More detailed definitions appear in Section 16.4.

THEOREM 16.2.2 (Optimal $(1,n-1)$-inequality). *Every compact, orientable n-manifold $(M,\mathcal{G})$ with positive Betti number $b = b_1(M,\mathbb{R})$ satisfies the inequality*

$$\operatorname{confsys}_1(\mathcal{G}) \operatorname{sys}_{n-1}(\mathcal{G}) \le \gamma_b' \operatorname{vol}_n(\mathcal{G})^{\frac{n-1}{n}}, \tag{16.2.1}$$

where γ_b' is the Bergé-Martinet constant (5.3.3).

The definition (5.3.3) of the Bergé-Martinet constant γ_b' and the associated dual-critical lattices, *cf.* Definition 5.3.4, implies that γ_b' is precisely the optimal constant in inequality (16.2.1) in the class of flat tori $(\mathbb{T}^b,\mathcal{G})$. We also generalize Hebda's discussion of equality in inequality (16.4.3) as follows.

THEOREM 16.2.3 (Characterisation of equality). *Under the hypotheses of Theorem 16.2.2, equality is attained in* (16.2.1) *if and only if there exists a dual-critical lattice $L \subset \mathbb{R}^b$ and a Riemannian submersion of M onto the flat torus $\mathbb{R}^b/L$, such that all fibers are connected minimal submanifolds of M.*

In particular, equality in (16.2.1) can only occur if $b_1(M,\mathbb{R}) \le n$.

REMARK 16.2.4. An alternative way of stating the conclusion of Theorem 16.2.3 is as follows: we have equality in (16.2.1) if and only if the deck transformations of the Jacobi variety (16.5.2) form a dual-critical lattice, while the Abel-Jacobi map (16.5.3) is a Riemannian submersion with connected minimal fibers.

COROLLARY 16.2.5. *If one has* $\dim(M) = b_1(M)+1$, *then equality in* (16.2.1) *is attained if and only if the deck transformations of the Jacobi variety form a dual-critical lattice, while the Abel-Jacobi map is a Riemannian submersion with geodesic fibers of constant length.*

EXAMPLE 16.2.6. Let $(M,\mathcal{G})$ be a Riemannian 4-manifold with first Betti number $b_1(M)=3$. For instance, M could be the product of a circle by the nilmanifold given by the total space of the circle bundle over $\mathbb{T}^2$ with nonzero Euler number. Then inequality (16.2.1) takes the form

$$\operatorname{confsys}_1(\mathcal{G}) \operatorname{sys}_3(\mathcal{G}) \le \sqrt{\tfrac{3}{2}}\ \operatorname{vol}_4(\mathcal{G})^{\frac{3}{4}},$$

with equality possible if and only if the deck transformations of the Jacobi torus of M form a cubic face-centered lattice, *cf.* Example 5.3.6. The possible topology of M was clarified in [**NV04**], *cf.* Remark 16.2.7 below.

Note that M admits a Riemannian submersion with minimal fibers to a flat b-dimensional torus if and only if all of its harmonic one-forms have pointwise constant norm, *cf.* [**EeS64**] and Section 16.7. In Theorem 16.14.2 we characterize the metrics which attain the optimal bound (16.2.1). We employ a multi-parameter generalisation of the method of J. Moser [**Mo65**] of constructing volume-preserving flows to determine all such metrics.

We study the relation between harmonic one-forms of constant norm and Riemannian submersions to a flat torus with minimal fibers, *cf.* Section 16.7. Such submersions are special cases of foliations with minimal leaves. The latter have been studied in the theory of foliations, *cf.* [**Rum79**], [**Su79**], [**KamT82**], and the survey by V. Sergiescu [**Serg88**]. However, the metrics in our context are much more special.

In the class of flat tori $\mathbb{T}^b$, the discussion of equality reduces to an extremal problem for Euclidean lattices, studied by A.-M. Bergé and J. Martinet [**BeM89**]. Namely, a flat torus $\mathbb{T}^b$ realizes equality if its deck transformations form a so-called dual-critical lattice of rank b, *cf.* Definition 5.3.4. Thus, if $b = n$, then Theorem 16.2.3 implies that $(M, \mathcal{G})$ is a flat torus $\mathbb{R}^n/L$, for a suitable dual-critical lattice $L \subset \mathbb{R}^n$.

If $\dim(M) = b_1(M)$ and all harmonic 1-forms have constant norm, G. Paternain [**Pat01**, Corollary 2] proved that M is flat. The relation of the present work to [**Pat01**] is further clarified in Remark 16.7.4. D. Kotschick [**Kot01**] defines the concept of *formal* Riemannian metric and proves, in particular, that for such a metric all harmonic forms (of arbitrary degree) have constant norm. If $n = b$, then formal Riemannian manifolds are just flat tori by a result of J. Lafontaine [**Laf79**], *cf.* [**Kot01**, Theorem 7]. In general one has $b \leq n$ [**Kot01**, Theorem 6]. Both of these statements parallel our discussion of equality in (16.2.1).

REMARK 16.2.7. P.-A. Nagy and C. Vernicos [**NV04**] study Riemannian manifolds $(M, \mathcal{G})$ with all harmonic one-forms of constant norm in the particular case of $b_1(M) = \dim M - 1$. Assuming that M is compact, orientable and connected, they prove that M is diffeomorphic to a two-step nilmanifold.

16.3. Norms on (co-)homology

In this section we assume that M is a compact, oriented and connected manifold, $\dim M = n$, with Riemannian metric $\mathcal{G}$. Let p be a real number satisfying $1 \leq p \leq \infty$. We will define the L^p-norm $\| \, \|_p^*$ on $H^k(M, \mathbb{R})$ and the dual norms $\| \, \|_p$ on the dual space $H^k(M, \mathbb{R})^* = H_k(M, \mathbb{R})$. Then the k-systoles of $(M, \mathcal{G})$ will be defined as the minima of such norms on non-zero elements in the integer lattice $H_k(M, \mathbb{Z})_{\mathbb{R}}$ in $H_k(M, \mathbb{R})$.

The L^p-norms on $H^k(M, \mathbb{R})$, $k \in \{0, \ldots, n\}$, are quotient norms of the corresponding norms on closed differential forms in $\Omega^k(M)$, *cf.* Section 14.2. We let $\omega \in \Omega^k M$, and let $1 \leq p < \infty$. We set

$$\|\omega\|_p = \left(\int_M \|\omega_x\|^p dvol_n(x) \right)^{\frac{1}{p}},$$

where

$$\|\omega_x\| = \max \left\{ \omega_x(v_1, \ldots, v_k) \big| \, |v_i| \leq 1 \text{ for } 1 \leq i \leq k \right\}$$

is the pointwise comass norm of ω_x. For $p = \infty$, we set

$$\|\omega\|_\infty = \max_{x \in M} \|\omega_x\|.$$

REMARK 16.3.1. In the cases $k = 1$ and $k = n - 1$ which are mainly of interest in the present chapter, the comass norm coincides with the usual (Euclidean) norm $|\omega_x|$ induced by $\mathcal{G}$. Thus, for 1-forms or $(n-1)$-forms ω, we will often use the symbol $|\omega|_p$ instead of $\|\omega\|_p$.

If $\alpha \in H^k(M, \mathbb{R})$ and $1 \leq p \leq \infty$, we define

$$\|\alpha\|_p^* = \inf\{\|\omega\|_p \mid \omega \in \alpha\},$$

where the expression $\omega \in \alpha$ means that ω is a smooth closed k-form representing α. For $p = 2$, Hodge theory implies that every cohomology class α contains a unique harmonic form ω. If $k \in \{1, n-1\}$, this form ω is characterized by the equality $|\omega|_2 = \|\alpha\|_2^*$. In particular, the norm $\| \ \|_2^*$ on $H^1(M, \mathbb{R})$ and $H^{n-1}(M, \mathbb{R})$ is induced by the L^2 scalar product $\langle , \rangle$, on harmonic forms.

The non-degenerate Kronecker pairing

$$(\ , \ ) : H_k(M, \mathbb{R}) \times H^k(M, \mathbb{R}) \to \mathbb{R}$$

induced by integration of closed forms over cycles, allows us to define a norm $\| \ \|_p$ on $H_k(M, \mathbb{R})$ dual to the L^p-norm $\| \ \|_p^*$ on $H^k(M, \mathbb{R})$. Thus, given a homology class $h \in H_k(M, \mathbb{R})$, we set

$$\|h\|_p = \max\left\{(h, \alpha) \,\middle|\, \alpha \in H^k(M, \mathbb{R}), \|\alpha\|_p^* \leq 1\right\}.$$

16.4. Definition of conformal systoles

In Section 15.9, we defined two flavors of the middle dimensional conformal systole. More generally, the conformal k-systole is defined as follows. Recall that $\lambda_1(L)$ is the first successive minimum of a lattice L, *cf.* Definition 4.4.1. For $p = \frac{n}{k}$, the norms $\| \ \|_p^*$ on $H^k(M, \mathbb{R})$ and $\| \ \|_p$ on $H_k(M, \mathbb{R})$ are conformally invariant.

DEFINITION 16.4.1. The conformally invariant quantity

$$\operatorname{Confsys}_k(\mathcal{G}) = \lambda_1\left(H_k(M, \mathbb{Z})_{\mathbb{R}}, \| \ \|_{\frac{n}{k}}\right) \tag{16.4.1}$$

is called the *Conformal* k-systole of $(M, \mathcal{G})$.

If $h \in H_k(M, \mathbb{R})$ then the quantity $\|h\|_{\frac{n}{k}}$ is the supremum of the stable norms of h with respect to all metrics that are conformal to $\mathcal{G}$ and have unit volume, *cf.* [**Gro83**, 7.4.A]. In particular, one has

$$\operatorname{stsys}_k(\mathcal{G}) \leq \operatorname{Confsys}_k(\mathcal{G}) \operatorname{vol}_n(\mathcal{G})^{\frac{k}{n}}, \tag{16.4.2}$$

cf. (12.1.7). However, the supremum of $\operatorname{stsys}_k(\tilde{\mathcal{G}})$ over all unit volume metrics $\tilde{\mathcal{G}}$ conformal to $\mathcal{G}$ may be smaller than $\operatorname{Confsys}_k(\mathcal{G})$. Indeed, the usual systole and the conformal systole obey dramatically different asymptotic bounds even in the case of surfaces. Thus, P. Buser and P. Sarnak [**BusS94**] construct a sequence of surfaces Σ_g of genus $g \to \infty$, satisfying an asymptotically *infinite* lower bound

$$\operatorname{confsys}_1(\Sigma_g)^2 \geq c \log g \text{ as } g \to \infty,$$

for a suitable constant $c > 0$, *cf.* [**KatzSV07**]. In contrast, an asymptotically *vanishing* upper bound

$$\frac{\operatorname{sys}_1(\Sigma_g)^2}{\operatorname{area}(\Sigma_g)} \leq C \frac{(\log g)^2}{g}$$

appears in [**Gro83, Gro96**], *cf.* Chapter 11.

Our Theorem 16.2.2 generalizes, in addition to the Loewner inequality (5.4.1), the following inequality due to J. Hebda [**Heb86**, Theorem A].

THEOREM 16.4.2. *Let M be a compact, orientable, n-dimensional manifold with first Betti number $b_1(M,\mathbb{R}) = 1$. Then every Riemannian metric $\mathcal{G}$ on M satisfies*

$$\operatorname{stsys}_1(\mathcal{G})\operatorname{sys}_{n-1}(\mathcal{G}) \leq \operatorname{vol}_n(\mathcal{G}) \tag{16.4.3}$$

with equality if and only if $(M,\mathcal{G})$ admits a Riemannian submersion with connected minimal fibers onto a circle.

Note that inequality (16.2.1) is a generalisation of (5.4.1), while (16.4.3) follows as a special case of inequality (16.4.2). Indeed, the first Bergé-Martinet constant γ_1' equals one, while $\gamma_2' = \frac{2}{\sqrt{3}}$, *cf.* [**BeM89**, Proposition 2.13] and Section 5.3.

16.5. Jacobi variety and Abel-Jacobi map

Let $\pi = \pi_1(M)$. Let $f : \pi \to \pi^{ab}$ be its abelianisation map. Let $tor = tor(\pi^{ab})$ be the torsion subgroup of π^{ab}. Let

$$g : \pi^{ab} \to \pi^{ab}/tor$$

be the quotient by torsion. Clearly, $\pi^{ab}/tor = \mathbb{Z}^b$, where $b = b_1(M)$. Let

$$\phi = g \circ f : \pi \to \mathbb{Z}^b$$

be the composed homomorphism.

DEFINITION 16.5.1. The cover $\overline{M}$ of the manifold M corresponding the subgroup $Ker(\phi) \subset \pi$ is called the universal (or maximal) free abelian cover.

Let E be the space of harmonic 1-forms on M, with dual E^* canonically identified with $H_1(M,\mathbb{R})$. By integrating an integral harmonic 1-form along paths from a basepoint $x_0 \in M$, we obtain a map to the circle $\mathbb{R}/\mathbb{Z} = S^1$.

Similarly, in order to define a map $M \to H_1(M,\mathbb{R})/H_1(M,\mathbb{Z})_{\mathbb{R}}$ without choosing a basis for cohomology, we argue as follows. Let x be a point in the universal cover $\widetilde{M}$ of M. Thus x is represented by a point of M together with a path c from x_0 to it. By integrating along the path c, we obtain a linear form, $h \to \int_c h$, on E. We thus obtain a map $\widetilde{M} \to E^* = H_1(M,\mathbb{R})$, which, furthermore, descends to a map

$$\overline{\mathcal{A}}_M : \overline{M} \to E^*, \quad c \mapsto \left(h \mapsto \int_c h\right), \tag{16.5.1}$$

where $\overline{M}$ is the universal free abelian cover.

DEFINITION 16.5.2. The Jacobi variety (Jacobi torus) of M is the torus

$$J_1(M) = H_1(M,\mathbb{R})/H_1(M,\mathbb{Z})_{\mathbb{R}}. \tag{16.5.2}$$

DEFINITION 16.5.3. The *Abel-Jacobi map*

$$\mathcal{A}_M : M \to J_1(M), \tag{16.5.3}$$

is obtained from the map (16.5.1) by passing to quotients.

The Abel-Jacobi map is unique up to translations of the Jacobi torus.

REMARK 16.5.4. The flat torus $\mathbb{R}^b/L$ in Theorem 16.2.3 is isometric to the Jacobi torus (16.5.2), while the Riemannian submersion is the Abel-Jacobi map (16.5.3), induced by the harmonic one-forms on M, originally introduced by A. Lichnerowicz [**Lich69**], *cf.* [**Gro99**, 4.21], *cf.* [**Mir95**, p. 249].

16.6. Summary of the proofs

The proof of Theorem 16.2.2, which will be completed in Section 16.10, depends on two results. First we prove that for conjugate exponents $p \in [1, \infty]$ and q, the Poincaré duality map

$$\mathrm{PD}_{\mathbb{R}} : (H^1(M, \mathbb{R}), \| \ \|_p^*) \to (H_{n-1}(M, \mathbb{R}), \| \ \|_q)$$

is an isometry, see Proposition 16.8.1. On the other hand, the Hölder inequality implies a chain of inequalities for the norms $\| \ \|_p^*$ on $H^1(M, \mathbb{R})$, see Proposition 16.9.1. Dually, we obtain the reverse inequalities for the norms $\| \ \|_p$ on $H_1(M, \mathbb{R})$. This allows us to reduce inequality (16.2.1) to the case $p = 2$ where the norms are Euclidean, so that the definition (5.3.3) of γ_b' applies. If one has equality in (16.2.1), and if $\alpha \in H^1(M, \mathbb{Z})_{\mathbb{R}}$ is a nonzero element of minimal L^2-norm, then the chain of inequalities for $\|\alpha\|_p^*$ reduces to equality for all p, and we conclude that the harmonic representative of α has constant norm.

Similarly, the L^2-dual of every nonzero homology class $h \in H_1(M, \mathbb{Z})_{\mathbb{R}}$ of minimal dual L^2-norm is represented by a harmonic one-form of constant norm, as well. Using a result of Bergé and Martinet [**BeM89**] on dual-critical lattices, we then prove that equality in (16.2.1) implies that *all* harmonic one-forms have constant norm. Finally, the Riemannian submersion is the Abel-Jacobi map $\mathcal{A}_M$ of formula (16.5.3), see Section 16.7.

In the case $b_1(M) = 1$, J. Hebda [**Heb86**] remarked already that equality in his inequality (16.4.3) implies the existence of a nonzero harmonic one-form with constant norm, and hence of a Riemannian submersion with minimal fibers over a circle. Although Riemannian submersions over flat tori with minimal fibers are relatively simple objects, it is not immediately clear what choices one has in constructing all of them. In the case $b_1(M) = 1$, J. Hebda [**Heb86**] describes the simplest class of examples, namely local Riemannian products. In Theorem 16.14.2, we present a construction that starts with a submersion $F : M \to \mathbb{T}^m$ over a flat torus, and characterizes all Riemannian metrics on M for which F is a Riemannian submersion with minimal fibers.

More specifically, we can prescribe arbitrary metrics of fixed volume on the fibers. To determine the metric, one additionally has to choose a suitable horizontal distribution. This choice depends on the choice of a linear map from $\mathbb{R}^m$ into the Lie algebra of vertical vector fields on M that preserve the volume elements of the fibers. If the fiber dimension, $\dim M - m$, is greater than one, this Lie algebra is infinite-dimensional. In particular, in this case the Riemannian metrics satisfying the case of equality in (16.2.1) are much more flexible than local Riemannian products.

REMARK 16.6.1. There are compact manifolds that admit a submersion to the circle, but do not carry a Riemannian metric possessing a nontrivial local product structure, *e.g.* hyperbolic manifolds fibering over a circle. Thus, our construction yields new topological types of examples even in the case $b_1(M) = 1$.

16.7. Harmonic one-forms of constant norm and flat tori

Most of the material of this section appears in [**EeS64**, pp. 127-128]. We include the proofs of Lemma 16.7.1 and Proposition 16.7.3 for the sake of completeness. The proofs are mostly straightforward calculations in local Riemannian geometry.

First we give a geometric characterisation of the existence of a nonzero harmonic one-form of constant (pointwise) norm on a Riemannian manifold $(M, \mathcal{G})$.

LEMMA 16.7.1. *If $\theta \neq 0$ is a harmonic one-form of constant norm, then the leaves of the distribution $\ker\theta$ are minimal, and the vector field V_θ that is $\mathcal{G}$-dual to θ has geodesic flow lines orthogonal to $\ker\theta$.*

Conversely, given a transversely oriented foliation of M by minimal hypersurfaces, such that the orthogonal foliation consists of geodesics, there exists a unique harmonic one-form θ with $|\theta| = 1$, such that $\ker\theta$ is everywhere tangent to the foliation by minimal hypersurfaces, and such that θ is positive on the oriented normals to the leaves.

PROOF. Since $d\theta = 0$, we can locally find a primitive f of θ, *i.e.* $df = \theta$ and hence $\text{grad} f = V_\theta$. The condition $d^*\theta = 0$ translates into $\text{div}(\text{grad} f) = \Delta f = 0$. Since $|V_\theta| = |\theta|$ is constant, for all vector fields W on M we have

$$0 = \mathcal{G}(\nabla_W V_\theta, V_\theta) = \text{Hess}(f)(W, V_\theta) = \mathcal{G}\left(W, \nabla_{V_\theta} V_\theta\right).$$

Hence $\nabla_{V_\theta} V_\theta = 0$, *i.e.* the flow lines of V_θ are geodesics. Since V_θ is normal to the leaves, while $|V_\theta|$ is constant and $\nabla_{V_\theta} V_\theta = 0$, the condition $\text{div}(V_\theta) = 0$ is equivalent to the fact that the mean curvature of the leaves of $\ker(\theta)$ vanishes.

Conversely, let θ be the one-form with $|\theta| = 1$ that defines the given foliation by minimal hypersurfaces and is positive on the oriented normals. We want to prove that θ is closed. Since the orthogonal foliation is geodesic by assumption, we have $\nabla_{V_\theta} V_\theta = 0$. If h is a locally defined regular function whose level sets are leaves of $\ker(\theta)$, then $\text{grad} h = \lambda V_\theta$ for some nowhere vanishing function λ. Then for all vector fields W, Z we have

$$\text{Hess}(h)(W, Z) = W(\lambda)\mathcal{G}(V_\theta, Z) + \lambda\mathcal{G}(\nabla_W V_\theta, Z).$$

Hence the bilinear form

$$(W, Z) \to \mathcal{G}(\nabla_W V_\theta, Z)$$

is symmetric if W and Z are orthogonal to V_θ. Since $\nabla_{V_\theta} V_\theta = 0$ and

$$0 = W(|V_\theta|^2) = 2\mathcal{G}(\nabla_W V_\theta, V_\theta),$$

we conclude that

$$\mathcal{G}(\nabla_W V_\theta, Z) = \mathcal{G}(\nabla_Z V_\theta, W)$$

for all vector fields W, Z. This implies $d\theta = 0$. As above, we see that the minimality of the leaves, together with the conditions $|V_\theta| = 1$ and $\nabla_{V_\theta} V_\theta = 0$, imply the equality $\text{div}(V_\theta) = 0$, and hence $d^*\theta = 0$. □

REMARK 16.7.2. If $\dim M = 2$ and if M carries a nonzero harmonic one-form θ of constant norm, then M is flat and, in particular, every harmonic one-form has constant norm. This can be seen as follows. According to Lemma 16.7.1, the form θ induces two orthogonal foliations by geodesics. So the Gauss-Bonnet formula applied to small geodesic squares implies that M is flat. Then the lift of a harmonic one-form on M to the universal cover is constant.

Recall that a closed one-form is called *integral* if its integral over an arbitrary loop is an integer.

PROPOSITION 16.7.3. *The following two properties of a Riemannian manifold $(M, \mathcal{G})$ are equivalent:*

(i) *There exists a flat torus $(T^m, \overline{\mathcal{G}})$ and a Riemannian submersion F with minimal fibers, $F : (M, \mathcal{G}) \to (T^m, \overline{\mathcal{G}})$;*
(ii) *The set of harmonic one-forms of constant norm contains an m-dimensional vector space generated by integral one-forms.*

PROOF OF THE IMPLICATION $(i) \Rightarrow (ii)$. Since $\overline{\mathcal{G}}$ is flat, every harmonic one-form $\overline{\theta}$ on $(T^m, \overline{\mathcal{G}})$ has constant norm. Integral forms pull back to integral forms, and so it suffices to prove that $F^*\overline{\theta}$ is a harmonic one-form of norm $|F^*\overline{\theta}| = |\overline{\theta}|$. This equality is true since F is a Riemannian submersion. To show that $F^*\overline{\theta}$ is harmonic, note that $F^*\overline{\theta}$ is closed and of constant norm. It follows that the foliation orthogonal to $\ker\left(F^*\overline{\theta}\right)$ is geodesic if $\overline{\theta} \neq 0$, *cf.* the proof of Lemma 16.7.1.

By Lemma 16.7.1, it remains to show that the leaves of the foliation $\ker\left(F^*\overline{\theta}\right)$ are minimal. This is a consequence of the minimality of the fibers of F, together with the fact that the leaves of $\ker\left(F^*\overline{\theta}\right)$ are geodesic in horizontal directions. Namely, if $v \in \ker\left(F^*\overline{\theta}|_x\right)$ is orthogonal to the fiber $F_x = F^{-1}(F(x))$ through x and if c is the geodesic in M with $\dot{c}(0) = v$, then

$$\dot{c}(t) \in \ker\left(F^*\overline{\theta}|_{c(t)}\right)$$

for all t. □

PROOF OF THE IMPLICATION $(ii) \Rightarrow (i)$. Let $\theta_1, \ldots, \theta_m$ be linearly independent integral harmonic one-forms on M, such that all linear combinations $\sum_{i=1}^{m} r_i\theta_i$ with $r_i \in \mathbb{R}$ have constant norm. Note that this last property is equivalent to the constancy of all scalar products $\mathcal{G}^*(\theta_i, \theta_j)$ with respect to the dual metric $\mathcal{G}^*$. Let $\left(\overline{g}_{ij}\right)$ denote the matrix inverse to $\left(\mathcal{G}^*(\theta_i, \theta_j)\right)$. Since the θ_i are integral, there exist functions $F_i : M \to \mathbb{R}/\mathbb{Z}$ such that $dF_i = \theta_i$. Then the map

$$F = (F_1, \ldots, F_m) : M \to \mathbb{R}^m/\mathbb{Z}^m$$

is a Riemannian submersion from $(M, \mathcal{G})$ to $\mathbb{R}^m/\mathbb{Z}^m$ endowed with the constant metric

$$\overline{\mathcal{G}} = \sum \overline{g}_{ij} dx_i \otimes dx_j.$$

It remains to prove that the fibers of F are minimal. We can find a global orthonormal frame field $V_1, \ldots, V_m$ for the horizontal bundle

$$\mathcal{H} = \ker(DF)^\perp \subseteq TM$$

by taking the $\mathcal{G}$-dual of the appropriate constant linear combinations of the m forms $\theta_1, \ldots, \theta_m$. Locally, we can complete $V_1, \ldots, V_m$ to an orthonormal frame field $V_1, \ldots, V_n$ for TM. Since $V_1, \ldots, V_m$ are $\mathcal{G}$-dual to harmonic one-forms, we know that

$$\sum_{j=1}^{n} \mathcal{G}\left(\nabla_{V_j} V_i, V_j\right) = \operatorname{div} V_i = 0$$

$$\begin{array}{ccc} M_0 & \xrightarrow{\quad I \quad} & M_1 \\ \Big\downarrow {\scriptstyle F_0} & & \Big\downarrow {\scriptstyle F_1} \\ (\mathbb{T}^m, \bar{g}_0) & \overset{\overline{I}}{\dashrightarrow} & (\mathbb{T}^m, \bar{g}_1) \end{array}$$

FIGURE 16.7.1. Construction of isometry $\overline{I}$

for $1 \leq i \leq m$, *cf.* the proof of Lemma 16.7.1. On the other hand, Lemma 16.7.1 implies $\nabla_{V_j} V_j = 0$ for $1 \leq j \leq m$, and hence

$$\mathcal{G}\left(\nabla_{V_j} V_i, V_j\right) = V_j(\mathcal{G}(V_i, V_j)) = 0$$

for $1 \leq j \leq m$, $1 \leq i \leq n$. The preceding equations imply that

$$\sum_{j=m+1}^{n} \mathcal{G}\left(\nabla_{V_j} V_i, V_j\right) = 0$$

for $1 \leq i \leq m$. Since $V_1, \ldots, V_m$ are orthogonal to the fibers of F, this says that the mean curvature vector of the fibers of F vanishes. $\square$

REMARK 16.7.4. In the case $m = \dim(M)$, the statement of Proposition 16.7.3 also follows from the arguments employed by G. Paternain to prove [**Pat01**, Corollary 2]. These arguments are global and use the solution of the E. Hopf conjecture by D. Burago and S. Ivanov. They do not apply in the case $m < \dim(M)$.

In the corollary below, we show that a Riemannian submersion F as in Proposition 16.7.3(i) is uniquely determined by the metric $\mathcal{G}$ on M and by the induced map $F_* : H_1(M, \mathbb{Z}) \to H_1(\mathbb{T}^m, \mathbb{Z})$.

COROLLARY 16.7.5. *Suppose $I : (M_0, g_0) \to (M_1, g_1)$ is an isometry and for $i = 0, 1$ we are given Riemannian submersions F_i with minimal fibers from (M_i, g_i) to flat tori $(\mathbb{T}^m, \bar{g}_i)$. If there exists an automorphism A of $H_1(\mathbb{T}^m, \mathbb{Z})$ such that $(F_1 \circ I)_* = A \circ (F_0)_*$ then there exists an isometry $\overline{I} = (\mathbb{T}^m, g_0) \to (\mathbb{T}^m, \bar{g}_1)$ such that $\overline{I} \circ F_0 = F_1 \circ I$.*

PROOF. Let $L_i \subseteq \mathbb{R}^m$ be the lattice of deck transformations of $(\mathbb{T}^m, \bar{g}_i)$. Then we can identify A with a linear map $J : \mathbb{R}^m \to \mathbb{R}^m$ such that $J(L_0) = L_1$. We denote by

$$\overline{J} : \mathbb{R}^m / L_0 \to \mathbb{R}^m / L_1$$

the diffeomorphism induced by J. Our assumption implies that

$$(F_1 \circ I)_* = \left(\overline{J} \circ F_0\right)_* .$$

Hence, if θ is a constant one-form on $\mathbb{R}^m / L_1$, then the closed one-forms $(F_1 \circ I)^* \theta$ and $\left(\overline{J} \circ F_0\right)^* \theta$ are cohomologous. From the proof of Proposition 16.7.3, we know that $(F_1 \circ I)^* \theta$ and $\left(\overline{J} \circ F_0\right)^* \theta$ are harmonic with respect to g_0. By the uniqueness of the harmonic representative of a cohomology class, we conclude that

$$(F_1 \circ I)^* \overline{\theta} = \left(\overline{J} \circ F_0\right)^* \overline{\theta}$$

for every constant one-form on $\mathbb{R}^m / L_1$. This implies the existence of $v \in \mathbb{R}^m / L_1$ such that

$$F_1 \circ I = \overline{J} \circ F_0 + v.$$

If we define $\overline{I} : \mathbb{R}^m/L_0 \to \mathbb{R}^m/L_1$ by $\overline{I}(y) = \overline{J}(y)+v$, then $\overline{I}$ satisfies $F_1 \circ I = \overline{I} \circ F_0$, as illustrated in Figure 16.7.1. The map $\overline{I}$ is a diffeomorphism and a (local) isometry since I is an isometry, and F_0 and F_1 are Riemannian submersions. □

16.8. Norm duality and the cup product

In this section, we assume that $(M, \mathcal{G})$ is a compact, n-dimensional, oriented, and connected Riemannian manifold. For $k \in \{1, n-1\}$ we consider the norms $\| \ \|_p^*$ on $H^k(M,\mathbb{R})$ and $\| \ \|_p$ on $H_k(M,\mathbb{R})$ defined in Section 16.3.

In Proposition 16.8.1, we prove that for every pair of conjugate exponents $p \in [1,\infty]$ and q, the Poincaré duality map is an isometry between the pair of normed spaces $\left(H^1(M,\mathbb{R}), \| \ \|_p\right)$ and $(H_{n-1}(M,\mathbb{R}), \| \ \|_q)$; respectively, it is an isometry between $\left(H^{n-1}(M,\mathbb{R}), \| \ \|_p\right)$ and $(H_1(M,\mathbb{R}), \| \ \|_q)$.

The results of the present section will be used in Sections 16.10 and 16.12 to prove the systolic inequality in Theorem 16.2.2 and to analyze the case of equality. For completeness and future reference, some of them are proved in greater generality than necessary for these applications.

PROPOSITION 16.8.1. *Let $p \in [1,\infty]$ and q be conjugate exponents. Then the normed spaces $\left(H^1(M,\mathbb{R}), \| \ \|_p^*\right)$ and $\left(H^{n-1}(M,\mathbb{R}), \| \ \|_q^*\right)$ are dual to each other with respect to the cup product.*

REMARK 16.8.2. The claim of Proposition 16.8.1 is equivalent to either of the following two statements:

(1) The Poincaré duality map
$$\mathrm{PD}_{\mathbb{R}} : \left(H^1(M,\mathbb{R}), \| \ \|_p^*\right) \to \left(H_{n-1}(M,\mathbb{R}), \| \ \|_q\right)$$
is an isometry.

(2) The Poincaré duality map
$$\mathrm{PD}_{\mathbb{R}} : \left(H^{n-1}(M,\mathbb{R}), \| \ \|_q^*\right) \to \left(H_1\ (M,\mathbb{R}), \| \ \|_p\right)$$
is an isometry.

PROOF OF PROPOSITION 16.8.1. A key idea is the application of the Hodge star operator in a suitable measurable context, as in formula (16.8.3). Consider cohomology classes $\alpha \in H^1(M,\mathbb{R})$ and $\beta \in H^{n-1}(M,\mathbb{R})$. If ω (respectively, π) is a closed form representing α (respectively, β), then the cup product $\alpha \cup \beta$ satisfies
$$\mathrm{PD}_{\mathbb{R}}(\alpha \cup \beta) = \int_M \omega \wedge \pi.$$
Since one-forms and $(n-1)$-forms take values in the set of simple covectors, we can use [**Fed69**, p. 32] and the Hölder inequality to estimate
$$\left| \int_M \omega \wedge \pi \right| \leq \int_M |\omega_x| \ |\pi_x| d\,\mathrm{vol}_n(x) \leq |\omega|_p |\pi|_q.$$
This proves the bound
$$\mathrm{PD}_{\mathbb{R}}(\alpha \cup \beta) \leq \|\alpha\|_p^* \|\beta\|_q^*. \tag{16.8.1}$$
It remains to show that for every cohomology class $\alpha \in H^1(M,\mathbb{R}) \setminus \{0\}$, there exists a class $\beta \in H^{n-1}(M,\mathbb{R}) \setminus \{0\}$ such that
$$\mathrm{PD}_{\mathbb{R}}(\alpha \cup \beta) = \|\alpha\|_p^* \|\beta\|_q^*, \tag{16.8.2}$$

or, conversely, that for every $\beta \in H^{n-1}(M,\mathbb{R})\setminus\{0\}$, there exists $\alpha \in H^1(M,\mathbb{R})\setminus\{0\}$ such that (16.8.2) holds. Assume first that $2 \le p < \infty$. It is shown in the proof of [**Sib70**, Lemma 4.2] that there exists a one-form ω in the L^p-closure of α such that $|\omega|_p = \|\alpha\|_p^*$. It follows from [**Ham92**, Theorem 5.1] that the form

$$\pi = |\omega|^{p-2}(*\omega) \tag{16.8.3}$$

is weakly closed and determines a class $\beta \in H^{n-1}(M,\mathbb{R})$ such that $\|\beta\|_q^* = |\pi|_q$. This implies $0 \neq \alpha \cup \beta = \|\alpha\|_p^* \|\beta\|_q^*$.

If $1 < p < 2$ and hence $2 < q < \infty$, we can apply analogous arguments starting with an arbitrary $\beta \in H^{n-1}(M,\mathbb{R}) \setminus \{0\}$. We then obtain $\alpha \in H^1(M,\mathbb{R}) \setminus \{0\}$ such that (16.8.2) holds. Finally we treat the case $p = \infty$ and $q = 1$. The proof in the case $p = 1$ and $q = \infty$ is completely analogous. According to Remark 16.8.2, it suffices to prove that the map

$$\mathrm{PD}_{\mathbb{R}} : \left(H^{n-1}(M,\mathbb{R}), \|\ \|_1^*\right) \to \left(H_1(M,\mathbb{R}), \|\ \|_\infty\right)$$

is an isometry. From (16.8.1) we conclude that $\|\mathrm{PD}_{\mathbb{R}}(\beta)\|_\infty \le \|\beta\|_1^*$ for every class $\beta \in H^{n-1}(M,\mathbb{R})$. To prove the opposite inequality, we use some elementary facts from geometric measure theory.

DEFINITION 16.8.3. A closed normal one-current is a linear functional

$$T : \Omega^1 M \to \mathbb{R}$$

(*cf.* Section 14.2) with the following two properties:

(1) we have $T(df) = 0$ for every $f \in C^\infty(M,\mathbb{R})$;
(2) the mass $\mathbf{M}(T)$, defined by

$$\mathbf{M}(T) = \sup\left\{T(\omega) \mid \omega \in \Omega^1(M), |\omega|_\infty \le 1\right\}, \tag{16.8.4}$$

is finite.

Every closed normal one-current T determines a homology class $[T] \in H_1(M,\mathbb{R})$ such that $([T],\alpha) = T(\omega)$ whenever $\omega \in \alpha$.

Every closed $(n-1)$-form $\pi \in \Omega^{n-1}(M)$ defines a closed normal one-current T_π by setting

$$T_\pi(\omega) = \int_M \omega \wedge \pi.$$

Such one-currents are called smooth. One easily sees that

$$\mathbf{M}(T_\pi) = |\pi|_1$$

and

$$\mathrm{PD}_{\mathbb{R}}(\pi) = \pm[T_\pi]$$

Functional analysis implies that for every $h \in H_1(M,\mathbb{R})$, there exists a closed normal one-current T such that $[T] = h$ and $\mathbf{M}(T) = \|h\|_\infty$, *cf.* [**Fed74**, Section 3].

We can smooth the current T, *cf.* [**Fed74**, 4.1.18], and obtain a sequence of classes $\pi_i \in \mathrm{PD}_{\mathbb{R}}^{-1}(h)$ such that

$$\lim_{i\to\infty} |\pi_i|_1 = \lim_{i\to\infty} \mathbf{M}(T_{\pi_i}) = \mathbf{M}(T) = \|h\|_\infty.$$

This implies $\|\mathrm{PD}_{\mathbb{R}}^{-1}(h)\|_1^* \le \|h\|_\infty$. □

16.9. Hölder inequality in cohomology and case of equality

The Hölder inequality implies inequalities between the L^p norms $\| \ \|_p$ for different values of p. We will prove that the harmonic representative of a cohomology class $\alpha \in H^1(M, \mathbb{R})$ has constant norm if and only if for some (and hence for every) $p \neq 2$, the inequality relating the L^2-norm and the L^p-norm is an equality at α. Note that in the case $p = \infty$, this follows from [**Pat01**, Theorem C].

PROPOSITION 16.9.1. *For every $\alpha \in H^1(M, \mathbb{R})$, the function*

$$p \in [1, \infty] \to \|\alpha\|_p^* \operatorname{vol}_n(\mathcal{G})^{-\frac{1}{p}}$$

is weakly increasing. For $p \in [1, \infty] \setminus \{2\}$, the equality

$$\|\alpha\|_p^* \operatorname{vol}_n(\mathcal{G})^{-\frac{1}{p}} = \|\alpha\|_2^* \operatorname{vol}_n(\mathcal{G})^{-\frac{1}{2}}$$

holds if and only if the harmonic representative of α has constant norm.

PROOF. The monotonicity follows from Hölder inequality. Next, we assume that the harmonic representative ω of α has constant norm. Due to the monotonicity, it suffices to prove that

$$\|\alpha\|_\infty^* \leq \|\alpha\|_2^* \operatorname{vol}_n(\mathcal{G})^{-\frac{1}{2}} \leq \|\alpha\|_1^* \operatorname{vol}_n(\mathcal{G})^{-1}$$

The first inequality follows directly from the constancy of $|\omega_x|$. To prove the second inequality, let $(\omega_i)_{i \in \mathbb{N}}$ be a sequence of representatives of α such that

$$\lim_{i \to \infty} |\omega_i|_1 = \|\alpha\|_1^*.$$

Since $\omega_i - \omega$ is exact and ω is harmonic, we see that $\langle \omega_i - \omega, \omega \rangle = 0$. Since $|\omega_x|$ is constant, we conclude that

$$|\omega|_2^2 = \langle \omega_i, \omega \rangle_2 \leq |\omega_i|_1 |\omega|_2 \operatorname{vol}_n(\mathcal{G})^{-\frac{1}{2}}$$

and hence

$$\|\alpha\|_1^* \operatorname{vol}_n(\mathcal{G})^{-\frac{1}{2}} \geq \|\alpha\|_2^*.$$

Finally, we assume that $p \in [1, \infty] \setminus \{2\}$ and that

$$\|\alpha\|_p^* \operatorname{vol}_n(\mathcal{G})^{-\frac{1}{p}} = \|\alpha\|_2^* \operatorname{vol}_(n)^{-\frac{1}{2}}, \tag{16.9.1}$$

and prove that the harmonic representative ω of α has constant norm. Due to the monotonicity, it suffices to treat the cases $1 < p < 2$ and $2 < p < \infty$. If $1 < p < 2$, the discussion of equality in the Hölder inequality implies that ω has constant norm. If $2 < p < \infty$, then there exists a one-form ω in the L^p-closure of α such that $|\omega|_p = \|\alpha\|_p^*$, *cf.* [**Sib70**, proof of Lemma 4.2]. Now the Hölder inequality and (16.9.1) imply

$$\begin{aligned} |\omega|_2 \operatorname{vol}_n(\mathcal{G})^{-\frac{1}{2}} &\leq |\omega|_p \operatorname{vol}_n(\mathcal{G})^{-\frac{1}{p}} \\ &= \|\alpha\|_p^* \operatorname{vol}_n(\mathcal{G})^{-\frac{1}{p}} \\ &= \|\alpha\|_2^* \operatorname{vol}_n(\mathcal{G})^{-\frac{1}{2}}. \end{aligned}$$

Hence ω is the harmonic representative of α, the first inequality is an equality, and the norm of ω is constant. □

Let $\langle\ ,\ \rangle_2^*$ denote the scalar product on $H^1(M,\mathbb{R})$, with norm $\|\ \|_2^*$. Thus, if we identify $H^1(M,\mathbb{R})$ with the space of harmonic one-forms, then $\langle\ ,\ \rangle_2^*$ corresponds to the L^2-scalar product. We define an isomorphism $I : H_1(M,\mathbb{R}) \to H^1(M,\mathbb{R})$ by setting

$$\langle I(h), \alpha\rangle_{L^2}^* = (h, \alpha) \quad \text{for } h \in H_1(M,\mathbb{R}), \alpha \in H^1(M,\mathbb{R}). \tag{16.9.2}$$

Note that $\|h\|_2 = \|I(h)\|_2^*$ for all $h \in H_1(M,\mathbb{R})$. Proposition 16.9.1 implies the following corollary.

COROLLARY 16.9.2. *For every $h \in H_1(M,\mathbb{R})$, the function*

$$p \in [1,\infty] \to \|h\|_p \operatorname{vol}_n(\mathcal{G})^{\frac{1}{p}}$$

is (weakly) decreasing. For $p \in [1,\infty] \setminus \{2\}$, the equality

$$\|h\|_p \operatorname{vol}_n(\mathcal{G})^{\frac{1}{p}} = \|h\|_2 \operatorname{vol}_n(\mathcal{G})^{\frac{1}{2}}$$

holds if and only if the harmonic representative of $I(h)$ has constant norm.

Similarly, we can combine Propositions 16.8.1 and 16.9.1 to obtain the following corollary.

COROLLARY 16.9.3. *For every $k \in H_{n-1}(M,\mathbb{R})$, the function*

$$p \in [1,\infty] \to \|k\|_p \operatorname{vol}_n(\mathcal{G})^{\frac{1}{p}}$$

is (weakly) decreasing. For $p \in [1,\infty] \setminus \{2\}$, the equality

$$\|k\|_p \operatorname{vol}_n(\mathcal{G})^{\frac{1}{p}} = \|k\|_2 \operatorname{vol}_n(\mathcal{G})^{\frac{1}{2}}$$

holds if and only if the harmonic representative of the class $\mathrm{PD}_{\mathbb{R}}^{-1}(k)$ has constant norm.

16.10. Proof of optimal $(1, n-1)$-inequality

We assume that $(M,\mathcal{G})$ is a compact, oriented, and connected Riemannian manifold with $b = b_1(M,\mathbb{R}) \geq 1$. We first show how the results proved in Sections 16.8 and 16.9 imply Theorem 16.2.2:

$$\operatorname{confsys}_1(\mathcal{G}) \operatorname{sys}_{n-1}(\mathcal{G}) \leq \gamma_b' \operatorname{vol}_n(\mathcal{G})^{\frac{n-1}{n}}, \tag{16.10.1}$$

where $n = \dim(M)$ and γ_b' is the Bergé-Martinet constant. Note that inequality (16.10.1) is true also for disconnected manifolds, once we have established it for connected ones.

PROOF OF THEOREM 16.2.2. Consider the following pair of dual Euclidean vector spaces, $(H_1(M,\mathbb{R}), \|\ \|_2)$ and $(H_{n-1}(M,\mathbb{R}), \|\ \|_2)$. These spaces are dual to each other with respect to the intersection pairing defined by

$$h \cdot k = (h, \mathrm{PD}_{\mathbb{R}}^{-1}(k)).$$

This follows from the fact that the map

$$\mathrm{PD}_{\mathbb{R}} : (H^1(M,\mathbb{R}), \|\ \|_2^*) \to (H_{n-1}(M,\mathbb{R}), \|\ \|_2)$$

is an isometry, *cf.* Remark 16.8.2. Moreover, the corresponding pair of integer lattices, $H_1(M,\mathbb{Z})_{\mathbb{R}} \subset H_1(M,\mathbb{R})$ and $H_{n-1}(M,\mathbb{Z})_{\mathbb{R}} \subset H_{n-1}(M,\mathbb{R})$, are dual to each other with respect to this pairing, *cf.* Section 15.4. Hence, formula (5.3.3) for the Bergé-Martinet constant γ_b' implies that

$$\lambda_1\left(H_1(M,\mathbb{Z})_{\mathbb{R}}, \|\ \|_2\right) \lambda_1\left(H_{n-1}(M,\mathbb{Z})_{\mathbb{R}}, \|\ \|_2\right) \leq \gamma_b',$$

where $b = b_1(M)$. Invoking Corollaries 16.9.2 and 16.9.3, we obtain the inequality

$$\lambda_1\left(H_1\left(M,\mathbb{Z}\right)_{\mathbb{R}}, \| \ \|_p\right) \lambda_1\left(H_{n-1}(M,\mathbb{Z})_{\mathbb{R}}, \| \ \|_{p'}\right) \leq \gamma_b' \operatorname{vol}_n(\mathcal{G})^{1-\frac{1}{p}-\frac{1}{p'}} \tag{16.10.2}$$

valid for all $p, p' \in [2, \infty]$. Now we specify the parameters to the following values: $p = n$ and $p' = \infty$. Then inequality (16.10.2) specifies to (16.10.1), *cf.* definition of $\operatorname{confsys}_1$ in formula (16.2.1). □

REMARK 16.10.1. Note that the inequality $\|\alpha\|_{L^2}^* (\operatorname{vol}_n(\mathcal{G}))^{\frac{1}{2}} \geq \| \operatorname{PD}_{\mathbb{R}} \alpha\|_\infty$ for an element $\alpha \in H^1(M,\mathbb{R})$ can also be proved more directly, as follows. Consider a nonzero element $\alpha \in H^1(M,\mathbb{Z})_{\mathbb{R}}$ of the integer lattice, and let $\omega \in \alpha$ be the harmonic 1-form. Then there exists a map $f : M \to S^1 = \mathbb{R}/\mathbb{Z}$ such that $df = \omega$. Using the Cauchy-Schwarz inequality and the coarea formula, *cf.* [**Fed69**, 3.2.11], [**Cha06**, p. 267], we obtain

$$\begin{aligned} \|\alpha\|_{L^2}^*(\operatorname{vol}_n(\mathcal{G}))^{1/2} &= |\omega|_2(\operatorname{vol}_n(\mathcal{G}))^{1/2} \\ &\geq \int_M |df| dvol_n \\ &= \int_{S^1} \operatorname{vol}_{n-1}\left(f^{-1}(t)\right) dt \\ &\geq \| \operatorname{PD}(\alpha)\|_\infty, \end{aligned} \tag{16.10.3}$$

since the hypersurface $f^{-1}(t)$ is Poincaré dual to α for every regular value of t.

16.11. Consequences of equality, criterion of dual-perfection

We now collect some consequences of equality in (16.10.1). From Corollaries 16.9.2 and 16.9.3, combined with Remark 16.7.2 (if $n = 2$), we obtain Lemma 16.11.2 below.

DEFINITION 16.11.1. An element $v \in (L, \| \ \|)$ is *short* if $\|v\| = \lambda_1(L, \| \ \|)$.

LEMMA 16.11.2. *Assume equality holds in* (16.10.1). *Then*

(1) *if a cohomology class* $\alpha \in (H^1(M,\mathbb{Z})_{\mathbb{R}}, \| \ \|_2^*)$ *is short, then the harmonic representative of* α *has constant norm;*
(2) *if a homology class* $h \in (H_1(M,\mathbb{Z})_{\mathbb{R}}, \| \ \|_2)$ *is short, then the harmonic representative of* $I(h) \in H^1(M,\mathbb{R})$ *as in* (16.9.2) *has constant norm.*

Given a pair of dual lattices L and L^* in Euclidean space $(E, \| \ \|_2)$, consider the sets of short vectors

$$\begin{aligned} S(L) &= \left\{\alpha \in L \,\middle|\, \|\alpha\|_2 = \lambda_1(L)\right\}, \\ S(L^*) &= \left\{\beta \in L^* \,\middle|\, \|\beta\|_2 = \lambda_1(L^*)\right\}, \end{aligned}$$

so that $S(L) \cup S(L^*) \subset E$. Let $Q(E)$ denote the vector space of symmetric bilinear forms on E, and let $\varphi : E \to Q(E)^*$ be the map defined by

$$\varphi(\gamma)(q) = q(\gamma, \gamma)$$

for $\gamma \in E$ and $q \in Q(E)$.

DEFINITION 16.11.3 ([**BeM89**]). A pair (L, L^*) of dual lattices in Euclidean space is called *dual-perfect* if the set $\varphi(S(L) \cup S(L^*))$ generates $Q(E)^*$ as a vector space.

The following was proved in [**BeM89**, item 3.10].

THEOREM 16.11.4. *Every dual-critical pair is dual-perfect.*

Note that a compact, oriented Riemannian manifold for which equality holds in (16.10.2) is necessarily connected.

PROPOSITION 16.11.5. *Let $(M, \mathcal{G})$ be a compact, oriented Riemannian manifold of dimension n, with positive first Betti number $b = b_1(M, \mathbb{R})$. Assume that*

$$\operatorname{confsys}_1(\mathcal{G}) \operatorname{sys}_{n-1}(\mathcal{G}) = \gamma'_b \operatorname{vol}_n(\mathcal{G})^{\frac{n-1}{n}}.$$

Then every harmonic one-form on $(M, \mathcal{G})$ has constant norm.

PROOF. By Hodge theory, the space E of harmonic 1-forms on M endowed with the L^2 scalar product is canonically isometric to the space $\left(H^1(M, \mathbb{R}), \| \|_2^*\right)$. In E, we have the pair of dual lattices $L = H^1(M, \mathbb{Z})_{\mathbb{R}}$ and $L^* = I\left(H_1(M, \mathbb{Z})_{\mathbb{R}}\right)$.

In the terminology of [**BeM89**], our assumption says that the pair (L, L^*) is dual-critical. By Theorem 16.11.4, the pair (L, L^*) is dual-perfect in the sense of Definition 16.11.3. We have a canonical map $p : M \to Q(E)$, $x \mapsto p_x$, defined by

$$p_x(\alpha, \beta) = \langle \alpha(x), \beta(x) \rangle$$

for $\alpha, \beta \in E$. By Lemma 16.11.2, every $\gamma \in S(L) \cup S(L^*)$ is a harmonic form of constant norm. This means that the linear form $\varphi(\gamma) \in Q(E)^*$ is constant on the image $p(M) \subseteq Q(E)$ of p. Since the elements $\varphi(\gamma)$, $\gamma \in S(L) \cup S(L^*)$ span $Q(E)^*$, it follows that every element of $Q(E)^*$ is constant on $p(M)$. In particular, if $\gamma \in E$ then

$$\varphi(\gamma)(p_x) = p_x(\gamma, \gamma) = |\gamma(x)|^2$$

does not depend on $x \in M$. □

REMARK 16.11.6. In terms of matrix traces, the argument above can be paraphrased as follows. The vectors $s \in S(L) \cup S(L^*)$ give rise to rank 1 symmetric matrices $ss^t \in Sym(\mathbb{R}^b)$. A quadratic form $q_A(x)$ defined by a symmetric matrix A gives rise to a linear map $Q : Sym(\mathbb{R}^b) \to \mathbb{R}$ by the formula

$$Q(xx^t) = q_A(x) = x^t A x = \operatorname{Tr}(x^t A x) = \operatorname{Tr}(A x x^t).$$

Therefore, the map Q is uniquely determined by its values solely on the elements $ss^t \in Sym(\mathbb{R}^b)$, which form a spanning set by dual-perfection. It follows that every harmonic 1-form on M has constant norm.

REMARK 16.11.7. The following question was asked by A. Katok in 2002. Does there exist an example of a Riemannian metric with a pair of harmonic one-forms of constant norm, whose pointwise scalar product is not constant? Such an example is, in fact, constructed in [**NV04**], showing that the Bergé-Martinet criterion (Theorem 16.11.4) cannot be bypassed in the proof of Theorem 16.2.3.

16.12. Characterisation of equality in $(1, n-1)$-inequality

To prove Theorem 16.2.3, assume first that there exists a Riemannian submersion F with connected, minimal fibers of a compact, oriented, n-dimensional Riemannian manifold $(M, \mathcal{G})$, with first Betti number $b \geq 1$, onto a flat torus $\mathbb{R}^b / L$ where L denotes a lattice in Euclidean space $(\mathbb{R}^b, | \, |)$. We will show that in such a situation, we necessarily have

$$\begin{aligned} \operatorname{stsys}_1(\mathcal{G}) &= \lambda_1(L), \\ \operatorname{confsys}_1(\mathcal{G}) &= \lambda_1(L) \operatorname{vol}_n(\mathcal{G})^{-\frac{1}{n}}, \\ \operatorname{sys}_{n-1}(\mathcal{G}) &= \lambda_1(L^*) \operatorname{vol}_n(\mathcal{G}). \end{aligned} \tag{16.12.1}$$

If L is dual-critical, formulas (16.12.1) imply that

$$\operatorname{stsys}_1(\mathcal{G}) \operatorname{sys}_{n-1}(\mathcal{G}) = \gamma_b' \operatorname{vol}_n(\mathcal{G})$$

and $\operatorname{confsys}_1(\mathcal{G}) \operatorname{sys}_{n-1}(\mathcal{G}) = \gamma_b' \operatorname{vol}_n(\mathcal{G})^{\frac{n-1}{n}}$. First note that the harmonic one-forms on M are precisely the pull-backs of the constant one-forms on $\mathbb{R}^b/L$, and hence all of them have constant norm, *cf.* Proposition 16.7.3.

Now Corollaries 16.9.2 and 16.9.3 imply that, up to the factor $\operatorname{vol}_n(\mathcal{G})^{\frac{1}{p}-\frac{1}{2}}$, the L^p-norms on $H_1(M,\mathbb{R})$ and on $H_{n-1}(M,\mathbb{R})$ coincide with the L^2-norms $\| \ \|_2$. In particular, we have

$$\operatorname{stsys}_1(\mathcal{G}) = \lambda_1\left(H_1\ (M,\mathbb{Z})_{\mathbb{R}}, \| \ \|_2\right) \operatorname{vol}_n(\mathcal{G})^{\frac{1}{2}}, \tag{16.12.2}$$

and

$$\operatorname{confsys}_1(\mathcal{G}) = \operatorname{stsys}_1(\mathcal{G}) \operatorname{vol}_n(\mathcal{G})^{-\frac{1}{n}}, \tag{16.12.3}$$

and furthermore,

$$\operatorname{sys}_{n-1}(\mathcal{G}) = \lambda_1\left(H_{n-1}(M,\mathbb{Z})_{\mathbb{R}}, \| \ \|_2\right) \operatorname{vol}_n(\mathcal{G})^{\frac{1}{2}}. \tag{16.12.4}$$

By Remark 16.8.2, the Poincaré duality map

$$\mathrm{PD}_{\mathbb{R}} : H^1(M,\mathbb{R}) \to H_{n-1}(M,\mathbb{R})$$

preserves L^2-norms. Since it maps $H^1(M,\mathbb{Z})_{\mathbb{R}}$ onto $H_{n-1}(M,\mathbb{Z})_{\mathbb{R}}$ (*cf.* Section 15.4), equation (16.12.4) is equivalent to

$$\operatorname{sys}_{n-1}(\mathcal{G}) = \lambda_1\left(H^1(M,\mathbb{Z})_{\mathbb{R}}, \| \ \|_2^*\right) \operatorname{vol}_n(\mathcal{G})^{\frac{1}{2}}. \tag{16.12.5}$$

Since the fibers of the submersion F are connected, the induced homomorphism between fundamental groups $\pi_1(M) \to \pi_1(\mathbb{R}^b/L)$ is onto, and hence we obtain an epimorphism $H_1(M,\mathbb{Z}) \to H_1(\mathbb{R}^b/L,\mathbb{Z})$. Since $b = b_1(M,\mathbb{R})$, the epimorphism $F_* : H_1(M,\mathbb{Z})_{\mathbb{R}} \to H_1(\mathbb{R}^b/L,\mathbb{Z})_{\mathbb{R}}$ is an isomorphism. By duality, the map $F^* : H^1(\mathbb{R}^b/L,\mathbb{Z})_{\mathbb{R}} \to H^1(M,\mathbb{Z})_{\mathbb{R}}$ is an isomorphism. Now $H^1(\mathbb{R}^b/L,\mathbb{Z})_{\mathbb{R}}$ can be identified with the dual lattice, L^*. An element $\theta \in L^*$ satisfies

$$\|F^*\theta\|_2^* = |\theta| \operatorname{vol}_n(\mathcal{G})^{\frac{1}{2}}.$$

Hence, equations (16.12.2), (16.12.3), and (16.12.5) imply our claim, *i.e.* equation (16.12.1).

Conversely, let us assume that $\operatorname{confsys}_1(\mathcal{G}) \operatorname{sys}_{n-1}(\mathcal{G}) = \gamma_b' \operatorname{vol}_n(\mathcal{G})^{\frac{n-1}{n}}$. Then Proposition 16.11.5 implies that every harmonic one-form on M has constant norm. Integrating harmonic one-forms whose classes form a basis for the integer lattice in $H^1(M,\mathbb{R})$, we obtain a Riemannian submersion F with minimal fibers from $(M,\mathcal{G})$ onto a flat torus $\mathbb{R}^b/L$, *cf.* (16.7.3). Recall that we obtain F from a basis of the integer lattice $H^1(M,\mathbb{Z})_{\mathbb{R}}$. Hence the map

$$F^* : H^1(\mathbb{R}^b/L,\mathbb{Z})_{\mathbb{R}} \to H^1(M,\mathbb{Z})_{\mathbb{R}}$$

is an isomorphism. It follows that F induces an epimorphism at the level of fundamental groups, and hence the fibers of F are connected. Note that the first part of the proof implies the identities

$$\operatorname{confsys}_1(\mathcal{G}) = \lambda_1(L) \operatorname{vol}_n(\mathcal{G})^{-\frac{1}{n}}$$

and $\operatorname{sys}_{n-1}(\mathcal{G}) = \lambda_1(L^*) \operatorname{vol}_n(\mathcal{G})$. Hence $\lambda_1(L)\lambda_1(L^*) = \gamma_b'$, and therefore L is dual-critical.

16.13. Construction of all extremal metrics

In this section, we will start with a submersion $F : M \to \mathbb{T}^m$ of a compact manifold M to a flat torus, and present a construction that yields all Riemannian metrics on M for which F is a Riemannian submersion with minimal fibers. If M admits a submersion to $\mathbb{T}^b$, where $b = b_1(M, \mathbb{R})$, this construction accounts for all metrics on M for which all harmonic one-forms have constant norm, *cf.* Proposition 16.7.3.

In this construction, we can prescribe arbitrary metrics of fixed volume on the fibers of F. This does not determine the metric on M uniquely. We additionally have to define a horizontal distribution $\mathcal{H}$ transverse to the vertical distribution

$$\mathcal{V} = \ker(dF).$$

The choice of $\mathcal{H}$ is restricted by the requirement that the horizontal lift of every flow on the base has to preserve the volume elements of the fibers. The necessity of this condition follows from the requirement that all fibers be minimal, *cf.* Lemma 16.13.3.

We start by providing a multiparameter variant of J. Moser's theorem [**Mo65**] on qthe existence of volume-preserving diffeomorphisms. This will allow us to find the horizontal distributions, starting with an arbitrary family of metrics on the fibers of constant volume.

We first explain the setting and fix some notation. Let $F : M \to M'$ denote a submersion between compact, connected and oriented manifolds M and M'.

DEFINITION 16.13.1. Let $\mathfrak{X}_F$ denote the set of *projectable vector fields on* M. Namely, a vector field V on M is in $\mathfrak{X}_F$ if there exists a vector field W on M' such that $DF \circ V = W \circ F$.

Such a vector field W on M' will be called the *projection of* $V \in \mathfrak{X}_F$ and it will be denoted by $W = F_*(V)$. In the literature on foliations, projectable vector fields are also called "basic" or "foliate". Note that $\mathfrak{X}_F$ is a Lie subalgebra of the Lie algebra of all vector fields on M. Furthermore, F_* is a homomorphism from the Lie algebra $\mathfrak{X}_F$ to the Lie algebra $\mathfrak{X}(M')$ of all vector fields on M'. In terms of the flow, Φ, of $V \in \mathfrak{X}_F$, and the flow, Ψ, of $F_*(V) = W$, the relation $DF \circ V = W \circ F$ is equivalent to

$$F \circ \Phi_t = \Psi_t \circ F \quad \text{for all } t \in \mathbb{R}.$$

In particular, if $y \in M'$ then diffeomorphism Φ_t maps the fiber $F^{-1}(y)$ to the fiber $F^{-1}(\Psi_t(y))$. We denote the diffeomorphism $\Phi_t|F^{-1}(y)$ by $\Phi_{t,y}$.

In addition, we assume that on every fiber $F^{-1}(y)$, we are given a volume element $\alpha^y \in \Omega^{n-m}(F^{-1}(y))$, where $\dim M = n$ and $\dim M' = m$, that varies smoothly with $y \in M'$. We say that the flow Φ of $V \in \mathfrak{X}_F$ *preserves* $\alpha = (\alpha^y)_{y \in M'}$ if

$$\Phi_{t,y}^* \alpha^{\psi_t(y)} = \alpha^y,$$

for all $(t, y) \in \mathbb{R} \times M'$. We denote by $\mathfrak{X}(\alpha)$ the Lie subalgebra of all projectable vector fields whose flows preserve α.

PROPOSITION 16.13.2. *The homomorphism* $F_* : \mathfrak{X}(\alpha) \to \mathfrak{X}(M')$ *is an epimorphism of Lie algebras.*

This multiparameter version of J. Moser's theorem [**Mo65**] can be proved along the lines of the original proof in [**Mo65**]. Now we specialize the preceding discussion to the case of a Riemannian submersion $F : M \to M'$ between Riemannian manifolds M and M', where $\dim M = n$, $\dim M' = m$. Actually, the discussion is purely local. Every flow Ψ on M' has a unique *horizontal lift* Φ to M, *i.e.* Φ is the flow on M whose flow lines are the horizontal lifts of the flow lines of Ψ. Equivalently, if Ψ is the flow of the vector field W on M', then Φ is the flow of the *horizontal lift* V of W defined by requiring that $V \in \mathcal{X}_F$, $F_*(V) = W$ and that V is everywhere horizontal, *i.e.* orthogonal to the vertical distribution $\mathcal{V} = \ker(dF)$. Obviously, the horizontal lift of any flow on M' maps fibers of F to fibers of F.

LEMMA 16.13.3. *Suppose $F : M \to M'$ is a Riemannian submersion between oriented Riemannian manifolds. Then all the fibers of F are minimal submanifolds if and only if the horizontal lift of every flow on M' maps the volume elements of the fibers of F to each other.*

PROOF. This follows from [**Rum79**, Proposition 1]. □

REMARK 16.13.4. Actually, the "if"-part in Lemma 16.13.3 is true under the following apparently weaker assumption. For every $y \in M'$ there are m vector fields on M' that are linearly independent at y and whose flows have horizontal lifts that preserve the volume elements of the fibers of F.

16.14. Submersions onto tori

Let us now assume that F is a submersion of a compact, oriented and connected n-manifold M onto a flat torus $\mathbb{R}^m/L$ where L is a lattice in Euclidean space $\mathbb{R}^m$. Moreover, we assume that for all $y \in \mathbb{R}^m/L$, we are given a volume element α^y on $F^{-1}(y)$ that depends smoothly on y.

We let $\mathcal{X}_c(\alpha)$ denote the set of all vector fields $V \in \mathcal{X}(\alpha)$ such that

$$dF(V) = (dF_1(V), \ldots, dF_m(V)) \text{ is a constant vector in } \mathbb{R}^m.$$

Obviously $\mathcal{X}_c(\alpha)$ is a linear subspace of $\mathcal{X}(\alpha)$, and the map

$$l_F : \mathcal{X}_c(\alpha) \to \mathbb{R}^m, \; l_F(V) = dF(V),$$

is a linear map. Proposition 16.13.2 implies

$$l_F : \mathcal{X}_c(\alpha) \to \mathbb{R}^m \text{ is onto.} \tag{16.14.1}$$

To define a horizontal distribution $\mathcal{H}$, we will choose a linear right inverse H to l_F, namely a map $H : \mathbb{R}^m \to \mathcal{X}_c(\alpha)$, and set

$$\mathcal{H}_x = \{H(v)|_x \mid v \in \mathbb{R}^m\} \subseteq TM_x$$

for $x \in M$. Note that we have much freedom in the choice of H if $n - m > 1$. The kernel of l_F consists of those vertical vector fields V that preserve each α^y, *i.e.* on every fiber $F^{-1}(y)$ one has a vector field $V^y = V|F^{-1}(y)$ whose flow preserves α^y and that varies smoothly with $y \in T^m$. In particular, $\ker(l_F) \subseteq \mathcal{X}_c(\alpha)$ is an infinite-dimensional Lie algebra if $n - m > 1$.

We are now in a position to describe our construction of metrics for which a given submersion onto a flat torus is Riemannian and with minimal fibers.

CONSTRUCTION 16.14.1. *Let M be a compact, oriented and connected manifold of dimension n. Let $F : M \to \mathbb{T}^m = \mathbb{R}^m/L$ be a submersion onto a flat torus.*

1) *Choose a smooth family g^y, $y \in \mathbb{T}^m$, of Riemannian metrics on the fibers $F^{-1}(y)$ of F, such that the total Riemannian volume of $(F^{-1}(y), g^y)$ does not depend on $y \in \mathbb{T}^m$.*

2) *Let* $\alpha = (\alpha_y)_{y \in M'}$ *denote the family of Riemannian volume forms on the fibers of* F. *Choose a linear right inverse* $H : \mathbb{R}^m \to \mathcal{X}_c(\alpha)$ *to* l_F *and define the subbundle* $\mathcal{H} \subseteq TM$ *by setting*

$$\mathcal{H}_x = \{H(v)|_x \mid v \in \mathbb{R}^m\}$$

for $x \in M$.

3) *Now define a Riemannian metric* $\mathcal{G}$ *on* M *by specifying the following three requirements:*
   (a) *The subbundles* $\mathcal{V} = \ker(dF)$ *and* $\mathcal{H}$ *are orthogonal to each other;*
   (b) *For all* $x \in M$ *the isomorphism* $dF_x|\mathcal{H}_x : \mathcal{H}_x \to \mathbb{R}^m$ *is an isometry;*
   (c) *For all* $y \in \mathbb{T}^m$ *the inclusion* $(F^{-1}(y), g^y) \to (M, \mathcal{G})$ *is isometric.*

THEOREM 16.14.2. *The Riemannian metrics* $\mathcal{G}$ *on* M *arising from Construction 16.14.1 are precisely those for which the map*

$$F : M \to \mathbb{T}^m = \mathbb{R}^m/L$$

is a Riemannian submersion with minimal fibers.

PROOF. Suppose $\mathcal{G}$ arises from the construction. Then properties (a) and (b) imply that F is a Riemannian submersion. The vector fields $V_i = H(e_i)$, $1 \leq i \leq m$, form an orthonormal frame field for $\mathcal{H} = \mathcal{V}^\perp$. By property (c), and since $\mathcal{X}_c(\alpha) \subseteq \mathcal{X}(\alpha)$, their flows preserve the volume elements induced by $\mathcal{G}$ on the fibers. It follows that the fibers are minimal submanifolds of $(M, \mathcal{G})$, *cf.* Remark 16.13.4. Conversely, suppose $\mathcal{G}$ is a Riemannian metric on M such that $F : M \to \mathbb{R}^m/L$ is a Riemannian submersion with minimal fibers. We describe how the metrics on the fibers and the map H have to be chosen so that our construction yields $\mathcal{G}$.

For the metrics g^y, $y \in \mathbb{R}^m/L$, on the fibers we must obviously take the metrics induced by $\mathcal{G}$. If $v \in \mathbb{R}^m$ let $\overline{\theta}$ be the constant one-form on $\mathbb{R}^m/L$ that is dual to v with respect to the scalar product on $\mathbb{R}^m$ and let V be $\mathcal{G}$-dual to $F^*\overline{\theta}$. Now Lemma 16.13.3 implies that $V \in \mathcal{X}(\alpha)$, where $\alpha = (\alpha_y)_{y \in M'}$ denotes the volume elements induced by $\mathcal{G}$ on the fibers. Moreover, $l_F(V) = v$, i.e. the (linear) map that maps $v \in \mathbb{R}^m$ to $V \in \mathcal{X}_c(\alpha)$ is a right inverse to l_F which we take to be H. Then the horizontal distribution $\mathcal{H}$ induced by H is $\mathcal{G}$-orthogonal to $\mathcal{V} = \ker(dF)$ and our construction applied to these choices of $g^y, y \in \mathbb{R}^m/L$, and H yields the metric $\mathcal{G}$. □

Corollary 16.7.5 shows to what extent our construction leads to non-isometric metrics $\mathcal{G}$, and how it depends on the submersion F and the flat torus $\mathbb{R}^m/L$.

If M admits a metric such that all harmonic one-forms have constant norm, then there exists a submersion $F : M \to \mathbb{T}^b$ for $b = b_1(M, \mathbb{R})$, *cf.* Proposition 16.7.3. Starting from such submersions F and from arbitrary flat metrics on $\mathbb{T}^b$, our construction yields all metrics on M for which all harmonic one-forms have constant norm.

Similarly, the manifold M can support metrics $\mathcal{G}$ for which equality holds in inequality (16.2.1) only if M admits a submersion onto $\mathbb{T}^b$ for $b = b_1(M, \mathbb{R})$. Starting from such submersions and from flat metrics on $\mathbb{T}^b$ coming from dual-critical lattices $L \subseteq \mathbb{R}^b$, the construction yields all metrics on M satisfying equality in (16.2.1).

CHAPTER 17

Generalized degree and Loewner-type inequalities

17.1. Burago-Ivanov-Gromov inequality

Let M be a Riemannian n-manifold, and $\overline{M}$ its maximal free abelian cover, *cf.* Definition 16.5.1. Let $b = b_1(M)$. Let $\bar{f} : \overline{M} \to \mathbb{R}^b$ be the lift of an Abel-Jacobi map. Denote by $[\overline{F}_M] \in H_{n-b}(\overline{M})$ the homology class of a typical fiber of $\bar{f}$, *i.e.* the preimage of a regular point of $\mathbb{R}^b$ (more detailed definitions may be found in Section 17.2).

In 1983, M. Gromov [**Gro83**, Theorem 7.5.B] initiated the study of the geometric consequences of the nonvanishing of the class $[\overline{F}_M]$. Namely, he proved the first inequality of the type studied in the present chapter. The inequality that appeared in [**Gro83**, Theorem 7.5.B] was not optimal, but more recent techniques allow one to replace Gromov's inequality by an optimal one. Gromov himself eventually proved an optimal inequality in the case of the torus, as implied by the discussion in [**Gro99**, pp. 259-260], see below.

We generalize optimal inequalities of C. Loewner and M. Gromov, by proving lower bounds for the total volume in terms of the homotopy systole and the stable systole. Our main tool is the construction of an area-decreasing map to the Jacobi torus, streamlining and generalizing the construction of D. Burago and S. Ivanov [**BuraI95**]. It turns out that one can successfully combine this construction with the coarea formula, to obtain new optimal inequalities. These results first appeared in [**IvK04**].

An optimal generalisation of the Loewner inequality (5.4.1) is implied (*i.e.* proved but not stated) in [**Gro99**, pp. 259-260] (*cf.* [**CrK03**, inequality (5.14)]), based on the techniques of D. Burago and S. Ivanov [**BuraI94, BuraI95**]. We give below a slight generalisation of Gromov's idea.

DEFINITION 17.1.1. Given a map $f : M \to M'$ between closed manifolds of the same dimension, we denote by $\deg(f)$, either the algebraic degree of f when both manifolds are orientable, or the absolute degree, otherwise, *cf.* [**BroBS01**].

We denote by $\mathcal{A}_M$ the Abel-Jacobi map of M, *cf.* formula (17.2.1). Let γ_n be the Hermite constant (5.3.1). We denote by $\operatorname{stsys}_1(\mathcal{G})$ the stable 1-systole of a metric $\mathcal{G}$, *cf.* formula (12.1.7).

THEOREM 17.1.2 (Burago-Ivanov-Gromov inequality). *Let M^n be a compact manifold of equal dimension and first Betti number:* $\dim(M) = b_1(M) = n$. *Then every metric $\mathcal{G}$ on M satisfies the following optimal inequality:*

$$\deg(\mathcal{A}_M) \operatorname{stsys}_1(\mathcal{G})^n \le (\gamma_n)^{\frac{n}{2}} \operatorname{vol}_n(\mathcal{G}). \tag{17.1.1}$$

The bound (17.1.1) is attained by flat tori whose group of deck transformations is a critical lattice in $\mathbb{R}^n$.

Note that the inequality is nonvacuous in the orientable case, only if the cup-length of M is n, *i.e.* the Abel-Jacobi map $\mathcal{A}_M$ is of nonzero algebraic degree. Recall that a critical lattice (*i.e.* one attaining the value of the Hermite constant (5.3.1)) is in particular extremal, and being extremal for a lattice implies perfection and eutaxy [**Bar57**].

Generalisations of the Burago-Ivanov-Gromov inequality, or BIG inequality for short, are described in Sections 17.2 and 17.3.

Historical remarks and a discussion of the related systolic literature can be found at the end of the next section.

17.2. Generalized degree and BIG(n, b) inequality

Let $(M, \mathcal{G})$ be a Riemannian manifold. Let $n = \dim(M)$ and $b = b_1(M)$. The Loewner inequality can be generalized by optimal inequalities in two different ways, as illustrated in diagram (15.1.1), where the constants γ_n and γ_n' are defined in Section 5.3, while

$$\mathcal{A}_M : M \to \mathbb{T}^b \tag{17.2.1}$$

is the Abel-Jacobi map ([**Lich69, Gro96**]), *cf.* (16.5.3), inducing an isomorphism in 1-dimensional cohomology.

The map $\mathcal{A}_M : M \to \mathbb{T}^b$ lifts to a proper map

$$\overline{\mathcal{A}}_M : \overline{M} \to \mathbb{R}^b, \tag{17.2.2}$$

where $\mathbb{R}^b$ is the universal cover of the Jacobi torus. The corresponding fibers of $\mathcal{A}_M$ and $\overline{\mathcal{A}}_M$ (*i.e.* fibers which project to the same point of $\mathbb{T}^b$) are homeomorphic. On the other hand, the nonvanishing of the homology class $[\overline{F}_M]$ of the typical fiber of $\overline{\mathcal{A}}_M$ in $H_{n-b}(\overline{M})$ is far weaker than the nonvanishing of the homology class of the typical fiber of $\mathcal{A}_M$ in $H_{n-b}(M)$ (and thus the resulting theorem is more interesting). For instance, for the standard nilmanifold of the Heisenberg group (*cf.* Remark 17.2.6) we have $[\overline{F}_M] \neq 0$, but the homology class in M of the fiber of $\mathcal{A}_M$ is trivial. The latter is true whenever $n = b + 1$.

In the theorems below, to obtain a nonvacuous inequality, we replace the condition of nonvanishing degree of the BIG inequality (17.1.1), by the nonvanishing of the homology class $[\overline{F}_M]$ of the lift, of the typical fiber of $\mathcal{A}_M$, to the universal abelian cover $\overline{M}$ of M, *cf.* [**Gro83**, p. 101] and equation (17.2.2) below. In particular, we must have $b \leq n$ to obtain a nonvacuous inequality. Following M. Gromov [**Gro83**, p. 101], we introduce the following notion of generalized degree.

DEFINITION 17.2.1. Denote by $\deg(\mathcal{A}_M)$ the infimum of $(n-b)$-volumes of integral cycles representing the class $[\overline{F}_M] \in H_{n-b}(\overline{M})$, if M is orientable, and infimum of volumes of $\mathbb{Z}_2$-cycles, otherwise.

This quantity is denoted 'deg' in [**Gro83**, p. 101]. When the dimension and the first Betti number coincide, this quantity is a topological invariant. In general, of course, it is not. Yet it is remarkable that the nonvanishing of this quantity suffices to produce a nonvacuous volume lower bound of Theorem 17.2.2, generalizing the BIG inequality (17.1.1).

THEOREM 17.2.2 (BIG(n, b)-inequality). *Let M be a compact manifold of dimension $n = \dim(M)$. Let $b = b_1(M)$, and assume $b \geq 1$. Then every metric $\mathcal{G}$ on M satisfies the inequality*

$$\deg(\mathcal{A}_M)\, \mathrm{stsys}_1(\mathcal{G})^b \leq (\gamma_b)^{\frac{b}{2}} \mathrm{vol}_n(\mathcal{G}). \tag{17.2.3}$$

The proof of Theorem 17.2.2 appears in Section 17.7. Note that the condition $[\overline{F_M}] \neq 0$ for a 3-manifold with Betti number 2 is implied by the non-vanishing of the Lescop invariant, *i.e.* the self-linking number of a typical fiber of its Jacobi map, *cf.* calculation (18.2.1).

Recall that we allow a systole of $(\overline{M}, \overline{\mathcal{G}})$ to participate in Definition 12.3.3 of systolic category, *cf.* Definition 12.3.1. By hypothesis, the class of the fiber is nontrivial, and therefore $\operatorname{sysh}_{n-b}(\overline{M}, \overline{\mathcal{G}}) \leq \deg(\mathcal{A}_M)$, proving the following corollary.

COROLLARY 17.2.3. *Let M be an n-manifold with $b_1(M) = b$. Then the following lower bound for the systolic category holds: if $[\overline{F_M}] \neq 0$ then* $\operatorname{cat}_{\mathrm{sys}}(M) \geq b+1$.

QUESTION 17.2.4. Does a manifold with $[\overline{F_M}] \neq 0$ satisfy a similar inequality for the Lusternik-Schnirelmann category,

$$\operatorname{cat}_{\mathrm{LS}}(M) \geq b_1(M) + 1 \ ? \tag{17.2.4}$$

For $n = b+1$, we obtain the following corollary of Theorem 17.2.2.

COROLLARY 17.2.5. *Let M be a compact manifold. Let $b = b_1(M)$. Assume that* $\dim(M) = b+1$, *and* $[\overline{F_M}] \neq 0$. *Then every metric $\mathcal{G}$ on M satisfies the following optimal inequality:*

$$\operatorname{stsys}_1(\mathcal{G})^b \operatorname{sys}\pi_1(\mathcal{G}) \leq (\gamma_b)^{\frac{b}{2}} \operatorname{vol}_{b+1}(\mathcal{G}), \tag{17.2.5}$$

where $\operatorname{sys}\pi_1(\mathcal{G})$ *denotes the least length of a shortest noncontractible loop for the metric $\mathcal{G}$.*

REMARK 17.2.6. To give an example where equality is attained in inequality (17.2.5), it suffices to take a Riemannian fibration by circles of constant length, over a flat torus whose group of deck tranformations is critical (choose the circles sufficiently short, so as to realize the value of the invariant $\operatorname{sys}\pi_1(\mathcal{G})$). This can be realized, for instance, by compact quotients of left invariant metrics on the 3-dimensional Heisenberg group (here $b = 2$), *cf.* [**Gro83**, p. 101].

The results of the present chapter are further generalized in Chapter 18. Note that a different sharp generalisation of the Loewner inequality is studied in Chapter 15:

$$\operatorname{stsys}_1(\mathcal{G}) \operatorname{sys}_{n-1}(\mathcal{G}) \leq \gamma'_b \operatorname{vol}_n(\mathcal{G}), \tag{17.2.6}$$

where γ'_b is the Bergé-Martinet constant (5.3.3), *cf.* diagram (15.1.1). In addition, Chapter 16 studies the boundary case of equality of a further generalisation of inequality (17.2.6).

We would like to point out an interesting difference between inequalities (17.2.5) and (17.2.6). Namely, inequalities for the 1-systole, such as (17.2.5), tend to be satified even by the ordinary (*i.e.* unstable) systole, *cf.* [**Gro96**, $3\mathrm{C}_1$] and [**CrK03**, (3.5)], albeit with a nonsharp constant. Meanwhile, inequality (17.2.6) for ordinary systoles is definitely violated by a suitable sequence of metrics, no matter what the constant, *cf.* [**BabK98**].

Volume lower bounds in terms of conformal 2-systoles and the study of the associated constants for 4-manifolds appear in Chapter 20.

17.3. Pu's inequality and generalisations

To state the next theorem, we need a generalisation of Pu's inequality (5.2.2). Let Σ be a nonorientable surface, and let $\phi : \pi_1(\Sigma) \to \mathbb{Z}_2$ be an epimorphism from its fundamental group to $\mathbb{Z}_2$, corresponding to a map $\hat{\phi} : \Sigma \to \mathbb{RP}^2$ of absolute degree $+1$. We define the "1-systole relative to ϕ", denoted $\mathrm{sys}_1(\mathcal{G}, \phi)$, of a metric $\mathcal{G}$ on Σ, by minimizing length over loops γ which are not in the kernel of ϕ, *i.e.* loops whose image under ϕ is not contractible in the projective plane:

$$\mathrm{sys}_1(\mathcal{G}, \phi) = \min\left\{\mathrm{length}(\gamma) \mid \phi([\gamma]) \neq 0 \in \mathbb{Z}_2\right\}. \tag{17.3.1}$$

Does every nonorientable surface $(\Sigma, \mathcal{G})$ and map $\hat{\phi} : \Sigma \to \mathbb{RP}^2$ of absolute degree one, satisfy the following relative version of Pu's inequality (*cf.* [**CrK03**, conjecture 2.7]):

$$\mathrm{sys}_1(\mathcal{G}, \phi)^2 \leq \tfrac{\pi}{2}\,\mathrm{area}(\mathcal{G}) \quad ? \tag{17.3.2}$$

This inequality is related to Gromov's (non-sharp) inequality $(*)_{inter}$ from [**Gro96**, 3.C.1], see also [**Gro99**, Theorem 4.41]. Let

$$\sigma_2 = \sup_{(\Sigma, \mathcal{G})} \frac{\mathrm{sys}_1(\mathcal{G}, \phi)^2}{\mathrm{area}(\mathcal{G})}, \tag{17.3.3}$$

where the supremum is over all nonorientable surfaces Σ, metrics $\mathcal{G}$ on Σ, as well as maps ϕ as above. Thus we ask whether $\sigma_2 = \frac{\pi}{2}$. Calculating σ_2 depends on calculating the filling area of the Riemannian circle.

PROPOSITION 17.3.1. *We have the following estimate:* $\sigma_2 \in [\frac{\pi}{2}, 2]$.

PROOF. If we open up a surface Σ as above along a shortest essential loop $\gamma \subset \Sigma$ (in the sense that $\phi([\gamma]) \neq 0$), we obtain a $\mathbb{Z}_2$-filling Σ' (possibly nonorientable) of a circle of length $2\mathrm{length}(\gamma)$. It is clear that the boundary circle is imbedded in Σ' isometrically as a metric space. Thus it suffices to prove that that the filling area of a circle of length 2π is at least $\frac{\pi^2}{2}$. Choose two points on the boundary circle at distance $\frac{\pi}{2}$ from each other. Consider the map $\Sigma' \to \mathbb{R}^2$ whose coordinate functions are the distances to these points. The map is area-decreasing, and its image contains a square of area $\frac{\pi^2}{2}$ (encircled by the image of the boundary), namely the Pythagorean "diamond" inside the square $[0, \pi] \times [0, \pi] \subset \mathbb{R}^2$. □

17.4. A Pu times Loewner inequality

THEOREM 17.4.1. *Let M be a compact nonorientable manifold, and let $b = b_1(M)$. Assume* $\dim(M) = b + 2$ *and* $\iota[\overline{F}_M] \neq 0$, *where*

$$\iota : H_2(\overline{M}, \mathbb{Z}_2) \to H_2(K, \mathbb{Z}_2)$$

is the homomorphism induced by a map to some aspherical space K. Then every metric $\mathcal{G}$ on M satisfies the following inequality:

$$\mathrm{stsys}_1(\mathcal{G})^b \,\mathrm{sys}\pi_1(\mathcal{G})^2 \leq \sigma_2 \gamma_b^{\frac{b}{2}}\, \mathrm{vol}_{b+2}(\mathcal{G}), \tag{17.4.1}$$

where σ_2 is the optimal systolic ratio from (17.3.3).

The proof appears at the end of Section 17.7.

REMARK 17.4.2. If, as conjectured, we have $\sigma_2 = \frac{\pi}{2}$, then the bound (17.4.1) is attained by Riemannian submersions over a flat critical torus, with minimal fibers isometric to a real projective plane with a metric of constant Gaussian curvature.

EXAMPLE 17.4.3. For $M = \mathbb{RP}^2 \times \mathbb{T}^2$, we obtain the following inequality:

$$\operatorname{stsys}_1(\mathcal{G})^2 \operatorname{sys}\pi_1(\mathcal{G})^2 \leq \sigma_2 \gamma_2 \operatorname{vol}_4(\mathcal{G}),$$

which can be thought of as a "Pu-times-Loewner" inequality, *cf.* (5.4.1) and (5.2.2) (particularly if one succeeds in proving that $\sigma_2 = \frac{\pi}{2}$).

17.5. A decomposition of the John ellipsoid

A proof of the following lemma may be found in [**BuraI94**].

DEFINITION 17.5.1. The *John ellipsoid* of a convex set in Euclidean space is the unique ellipsoid of largest volume inscribed in it, *cf.* [**MilmS86**, p. 10].

LEMMA 17.5.2. *Let $(V^d, \| \ \|)$ be a Banach space. Let $\| \ \|_E$ be the Euclidean norm determined by the John ellipsoid of the unit ball of $\| \ \|$. Then there exists a decomposition of $\| \ \|_E^2$ into rank-1 quadratic forms L_i:*

$$\|x\|_E^2 = \sum_{i=1}^{N} \lambda_i L_i^2(x),$$

with the following four properties:

- $N \leq \frac{d(d+1)}{2}$,
- $\lambda_i > 0$ *for all* i,
- $\sum \lambda_i = d$,
- $L_i : V \to \mathbb{R}$ *are linear functions with* $\|L_i\|^* = 1$, *where* $\| \ \|^*$ *is the dual norm to* $\| \ \|$.

Note that the bound $N \leq \frac{b(b+1)}{2} + 1$ appears in [**BuraI94**], but one easily sees that it can be reduced by one.

17.6. An area-nonexpanding map

Let M be a compact Riemannian manifold. Let M' a topological space, and let $\varphi : M \to M'$ be a continuous map inducing an epimorphism in one-dimensional real homology. Then one defines the *relative stable norm* $\| \ \|_{st/\varphi}$ on $H_1(M'; \mathbb{R})$ as follows. Given $\alpha \in H_1(M', \mathbb{R})$, we set

$$\|\alpha\|_{st/\varphi} = \inf \left\{ \|\beta\|_{st} \middle| \beta \in H_1(M; \mathbb{R}), \ \varphi_*(\beta) = \alpha \right\}.$$

where $\| \ \|_{st}$ is the ordinary ("absolute") stable norm. The stable norm itself may be thought of as the relative stable norm defined by the Abel-Jacobi map to the torus $H_1(M, \mathbb{R})/H_1(M, \mathbb{Z})_{\mathbb{R}}$.

DEFINITION 17.6.1. We will say that a Lipschitz map $\mathcal{A} : M \to M'$ between Riemannian manifolds is "non-expanding on all d-dimensional areas" if for every smooth d-dimensional submanifold $\Sigma \subset M$, one has $\operatorname{vol}_d(\mathcal{A}(\Sigma)) \leq \operatorname{vol}_d(\Sigma)$.

Equivalently, $Jac(\mathcal{A}|_\Sigma) \leq 1$ wherever $\mathcal{A}|_\Sigma$ is differentiable.

Let M^n be a compact Riemannian manifold, V^d a vector space, and Γ a lattice in V. We will identify V and $H_1(V/\Gamma; \mathbb{R})$.

PROPOSITION 17.6.2. *Let $\varphi : M \to V/\Gamma$ be a continuous map inducing an epimorphism of the fundamental groups. Let $\| \ \|_E$ denote the Euclidean norm on V defined by the John ellipsoid of the relative stable norm $\| \ \|_{st/\varphi}$. Then there exists a Lipschitz map $\mathcal{A} : M \to (V/\Gamma, \| \ \|_E)$ which is homotopic to φ and non-expanding on all d-dimensional areas, where $d = \dim(V)$.*

REMARK 17.6.3. More generally, this map expands k-dimensional areas, for all $1 \le k \le d$, by at most a factor of $(d/k)^{k/2}$.

The proof of Propostion 17.6.2 appears at the end of this section. There is a natural isomorphism $\Gamma \simeq \pi_1(V/\Gamma) \simeq \pi_1(M)/\ker(\varphi_*)$. Consider a covering space $\overline{M}$ of M defined by the subgroup $\ker(\varphi_*)$ of $\pi_1(M)$. Then Γ acts on $\overline{M}$ as the deck group $\pi_1(M)/\ker(\varphi_*)$. This action will be written additively, as in (17.6.1) below. It is sufficient to construct a Lipschitz map $\overline{M} \to V$ which is Γ-equivariant and does not expand d-dimensional areas. We need the following lemma.

LEMMA 17.6.4. *For every linear function $L : V \to \mathbb{R}$ such that $\|L\|^*_{st/\varphi} = 1$ there exists a distance-nonincreasing function $f : \overline{M} \to \mathbb{R}$ such that*

$$f(x + v) = f(x) + L(v) \tag{17.6.1}$$

for all $x \in \overline{M}$ and $v \in \Gamma$.

PROOF. Fix a point $x_0 \in \overline{M}$. We consider the orbit

$$\overline{M}_0 = \{x_0 + v : v \in \Gamma\},$$

and define a function $f_0 : \overline{M}_0 \to \mathbb{R}$ by $f_0(x_0 + v) = L(v)$. Note that f_0 satisfies (17.6.1) for $x \in \overline{M}_0$. For every $v \in \Gamma$ and $x \in \overline{M}$, one has $L(v) \le \|v\|_{st/\varphi}$ and $\|v\|_{st/\varphi}$ is no greater than the distance between x and $x + v$. Hence f_0 is distance-nonincreasing. Every distance-nonincreasing function defined on a subset of a metric space admits a distance-nonincreasing extension to the whole space, by the triangle inequality. Moreover, an extension can be chosen so that the equivariance (17.6.1) is preserved. For example, we can set

$$f(x) = \inf\{f_0(y) + |xy| : y \in \overline{M}_0\},$$

where $|xy|$ denotes the distance. □

PROOF OF PROPOSITION 17.6.2. Applying Lemma 17.5.2 to the norm $\| \; \|_{st/\varphi}$ yields a decomposition

$$\| \; \|^2_E = \sum_{i=1}^{N} \lambda_i L_i^2,$$

where $\lambda_i > 0$, $\sum \lambda_i = d$, $L_i \in V^*$ and $\|L_i\|^*_{st/\varphi} = 1$. Then a linear map $L : V \to \mathbb{R}^N$ defined by

$$L(x) = (\lambda_1^{1/2} L_1(x), \lambda_2^{1/2} L_2(x), \ldots, \lambda_N^{1/2} L_N(x))$$

is an isometry from $(V, \| \; \|_E)$ onto a subspace $L(V)$ of $\mathbb{R}^N$, equipped with the restriction of the standard coordinate metric of $\mathbb{R}^N$.

By Lemma 17.6.4, for every $i = 1, 2, \ldots, N$ there exists a distance-nonincreasing function $f_i : \overline{M} \to \mathbb{R}$ such that $f_i(x + v) = f_i(x) + L_i(v)$ for all $x \in \overline{M}$ and $v \in \Gamma$. Define a map $F : \overline{M} \to \mathbb{R}^N$ by

$$F(x) = (\lambda_1^{1/2} f_1(x), \lambda_2^{1/2} f_2(x), \ldots, \lambda_N^{1/2} f_N(x)). \tag{17.6.2}$$

Observe that both L and F are Γ-equivariant with respect to the following action of Γ on $\mathbb{R}^N$:

$$\Gamma \times \mathbb{R}^N \to \mathbb{R}^N, \quad (v, x) \mapsto x + L(v).$$

Now let $\mathrm{Pr}_{L(V)} : \mathbb{R}^N \to L(V)$ be the orthogonal projection to $L(V)$. Then the composition $L^{-1} \circ \mathrm{Pr}_{L(V)} \circ F$ is a Γ-equivariant map from $\overline{M}$ to V. Since the projection is nonexpanding and L is an isometry, it suffices to prove that the map F of (17.6.2) is nonexpanding on d-dimensional areas.

Let Σ be a smooth d-dimensional submanifold of $\overline{M}$. Since F is Lipschitz, the restriction $F|_\Sigma$ is differentiable a.e. on Σ. Let $y \in \Sigma$ be such that $F|_\Sigma$ is differentiable at y, and let $A = d(F|_\Sigma) : T_y\Sigma \to \mathbb{R}^N$. Then we obtain

$$\mathrm{Tr}(A^*A) = \sum_i \lambda_i \left|d\left(f_i\,|_\Sigma\right)\right|^2 \leq \sum_i \lambda_i = d$$

since the functions f_i are distance-nonincreasing. By the inequality of geometric and arithmetic means, we have

$$\begin{aligned} Jac(F|_\Sigma)(x) &= \det(A^*A)^{1/2} \\ &\leq \left(\tfrac{1}{d}\mathrm{Tr}(A^*A)\right)^{d/2} \\ &\leq 1, \end{aligned}$$

proving the proposition. □

17.7. Proof of BIG(n, b)-inequality and Theorem 17.4.1

To prove Theorem 17.2.2, we proceed in the following way. Consider the Jacobi torus $J_1(M) = H_1(M, \mathbb{R})/H_1(M, \mathbb{Z})_\mathbb{R}$ and the Abel-Jacobi map $\mathcal{A}_M : M \to J_1(M)$ constructed in Proposition 17.6.2.

REMARK 17.7.1. The map $\mathcal{A}_M$ can be replaced by a smooth one, by an arbitrarily small perturbation, in such a way as to expand d-dimensional areas at most by a factor $1 + \epsilon$. Our main tool will be the coarea formula, see below. We can also avoid the above smoothing argument, and use instead the current-theoretic version of the coarea formula, relying on H. Federer's theory of "slicing". Given an integral current T in a smooth oriented manifold M (*e.g.* in our case $T = [M]$), and a Lipschitz map $f : M \to N$, one can in a sense decompose T by f, obtaining currents (slices) $\langle T, f, y\rangle$ supported in the fibers $f^{-1}(y)$, for a.e. $y \in N$. If T is a cycle then so are the slices; if T is a submanifold and f is smooth, then the slices at regular values of f are just (integration over) the typical fibers of the map. The main properties of this operation are given in [**Fed69**, Thm 4.3.2, p. 438]. In particular, item 4.3.2(2) is a version of the coarea formula. In section 4.4, Federer shows that the usual homology groups of a reasonably good space (*e.g.* manifold in our case) coincide with the ones defined via currents.

PROOF OF THEOREM 17.2.2. We will exploit the coarea formula, *cf.* [**Fed69**, 3.2.11], [**Cha06**, p. 267] as in [**Gro83**, Theorem 7.5.B].

Away from the negligible singular set, the smooth map is a submersion. Therefore the metric on M can be modified by a volume-preserving deformation so that the map actually becomes *distance* decreasing, up to an arbitrarily small amount. Note, however, that the coarea formula could be applied even without replacement by a short map. Formally we only need two facts:

(1) if the map is area-nonincreasing, then the volume is no smaller than the area of the image times the minimal area of a fiber,
(2) almost every fiber is "typical" and hence has area no less than the generalized degree $\deg(\mathcal{A}_M)$, *cf.* inequality (17.7.1).

Let $S = \mathcal{A}_M^{-1}(p)$ be the surface which is a regular fiber of least $(n-b)$-volume. (If there is none, choose S to be within $\epsilon > 0$ of the infimum, and then let $\epsilon \to 0$.) Then

$$\deg(\mathcal{A}_M)\,\mathrm{stsys}_1(\mathcal{G})^b \le \mathrm{vol}_{n-b}(S)\ \mathrm{stsys}_1(\mathcal{G})^b. \tag{17.7.1}$$

Note that $\mathcal{A}_M$ induces isometry in real one-dimensional homology, with respect to the stable norm of the metric $\mathcal{G}$. Hence

$$\deg(\mathcal{A}_M)\,\mathrm{stsys}_1(\mathcal{G})^b \le \mathrm{vol}_{n-b}(S)\ \mathrm{stsys}_1(J_1(M), \|\ \|)^b.$$

We now replace the stable norm $\|\ \|$ by the flat Euclidean metric $\|\ \|_E$ defined by the John ellipsoid of the stable norm:

$$\deg(\mathcal{A}_M)\,\mathrm{stsys}_1(\mathcal{G})^b \le \mathrm{vol}_{n-b}(S)\ \mathrm{stsys}_1(J_1(M), \|\ \|_E)^b.$$

By definition of the Hermite constant,

$$\deg(\mathcal{A}_M)\,\mathrm{stsys}_1(\mathcal{G})^b \le \mathrm{vol}_{n-b}(S)\ \gamma_b^{\frac{b}{2}}\ \mathrm{vol}_b(J_1(M), \|\ \|_E).$$

Now we apply the coarea formula to our map which is decreasing on b-dimensional volumes, to obtain the inequality $\mathrm{vol}_{n-b}(S)\,\mathrm{vol}_b(J_1(M), \|\ \|_E) \le \mathrm{vol}_n(M)$, completing the proof. □

PROOF OF THEOREM 17.4.1. We apply Theorem 17.2.2 together with the inequality $\mathrm{sys}\pi_1^2(\mathcal{G}) \le \sigma_2 \deg(\mathcal{A}_M)$. Here Pu's inequality does not suffice. Indeed, a typical fiber of $\mathcal{A}_M$ may not be diffeomorphic to $\mathbb{RP}^2$. An application of Pu's inequality (5.2.2) yields a suitably short loop which is essential in the typical fiber. However, a loop which is essential in the typical fiber, may not be essential in the ambient manifold M. Thus, we need a generalisation of Pu's inequality. The required generalisation is inequality (17.3.2) above, *cf.* [**CrK03**], applied to the composed map $\hat{\phi} : S \to \overline{M} \to K$. □

CHAPTER 18

Higher inequalities of Loewner-Gromov type

18.1. Introduction, conjectures, and some results

In the study of optimal systolic inequalities on a Riemannian manifold $(M, \mathcal{G})$, generalizing the Loewner inequality, two questions arise:

(1) Which stable systolic inequalities can be replaced by their conformal analogues?

(2) Calculate the values of the constants arising in the inequalities.

The results of this chapter first appeared in [**BaCIK07**]. The definitions of the systolic invariants $\operatorname{stsys}_1(\mathcal{G})$ and $\operatorname{sys}\pi_1(\mathcal{G})$ can be found in Chapter 12. See Definition 16.4.1 for the invariant $\operatorname{confsys}_1(\mathcal{G})$. The Abel-Jacobi map $\mathcal{A}_M$ was defined in [**Lich69, Gro96**], *cf.* (16.5.3).

The first such optimal inequality is implied in [**Gro99**, pp. 259-260] (*cf.* [**CrK03**, inequality (5.14)]) based on the techniques of D. Burago and S. Ivanov [**BuraI94, BuraI95**], *cf.* Theorem 17.1.2.

The following conformally invariant generalisation of the Burago-Ivanov-Gromov inequality (17.1.1) follows from Proposition 18.4.3, *cf.* inequality (18.3.3).

THEOREM 18.1.1 (Conformal BIG inequality). *Let M be a closed orientable manifold. Let $n = \dim(M)$ and assume $n = b_1(M) = \text{cup-length}_{\mathbb{R}}(M)$. Then every metric $\mathcal{G}$ on M satisfies the following optimal inequality:*

$$\operatorname{confsys}_1(\mathcal{G})^n \leq \deg(\mathcal{A}_M)^{-1} (\gamma_n)^{\frac{n}{2}} . \tag{18.1.1}$$

In connection with question 1 above, we aim for the following results. We show in Section 18.5 that if the L^p norm is equal to the comass norm, then the minimizing measurable 1-form is in fact continuous and has constant pointwise norm. Then we can apply a Voronoi-type argument to prove that in an extremal situation, in fact all cohomology classes are represented by 1-forms of constant norm. This allows us to prove that a manifold M such that $\dim(M) = b_1(M) = \text{cup-length}_{\mathbb{R}}(M)$, which attains (Gromov's) lower bound for the volume in terms of the stable 1-systole, must be a critical flat torus (Theorem 18.1.3). The proof passes via the conformally invariant strengthening of the Burago-Ivanov-Gromov inequality, namely Theorem 18.1.1.

Recall that a lattice $L \subset \mathbb{R}^b$ is called *critical* if it realizes the supremum in (5.3.1). A critical (or, more generally, extremal) lattice is necessarily perfect and eutactic [**Bar57**]. The following two theorems are proved in Section 18.9.

THEOREM 18.1.2. *The boundary case of equality in the conformally invariant inequality* (18.1.1), *occurs if and only if M is a torus and the metric $\mathcal{G}$ is conformal to a flat critical metric.*

THEOREM 18.1.3. *In the hypotheses of Theorem 17.1.2, equality in* (17.1.1) *is attained precisely when* M *is a torus, while* $\mathcal{G}$ *is a flat metric whose deck transformations form a critical lattice.*

In Section 18.2, we discuss a notion of degree of the Abel-Jacobi map when the dimension n of the manifold M exceeds its first Betti number b. Here we prove the positivity of the degree under a suitable topological hypothesis of homological nonvanishing. In Section 18.3, we extend Theorem 18.1.2 and Theorem 18.1.3 to a general pair $n \geq b$, and discuss the related results in the literature. In Section 18.4, we define the norms involved, and state a key proposition used in the proof of Theorems 18.1.2 and 18.1.3. In Sections 18.5 and 18.6, we collect analytical results pertaining to closed forms that minimize an L^p-norm. This constitutes the analytical backbone of this chapter. Section 18.7 contains a generalisation of a construction of [**BuraI94, BuraI95**].

In chapter 19, we prove a further generalisation in the context of L^p norms, see Section 19.2. The remaining sections are devoted to the proof of the results of Section 18.3.

18.2. Notion of degree when dimension exceeds Betti number

Let $(M, \mathcal{G})$ be a closed Riemannian manifold, $b = b_1(M)$. We denote by $\mathcal{A}_M$ the Abel-Jacobi map,

$$\mathcal{A}_M : M \to H_1(M, \mathbb{R})/H_1(M, \mathbb{Z})_{\mathbb{R}} \simeq \mathbb{T}^b,$$

inducing (the natural) isomorphism in 1-dimensional cohomology. Now let $\overline{M}$ be the maximal free abelian cover of M, *cf.* Definition 18.7.2. Denote by $\overline{\mathcal{A}}_M$ the proper map obtained as the lift of $\mathcal{A}_M$ to $\overline{M}$:

$$\overline{\mathcal{A}}_M : \overline{M} \to H_1(M, \mathbb{R}) \simeq \mathbb{R}^b.$$

If $\overline{M}$ is orientable we fix an orientation for $\overline{M}$. In the case that M is orientable, the following definition is due to M. Gromov [**Gro83**, p. 101].

DEFINITION 18.2.1. Let $[\overline{F}_M] \in H_{n-b}(\overline{M}, R)$ denote the homology class of the regular fibers $\overline{\mathcal{A}}_M^{-1}(y)$ of $\overline{\mathcal{A}}_M$, where we take $R = \mathbb{Z}$ if $\overline{M}$ is orientable, and $R = \mathbb{Z}_2$ otherwise. Then the geometric degree

$$\deg(\mathcal{A}_M)$$

of $\overline{\mathcal{A}}$ is the infimum of the $(n-b)$-volumes of all cycles representing $[\overline{F}_M]$.

REMARK 18.2.2. If $\overline{M}$ is orientable, then changing the orientation of $\overline{M}$ will not change $\deg(\mathcal{A}_M)$. If $n = b$ and M is orientable and connected, then $\deg(\mathcal{A}_M)$ is (the absolute value of) the topological degree of $\mathcal{A}_M$. If $n > b$ and $[\overline{F}_M] \neq 0$, then $\deg(\mathcal{A}_M)$ depends on the metric on M.

EXAMPLE 18.2.3. If M is the quotient of the 3-dimensional Heisenberg group by its subgroup of integral elements, then the fibers of $\mathcal{A}_M$ are homologically trivial in M, while $[\overline{F}_M] \neq 0$.

This example helps explain why one chooses to work in the cover $\overline{M}$ in Definition 18.2.1.

The following obvious consequence of Definition 18.2.1 will be crucial for our arguments.

LEMMA 18.2.4. *Suppose $f : M \to H_1(M,\mathbb{R})/H_1(M,\mathbb{Z})_\mathbb{R}$ is homotopic to $\mathcal{A}_M$, and y is a regular value of f. Then we have*

$$\mathrm{vol}_{n-b}(f^{-1}(y)) \geq \deg(\mathcal{A}_M).$$

PROOF. Let $\bar{y} \in H_1(M,\mathbb{R})$ be a preimage of y. Let $\bar{f} : \overline{M} \to H_1(M,\mathbb{R})$ be a lift of f. Then the covering projection $\overline{M} \to M$ maps the fiber $\bar{f}^{-1}(\bar{y})$ isometrically onto the fiber $f^{-1}(y)$, so that we have

$$\mathrm{vol}_{n-b}(\bar{f}^{-1}(\bar{y})) = \mathrm{vol}_{n-b}(f^{-1}(y)).$$

On the other hand, $\bar{f}$ is homotopic to $\overline{\mathcal{A}}_M$ by a proper map, and hence the fiber $\bar{f}^{-1}(\bar{y})$ represents the class $[\overline{F}_M]$. □

All the results discussed below are nonvacuous only if $\deg(\mathcal{A}_M)$ is positive. Thus, it is important to know when this is the case. First note that this is a topological property.

REMARK 18.2.5. If $[\overline{F}_M] \neq 0$ then $\deg(\mathcal{A}_M) > 0$.

This follows from the deformation theorem as stated in [**Fed69**, 4.2.9], complemented by [**Fed69**, $(4.2.9)^\nu$] for the case of $\mathbb{Z}_2$-coefficients, see also [**FedF60**], [**Wh99**], and [**Mo65**, Chapter 5] for a helpful picture. The deformation theorem implies that for every compact Riemannian manifold M, there exists $\epsilon > 0$ such that for every $q \in \mathbb{N}$ every q-cycle of q-volume smaller than ϵ can be deformed into the $(q-1)$-skeleton of a triangulation of M. Now, if z is a q-cycle in $\overline{M}$ of q-volume smaller than ϵ, then the projection of z to M can be deformed into the $(q-1)$-skeleton.

By the covering homotopy property, we can lift this deformation to the cover $\overline{M}$. We can therefore conclude that the homology class of z vanishes, regardless of the ring of coefficients.

Note that, if the minimizing rectifiable current in $[\overline{F}_M]$ happens to be nonsingular, then the work [**Lê93**] produces explicit lower bounds for $\deg(\mathcal{A}_M)$ in terms of the injectivity radius and an upper bound on the sectional curvature.

In analogy with the maximal real cup-length condition in Theorem 17.1.2, we note that $[\overline{F}_M] \neq 0$ if M is orientable and if

$$\alpha_1 \cup \ldots \cup \alpha_b \neq 0$$

for every basis $\alpha_1, \ldots, \alpha_b$ of $H^1(M,\mathbb{R})$. Even if the real cup product vanishes, for a 3-manifold M with Betti number 2, we have the following result, due to C. Lescop and A. Marin, see [**KatzL05**], generalized in [**CM05**].

PROPOSITION 18.2.6. *The nonvanishing of the self-linking number of the typical fiber $\mathcal{A}_M^{-1}(p)$ of $\mathcal{A}_M : M \to \mathbb{T}^2$ is a sufficient condition for the nonvanishing of the fiber class $[\overline{F}_M]$.*

PROOF. The argument goes roughly as follows. Consider the map $\overline{\mathcal{A}} : \overline{M} \to \mathbb{R}^2$. Let $a, b \in \mathbb{R}^2$. Let $\bar{F}_a = \overline{\mathcal{A}}^{-1}(a)$ and $\bar{F}_b = \overline{\mathcal{A}}^{-1}(b)$ be lifts of the corresponding fibers $F_a, F_b \subset M$. Choose a properly imbedded ray $r_b \subset \mathbb{R}^2$ joining $b \in \mathbb{R}^2$ to infinity while avoiding a (as well as its translates), and consider the complete surface

$$S = \overline{\mathcal{A}}^{-1}(r_b) \subset \overline{M}$$

with $\partial S = \bar{F}_b$. We have $S \cap g.\bar{F}_a = \emptyset$ for all $g \in G$, where G denotes the group of deck transformations of the cover $p_G : \overline{M} \to M$. If the link $\bar{F}_a$ is zero-homologous in $\overline{M}$, choose a compact surface $\Sigma \subset \overline{M}$ with

$$\partial \Sigma = \bar{F}_a.$$

The linking number $\ell_M(F_a, F_b)$ in M can therefore be computed as the *algebraic* intersection

$$\begin{aligned} \ell_M(F_a, F_b) &= p_G(\Sigma) \cap F_b \\ &= \sum_{g \in G} g.\Sigma \cap \bar{F}_b \\ &= \sum_{g \in G} \partial\left(g.\Sigma \cap S\right) \\ &= 0, \end{aligned} \tag{18.2.1}$$

where all sums and intersections are algebraic, and the intersection curve $\Sigma \cap S$ is compact by construction. The details of a rigorous proof may be found in [**KatzL05**]. The nonvanishing of the self-linking number is equivalent, in this Betti number, to the nonvanishing of C. Lescop's generalisation λ of the Casson-Walker invariant, *cf.* [**Les96**]. See T. Cochran and J. Masters [**CM05**] for generalisations. □

REMARK 18.2.7. Clearly, the condition $[\overline{F_M}] \neq 0$ implies that the corank of the fundamental group of M equals 1, *cf.* [**LeiR02**]. Recall that the corank of a group G is the greatest rank, k, of a free group F_k admitting an epimorphism $G \to F_k$ from G. This invariant is also known in the literature as the cut number of M. It is unknown whether or not the condition of unit corank is sufficient to guarantee that every typical fiber contains a noncontractible loop of M.

18.3. Conformal BIG(n, p)-inequality

The following two results were proved in Chapter 17.

THEOREM 18.3.1. *Let M be a closed Riemannian manifold. Let $n = \dim(M)$ and $b = b_1(M)$, and assume $n \geq b \geq 1$. Then every metric $\mathcal{G}$ on M satisfies the inequality*

$$\deg(\mathcal{A}_M)\, \mathrm{stsys}_1(\mathcal{G})^b \leq (\gamma_b)^{\frac{b}{2}} \, \mathrm{vol}_n(\mathcal{G}). \tag{18.3.1}$$

COROLLARY 18.3.2. *Let M be a closed orientable manifold. Let $b = b_1(M)$. Assume that* $\dim(M) = b + 1$, *and* $[\overline{F_M}] \neq 0$. *Then every metric $\mathcal{G}$ on M satisfies the following optimal inequality:*

$$\mathrm{stsys}_1(\mathcal{G})^b \, \mathrm{sys}\pi_1(\mathcal{G}) \leq (\gamma_b)^{\frac{b}{2}} \, \mathrm{vol}_{b+1}(\mathcal{G}). \tag{18.3.2}$$

The main tool in studying the case of equality of the above inequalities is Proposition 18.3.5 which concerns a kind of an L^p-systole, *cf.* Definition 19.2.1. Let $p \geq 1$.

DEFINITION 18.3.3. The norm $\| \ \|_p$ on $H_1(M; \mathbb{R})$ is the dual norm to the L^p-norm $\| \ \|_p^*$ on the cohomology space $H^1(M; \mathbb{R})$, *cf.* Chapter 15.

REMARK 18.3.4. Given a class $h \in H_k(M, \mathbb{R})$, the quantity $\|h\|_{\frac{n}{k}}$ is the supremum of the stable norms of h with respect to all metrics that are conformal to $\mathcal{G}$ and have unit volume, *cf.* [**Gro83**, 7.4.A]. In particular, one has

$$\operatorname{stsys}_k(\mathcal{G}) \leq \operatorname{Confsys}_k(\mathcal{G}) \operatorname{vol}_n(\mathcal{G})^{\frac{k}{n}}, \tag{18.3.3}$$

cf. Section 16.4, inequality (16.4.2).

Given a lattice L equipped with a norm $\| \ \|$, denote by $\lambda_1(L, \| \ \|)$ the least norm of a nonzero vector in L. The following proposition generalizes Theorem 18.3.1.

PROPOSITION 18.3.5. *Let M be a closed orientable Riemannian manifold of unit volume. Let $b = b_1(M)$. Let $p \geq \max\{b, 2\}$. Then the following inequality is satisfied:*

$$\deg(\mathcal{A}_M)\, \lambda_1\big(H_1(M, \mathbb{Z})_{\mathbb{R}}, \| \ \|_p\big)^b \leq (\gamma_b)^{\frac{b}{2}}. \tag{18.3.4}$$

The proof of Proposition 18.3.5 appears in Section 19.4. By choosing $p = n$ in Proposition 18.3.5, we obtain the following generalisation of Theorem 18.1.1 and Theorem 18.3.1, which could be called the conformal BIG(n, b)-inequality.

THEOREM 18.3.6. *Let M be a closed orientable manifold. Let $n = \dim(M)$ and $b = b_1(M)$. Then every metric $\mathcal{G}$ on M satisfies the inequality*

$$\deg(\mathcal{A}_M) \operatorname{confsys}_1(\mathcal{G})^b \leq (\gamma_b)^{\frac{b}{2}} \operatorname{vol}_n(\mathcal{G})^{n-b}. \tag{18.3.5}$$

REMARK 18.3.7. In the case $b = 2$, Theorem 18.3.6 follows by setting $p = 2$, and applying the coarea formula to Lichnerowicz's harmonic map [**Lich69**], see Remark 19.3.1. This proves inequality (18.3.4) for the "L^2-systole", and the monotonicity in p then proves (18.3.5). One can thus avoid in this case the extensions of the BI construction, described in Section 18.7 and Section 19.2. Thus our construction can be viewed as a generalisation, to $b \geq 3$, of the property of being area-decreasing (on the average) of Lichnerowicz's map in the case $b = 2$.

In analogy with our Theorem 18.1.3, we prove the following characterisation of the boundary case of equality of the conformal BIG(n, b) inequality.

THEOREM 18.3.8. *Let M be a closed orientable Riemannan manifold. Assume that* $\dim(M) > b_1(M)$. *Suppose equality is attained in* (18.3.5). *Then*

(1) *all harmonic one-forms on M have pointwise constant norm;*
(2) *M admits a harmonic Riemannian submersion onto a flat b-torus whose deck group is a critical lattice in $\mathbb{R}^b$;*
(3) *the submersion is given by the Abel-Jacobi map defined via harmonic one-forms;*
(4) *the fibers of the submersion are minimal surfaces of constant $(n - b)$-dimensional area.*

Note that Riemannian manifolds for which the Abel-Jacobi map is a harmonic Riemannian submersion onto a flat torus are studied in Chapter 16. Harmonic Riemannian submersions to flat tori are examples of harmonic morphisms [**BaiW03**].

As a consequence of Theorem 18.3.8, we obtain the following, see some related results in [**NV04, Nag06**].

COROLLARY 18.3.9. *In the hypotheses of Corollary 18.3.2, equality in* (18.3.2) *is attained precisely when M is a two-step nilpotent manifold with* 1*-dimensional center, while $\mathcal{G}$ is the metric of a Riemannian submersion with geodesic fibers of constant length.*

Let us briefly review related material. The case $n = b + 2$ is studied in Chapter 17 and also Chapter 8, which explores the closely related filling area conjecture. A different higher-dimensional generalisation of the Loewner inequality is studied in Chapters 15 and 16, *cf.* diagram 15.1.1. A proof of the Loewner inequality in genus 2 appears in Chapter 9, while the asymptotics of the optimal systolic ratio for large genus are studied in Chapter 11. Analogous asymptotics for the conformal 2-systole are studied in Chapter 20. Chapter 12 provides a topological framework for systolic geometry, in terms of a notion of *systolic category*, related to Lusternik-Schnirelmann category.

18.4. Stable norms and conformal norms

Let M and M' be compact smooth manifolds. Let

$$\varphi : M \to M' \tag{18.4.1}$$

be a continuous map inducing an epimorphism in one-dimensional real homology. Given a Riemannian metric on M (more generally, the structure of a locally simply connected length space), one defines the *relative stable norm* $\| \; \|_{\mathrm{st}/\varphi}$ on $H_1(M';\mathbb{R})$ by setting

$$\|\alpha\|_{\mathrm{st}/\varphi} = \inf \left\{ \|\beta\|_{\mathrm{st}} \;\middle|\; \beta \in H_1(M;\mathbb{R}),\ \varphi_*(\beta) = \alpha \right\}, \tag{18.4.2}$$

where $\| \; \|_{\mathrm{st}}$ is the ordinary ("absolute") stable norm of M. The stable norm itself may be thought of as the relative stable norm defined by the Abel-Jacobi map to the torus $H_1(M,\mathbb{R})/H_1(M,\mathbb{Z})_{\mathbb{R}}$.

Let M^n be a compact Riemannian manifold, V^b a vector space, and Γ a lattice in V. We will identify V and $H_1(V/\Gamma;\mathbb{R})$. Let $\varphi : M \to V/\Gamma$ be a continuous map inducing an epimorphism of the fundamental groups. Note that there is a natural isomorphism

$$\Gamma \simeq \pi_1(V/\Gamma) \simeq \pi_1(M)/ker(\varphi_*).$$

The following proposition was proved in Chapter 17. Denote by $\| \; \|_E$ the Euclidean norm on V, which is defined by the John ellipsoid, *cf.* [**Jo48**], of the relative stable norm $\| \; \|_{\mathrm{st}/\varphi}$.

PROPOSITION 18.4.1. *There exists a Lipschitz map*

$$M \to (V/\Gamma, \| \; \|_E)$$

homotopic to φ and non-expanding on b-dimensional areas, where $b = \dim V$.

The John ellipsoid was defined above, in Definition 17.5.1. The proof of following statement, already exploited in Chapter 17, may be found in [**BuraI94**].

LEMMA 18.4.2. *Let $(V^b, \| \; \|)$ be a Banach space. Let $\| \; \|_E$ be the Euclidean norm determined by the John ellipsoid of the unit ball of the norm $\| \; \|$. Then there exists a decomposition of $\| \; \|_E^2$ into rank-1 quadratic forms:*

$$\| \; \|_E^2 = \sum_{i=1}^{N} \lambda_i L_i^2$$

such that $N \leq \frac{b(b+1)}{2}$, $\lambda_i > 0$ for all i, $\sum \lambda_i = b$, and $L_i : V \to \mathbb{R}$ are linear functions with $\|L_i\|^* = 1$ where $\| \; \|^*$ is the dual norm to $\| \; \|$.

Note that the bound $N \leq \frac{b(b+1)}{2} + 1$ appears in [**BuraI94**], but one easily sees that it can be reduced by one.

Given an n-dimensional Riemannian manifold M, one defines a conformally invariant norm $\| \, \|_{\mathrm{conf}}$ on $H_1(M;\mathbb{R})$ as the dual norm to the L^n-norm $\| \, \|_n^*$ on the cohomology space $H^1(M;\mathbb{R})$, *cf.* Remark 18.3.4. If $\mathrm{vol}(M) = 1$, then we have an inequality $\| \, \|_n^* \leq \| \, \|^*$, where $\| \, \|^*$ is the comass norm in cohomology, dual to the stable norm in homology. Therefore dually, we have the inequality $\| \, \|_{\mathrm{conf}} \geq \| \, \|_{\mathrm{st}}$, *cf.* inequality (18.3.3). In particular, the volume form defined by the John ellipsoid of the conformal norm is greater than or equal to that of the stable norm.

For a continuous map $\varphi : M \to M'$ inducing an epimorphism in one-dimensional homology, one defines a relative conformal norm $\| \, \|_{\mathrm{conf}\,/\varphi}$ on $H_1(M';\mathbb{R})$ by setting

$$\|\alpha\|_{\mathrm{conf}\,/\varphi} = \inf \left\{ \|\beta\|_{\mathrm{conf}} \;\middle|\; \beta \in H_1(M;\mathbb{R}), \ \varphi_*(\beta) = \alpha \right\}. \tag{18.4.3}$$

We present an "integral" version of Proposition 18.4.1 for conformal norms.

PROPOSITION 18.4.3 (Construction of Abel-Jacobi map for conformal norms). *Let M^n be a compact Riemannian manifold, V^n a vector space, and let Γ be a lattice in V. Let $\varphi : M \to V/\Gamma$ be a continuous map inducing an epimorphism of the fundamental groups. Let $\| \, \|_E$ denote the Euclidean norm on V defined by the John ellipsoid of the relative conformal norm $\| \, \|_{\mathrm{conf}\,/\varphi}$. Then there exists a C^1 map*

$$f : M \to (V/\Gamma, \| \, \|_E)$$

which is homotopic to φ and satisfies

$$\int_M \mathrm{Jac}(f) \leq 1.$$

In particular, if φ has nonzero degree (or nonzero absolute degree), then we have the following upper bound for the volume of the torus V/Γ:

$$\mathrm{vol}(V/\Gamma, \| \, \|_E) \leq \frac{1}{\deg(\varphi)}.$$

An adaptation of the main construction [**BuraI94, BuraI95**] involved in the proof will be described in Section 18.7. We will then use it to prove Proposition 18.4.3 in Section 18.8.

18.5. Existence of L^p-minimizers in cohomology classes

In this section, we discuss the existence of a (weakly closed) L^p-form that minimizes the L^p-norm, $1 < p < \infty$, in a given cohomology class, used in the proof of Proposition 18.4.3 in Section 18.8.

Although this is known, we could not find a good reference. The proof consists in a straightforward application of the direct method of the calculus of variations. Actually C. Hamburger [**Ham92**] even proved Hölder continuity of the minimizers.

Note that the existence of a continuous 1-form minimizing the comass, and defining a geodesic lamination, is proved in [**FaS04**, Theorem 1.7].

If $E \to M$ is a Euclidean vector bundle over a compact, oriented Riemannian manifold M, and if $1 \leq p < \infty$, let $L^p(E)$ denote the vector space of L^p-sections s of E endowed with the Banach norm

$$|s|_p = \left(\int_M |s(x)|^p d\,\mathrm{vol}(x) \right)^{\frac{1}{p}}. \tag{18.5.1}$$

If $p = \infty$, let $L^\infty(E)$ denote the essentially bounded sections of E and let $|s|_\infty$ denote the essential supremum of the function $x \in M \to |s(x)|$.

In analogy with the case of functions $M \to \mathbb{R}$, we have:

LEMMA 18.5.1. *If $1 \le p < \infty$ and $\frac{1}{p} + \frac{1}{q} = 1$, then the natural map*

$$J_p : L^p(E) \to L^q(E)^*, \quad (J_p(s))(s') := \int_M \langle s, s' \rangle d\,\mathrm{vol}$$

for $s \in L^p(E)$, $s' \in L^q(E)$ is an isometric isomorphism between $L^p(E)$ and the dual space $L^q(E)^*$ of the Banach space $L^q(E)$. For $1 < p < \infty$ the Banach space $L^p(E)$ is reflexive.

PROOF. If we add a vector bundle E' to E such that $E \oplus E' \to M$ is trivial, we can easily reduce the claim to the function case. For this case see, for example, [**Alt99**, 4.12 and 6.10]. □

In the sequel, the bundle E will be the bundle $\Lambda^k M$ of alternating k-forms, where $0 \le k \le \dim M =: n$. The Euclidean structure on $\Lambda^k M$ will be the one induced by the Riemannian metric on M. For $k \in \{0, 1, n-1, n\}$ the corresponding Euclidean norm on the fibers of $\Lambda^k M$ coincides with the comass norm, *cf.* [**Fed69**, 1.8.1]. We assume that the total volume of M is normalized to one.

Denote by $\Omega^k M$ the space of smooth differential k-forms on M, *cf.* Section 14.2. Given a cohomology class $\alpha \in H^k(M, \mathbb{R})$, denote by α_{dR} the collection of smooth closed k-forms $\omega \in \Omega^k M$ that represent α. Then the formula

$$|\alpha|_p^* := \inf \left\{ |\omega|_p \,\middle|\, \omega \in \alpha_{\mathrm{dR}} \right\}$$

defines a norm on the finite-dimensional vector space $H^k(M, \mathbb{R})$. To see that one has $|\alpha|_p^* > 0$ if $\alpha \neq 0$, note that there exists $\beta \in H^{n-k}(M, \mathbb{R})$ such that the cup product $\alpha \cup \beta \in H^n(M, \mathbb{R})$ is non-zero. It follows that

$$\left| \int_M \omega \wedge \pi \right| = |(\alpha \cup \beta)([M])| > 0$$

whenever $\omega \in \alpha_{\mathrm{dR}}$ and $\pi \in \beta_{\mathrm{dR}}$. In view of the pointwise inequality

$$|\omega \wedge \pi| \le \binom{n}{k}^{\frac{1}{2}} |\omega|\, |\pi|,$$

cf. [**Fed69**, 1.7.5], we obtain

$$\int_M |\omega|\, |\pi| d\mathrm{vol} \ge \binom{n}{k}^{-\frac{1}{2}} |(\alpha \cup \beta)([M])|.$$

Hence the Hölder inequality implies

$$|\alpha|_p^*\, |\beta|_q^* \ge \binom{n}{k}^{-\frac{1}{2}} |(\alpha \cup \beta)([M])| > 0.$$

We let $\overline{\alpha}_{\mathrm{dR}}^p \subseteq L^p(\Lambda^k M)$ denote the closure of α_{dR} with respect to the L^p-norm. Note that

$$|\alpha|_p^* = \inf \left\{ |\omega|_p \,\middle|\, \omega \in \overline{\alpha}_{\mathrm{dR}}^p \right\} \tag{18.5.2}$$

Moreover, if $\beta \in H^{n-k}(M, \mathbb{R})$, $\frac{1}{p} + \frac{1}{q} = 1$, $\omega \in \overline{\alpha}_{\mathrm{dR}}^p$ and $\pi \in \overline{\beta}_{\mathrm{dR}}^q$, then

$$\int_M \omega \wedge \pi = [\alpha \cup \beta](M). \tag{18.5.3}$$

Since the cup product $\cup : H^k(M,\mathbb{R}) \times H^{n-k}(M,\mathbb{R}) \to \mathbb{R}$ is non-degenerate, we obtain the following consequence.

LEMMA 18.5.2. *Every smooth closed form in* $\overline{\alpha}^p_{\mathrm{dR}}$ *represents the class* α.

Next, we prove that $\overline{\alpha}^p_{\mathrm{dR}}$ contains a unique L^p-form with minimal p-norm if we have $1 < p < \infty$.

PROPOSITION 18.5.3. *Let* $1 < p < \infty$ *and* $k \in \{1, \ldots, \dim M - 1\}$. *For every* $\alpha \in H^k(M,\mathbb{R})$ *there exists a unique* $\omega \in \overline{\alpha}^p_{\mathrm{dR}}$ *such that*

$$|\omega|_p = |\alpha|^*_p.$$

PROOF. Let $(\omega_i)_{i\in\mathbb{N}}$ be a sequence in α_{dR} such that

$$\lim_{i\to\infty} |\omega_i|_p = |\alpha|^*_p.$$

Since $L^p(\Lambda^k M)$ is a reflexive Banach space, *cf.* Lemma 18.5.1, we can assume that the sequence $(\omega_i)_{i\in\mathbb{N}}$ converges weakly to a suitable element $\omega \in L^p(\Lambda^k M)$, *cf.* [**Alt99**, 6.9]. Since $\overline{\alpha}^p_{\mathrm{dR}}$ is a closed affine subspace of the space $L^p(\Lambda^k M)$, the Hahn-Banach theorem implies that $\overline{\alpha}^p_{\mathrm{dR}}$ is also weakly sequentially closed, *cf.* [**Alt99**, 6.12]. It follows that $\omega \in \overline{\alpha}^p_{\mathrm{dR}}$. Moreover, the weak convergence of $(\omega_i)_{i\in\mathbb{N}}$ to ω implies that

$$|\omega|_p \leq \lim_{i\to\infty} |\omega_i|_p = |\alpha|^*_p.$$

Using equality (18.5.2) and $\omega \in \overline{\alpha}^p_{\mathrm{dR}}$, we conclude that $|\omega|_p = |\alpha|^*_p$. The uniqueness of the minimizer ω follows from the fact that the equality $|\omega + \tilde{\omega}|_p = |\omega|_p + |\tilde{\omega}|_p$ can hold only if ω and $\tilde{\omega}$ are linearly dependent in $L^p(\Lambda^k M)$. □

18.6. Existence of harmonic forms with constant norm

We continue to assume that the volume of M^n is normalized to one. Then, as a direct consequence of the Hölder inequality, the function

$$p \in [1,\infty] \to |\alpha|^*_p$$

is (weakly) monotonely increasing for every $\alpha \in H^k(M,\mathbb{R})$, where $0 < k < n$.

PROPOSITION 18.6.1. *Let* $\alpha \in H^k(M,\mathbb{R})$, $0 < k < n$. *Assume there exist reals* $p < p'$ *in* $[1,\infty]$ *such that* $|\alpha|^*_p = |\alpha|^*_{p'}$. *Then the following holds:*

(1) *the function* $q \in [1,\infty] \to |\alpha|^*_q$ *is constant;*
(2) *the harmonic representative* ω *of* α *has constant (pointwise) norm;*
(3) *we have* $|\omega|_q = |\alpha|^*_q$ *for all* $q \in [1,\infty]$.

Conversely, if $p \in [1,\infty)$ *and if there exists a form* $\omega \in \bar{\alpha}^p_{\mathrm{dR}}$ *such that* $|\omega|_p = |\alpha|^*_p$ *and* ω *has constant norm almost everywhere, then*

$$|\omega|_p = |\omega|_q = |\alpha|^*_q$$

for all $q \in [1,\infty]$. *In particular,* ω *is harmonic.*

PROOF. Let $p < p'$ in $[1,\infty]$ satisfy $|\alpha|^*_p = |\alpha|^*_{p'}$. Since $q \mapsto |\alpha|^*_q$ is monotonic, we may assume that $1 < p < p' < \infty$. According to Proposition 18.5.3, there exists $\omega \in \overline{\alpha}^{p'}_{\mathrm{dR}}$ such that $|\omega|_{p'} = |\alpha|^*_{p'}$. First we show that ω has constant norm. Since $\overline{\alpha}^{p'}_{\mathrm{dR}} \subseteq \overline{\alpha}^p_{\mathrm{dR}}$, we see that

$$|\alpha|^*_p \leq |\omega|_p \leq |\omega|_{p'} = |\alpha|^*_{p'}.$$

Hence our assumption $|\alpha|_p^* = |\alpha|_{p'}^*$ implies that $|\omega|_p = |\omega|_{p'}$. Therefore the function $M \to \mathbb{R}$ sending $x \mapsto |\omega_x|$ is almost everywhere constant.

Next, we prove that ω is indeed harmonic. If $p \leq 2$, this follows directly from

$$|\alpha|_p^* = |\omega|_p = |\omega|_2 \geq |\alpha|_2^* \geq |\alpha|_p^*.$$

For general p, we will now show that ω is weakly closed and coclosed, *i.e.* that

$$\int_M \langle \omega, d^*\pi \rangle d\,\mathrm{vol} = 0 \text{ for all } \pi \in \Omega^{k+1}M, \tag{18.6.1}$$

and

$$\int_M \langle \omega, d\sigma \rangle d\,\mathrm{vol} = 0 \text{ for all } \sigma \in \Omega^{k-1}M. \tag{18.6.2}$$

To prove (18.6.1), consider a fixed $\pi \in \Omega^{k+1}M$ and the linear functional

$$\tilde{\omega} \in L^p(\Lambda^k M) \to \int_M \langle \tilde{\omega}, d^*\pi \rangle d\,\mathrm{vol}$$

on $L^p(\Lambda^k M)$. This functional vanishes on α_{dR} and hence, by continuity, on $\omega \in \overline{\alpha}_{\mathrm{dR}}^p$.

Equation (18.6.2) is a consequence of the minimality of $|\omega|_p$ in $\overline{\alpha}_{\mathrm{dR}}^p$ and of the constancy of $|\omega_x|$. Note that we may assume that $\alpha \neq 0$ and hence $|\omega_x| = r > 0$ for almost all $x \in M$. For every $\sigma \in \Omega^{k-1}M$ we have

$$\begin{aligned} 0 &= \frac{d}{ds}\bigg|_{s=0} \left(\int_M |\omega + s\, d\sigma|^p d\,\mathrm{vol} \right) \\ &= p \int_M |\omega|^{p-2} \langle \omega, d\sigma \rangle d\,\mathrm{vol} \\ &= p r^{p-2} \int_M \langle \omega, d\sigma \rangle d\,\mathrm{vol}. \end{aligned}$$

Since $r \neq 0$ this implies (18.6.2). Finally, we consider the Euclidean bundle

$$\Lambda^* M = \bigoplus_{k=0}^{n} \Lambda^k M$$

and the selfadjoint, first order linear differential operator $d + d^*$ acting on sections of $\Lambda^* M$. It is well-known that the operator $d + d^*$ is elliptic. Hence the standard regularity theory for linear elliptic operators implies that ω is smooth and one has $(d + d^*)(\omega) = 0$ in the strong sense. Since $(d + d^*)^2 = \Delta$, we conclude that ω is harmonic. From the fact that ω is a harmonic form of constant norm, we can easily conclude that $|\alpha|_q^*$ does not depend on q. This is proved in Proposition 16.9.1 for the case $k \in \{1, n-1\}$, and the proof carries over to the general case without any change.

Finally, we assume that $p \in [1, \infty)$, and that the form $\omega \in \overline{\alpha}_{\mathrm{dR}}^p$ has constant norm almost everywhere and satisfies $|\omega|_p = |\alpha|_p^*$. For every exponent $p' \in (p, \infty]$, we have $\omega \in \overline{\alpha}_{\mathrm{dR}}^{p'}$ and

$$|\alpha|_{p'}^* \leq |\omega|_{p'} = |\omega|_p = |\alpha|_p^* \leq |\alpha|_{p'}^*.$$

Hence the first part of the proof shows that $|\alpha|_q^*$ does not depend on q. Since $|\omega|_q$ is independent of q as well, we obtain

$$|\alpha|_p^* = |\omega|_p = |\omega|_q = |\alpha|_q^*$$

for all $q \in [1, \infty]$. □

18.7. The BI construction adapted to conformal norms

We now continue with the notation of Section 18.4. Consider the norm $\| \|_{\mathrm{conf}\,/\varphi}$ in V. We apply Lemma 18.4.2 to the norm $\| \|_{\mathrm{conf}\,/\varphi}$. This yields a decomposition

$$\| \|_E^2 = \sum_{i=1}^{N} \lambda_i L_i^2,$$

where $\lambda_i > 0$, while $\sum \lambda_i = n$, $L_i \in V^*$ and $\|L_i\|^*_{\mathrm{conf}\,/\varphi} = 1$. Then a linear map $L : V \to \mathbb{R}^N$ defined by

$$L(x) = (\lambda_1^{1/2} L_1(x), \lambda_2^{1/2} L_2(x), \ldots, \lambda_N^{1/2} L_N(x)) \tag{18.7.1}$$

is an isometry from $(V, \| \|_E)$ onto a subspace $L(V)$ of $\mathbb{R}^N$, equipped with the restriction of the standard coordinate metric of $\mathbb{R}^N$. (Note that the L_i are not the components of L.)

PROPOSITION 18.7.1. *For every L_i there exists a weakly closed 1-form ω_i on M from the pullback cohomology class, $\omega_i \in \varphi^*(L_i)$, satisfying*

$$\|\omega_i\|_n = \|L_i\|^*_{\mathrm{conf}\,/\varphi} = 1.$$

PROOF. The proof results from Proposition 18.5.3. Briefly, the issue is that the minimizing form ω_i will *a priori* only be an L^n-form, *i.e.* the L^n-limit of a sequence of smooth forms in the pullback class $\varphi^*(L_i)$. Then, it is a non-trivial result from nonlinear PDE that ω_i is Hölder. This can be found in [**To84**], and in [**Ham92**] for the case of forms of arbitrary degree. □

The definition below generalizes Definition 16.5.1.

DEFINITION 18.7.2. The *free abelian cover* of M is the covering space $\overline{M}$ of M defined by the subgroup $\ker(\varphi_*) \subset \pi_1(M)$. We denote the action of $\Gamma = \pi_1(M)/\ker(\varphi_*)$ on $\overline{M}$ by $(v, x) \mapsto x + v$.

COROLLARY 18.7.3. *There exists a C^1 function*

$$f_i : \overline{M} \to \mathbb{R} \text{ such that } df_i = \tilde{\omega}_i, \tag{18.7.2}$$

where $\tilde{\omega}_i$ is the lift of ω_i. Furthermore, f_i satisfies the relation

$$f_i(x + v) = f_i(x) + L_i(v) \tag{18.7.3}$$

for all $x \in \overline{M}$ and $v \in \Gamma$.

PROOF. The proof of Proposition 18.7.1 implies the existence of f_i with Hölder continuous first derivatives satisfying (18.7.2). Then formula (18.7.3) follows from the fact that $\omega_i \in \varphi^*(L_i)$. □

Now we use the functions $f_i : \overline{M} \to \mathbb{R}$ of formula (18.7.2) to define a vector valued map $F : \overline{M} \to \mathbb{R}^N$ as follows:

$$F(x) = (\lambda_1^{1/2} f_1(x), \lambda_2^{1/2} f_2(x), \ldots, \lambda_N^{1/2} f_N(x)). \tag{18.7.4}$$

Observe that both the map L of (18.7.1) and F of (18.7.4) are Γ-equivariant with respect to the following action of Γ on $\mathbb{R}^N$:

$$\Gamma \times \mathbb{R}^N \to \mathbb{R}^N, \quad (v, x) \mapsto x + L(v).$$

Now let $\mathrm{Pr}_{L(V)} : \mathbb{R}^N \to L(V)$ be the orthogonal projection to the image $L(V)$. Then the composition

$$L^{-1} \circ \mathrm{Pr}_{L(V)} \circ F : \overline{M} \to V \tag{18.7.5}$$

is a Γ-equivariant map covering a map $f : M \to V/\Gamma$. (Note that f_i are not the components of f.) We will exploit it in Section 18.8.

18.8. Abel-Jacobi map for conformal norms

PROOF OF PROPOSITION 18.4.3. The proof is similar to the one in Chapter 17, with some crucial differences. Therefore we present the modified proof here. The main difference is that the functions f_i in formula (18.7.2) were defined differently in Chapter 17. Consider the norm $\| \; \|_{\mathrm{conf}\,/\varphi}$. We apply the construction of Section 18.7 to this norm. Since the projection is nonexpanding and the map L is an isometry, it suffices to prove the proposition for the map F of formula (18.7.4). Let $A = dF : T_x M \to \mathbb{R}^N$. Then

$$\mathrm{trace}(A^*A) = \sum \lambda_i |df_i|^2.$$

By the inequality of geometric and arithmetic means, we have

$$\begin{aligned}\mathrm{Jac}(F)(x) &= \det(A^*A)^{1/2}\\ &\le \left(\tfrac{1}{n}\,\mathrm{trace}(A^*A)\right)^{n/2}.\end{aligned}$$

Thus, the map $F : \overline{M} \to \mathbb{R}^N$ satisfies

$$\mathrm{Jac}(F) \le \left(\tfrac{1}{n} \sum \lambda_i |df_i|^2\right)^{n/2}. \tag{18.8.1}$$

Therefore,

$$\begin{aligned}\mathrm{Jac}(F) &\le \left(\sum \tfrac{\lambda_i}{n} |\tilde{\omega}_i|^2\right)^{n/2}\\ &\le \sum \tfrac{\lambda_i}{n} |\tilde{\omega}_i|^n.\end{aligned}$$

The last inequality follows from Jensen's inequality applied to the function $t \mapsto t^{n/2}$, when $n \ge 2$. In the case $n = 1$, there is only one term in the ellipsoid decomposition, so Jensen is not needed.

We now integrate the inequality over M. Note that $\int_M |\omega_i|^n = 1$. This yields the desired inequality for the area of the corresponding map $M \to V/\Gamma$. □

18.9. Attaining the conformal BIG bound

This Section contains the proof of Theorem 18.1.2 as well as Theorem 18.1.3.

PROOF OF THEOREM 18.1.2. We apply the results of the preceding sections, to the case where the map φ of (18.4.1) is taken to be the Abel-Jacobi map $\mathcal{A}_M$.

Note that equality in (18.8.1) is possible only if F is conformal. The discussion of equality in (18.8.1) leads to the fact that equality in (18.1.1) implies that the map f is conformal at all points where the differential does not vanish. If $n \ge 3$, then by [**Fer77**], the map f is indeed a smooth local (conformal) diffeomorphism, hence a covering. Here we are assuming that the metric on M is smooth. Hence M is indeed a torus and f is a conformal diffeomorphism.

If $n = 2$, then M is a torus, since $b_1(M) = n$ by hypothesis. Hence f is a holomorphic map between 2-tori. It has degree one since it is homotopic to the Abel-Jacobi map. It follows that f is a conformal diffeomorphism. □

PROOF OF THEOREM 18.1.3. We apply Proposition 18.6.1 combined with Theorem 18.1.1, which is immediate from Proposition 18.4.3, and an argument based on the perfection of a critical lattice similar to the proof of Proposition 16.11.5. In more detail, let us normalize the metric to unit volume to fix ideas. The equality of the stable norm and the conformally invariant norm for each of the integral elements realizing $\lambda_1 (H_1(M, \mathbb{Z})_{\mathbb{R}}, \| \|)$ proves that the harmonic 1-form in each of those integral classes has pointwise constant norm. Next, perfection implies that in fact *every* harmonic 1-form on M has pointwise constant norm. It follows that the standard Abel-Jacobi map defined by those forms, is a Riemannian submersion onto a flat torus, without any ramification. Since the induced homomorphism of fundamental groups is surjective, the degree of the submersion is 1, proving the theorem. □

CHAPTER 19

Systolic inequalities for L^p norms

19.1. Case $n \geq b$ and L^p norms in homology

In this section, we generalize the techniques of Section 18.7.

DEFINITION 19.1.1. Let M^n and M'^b be Riemannian manifolds, and assume we have $n \geq b$. Let $f\colon M \to M'$ a smooth map. For every $x \in M$, we consider the derivative $d_x f$, and define

$$\operatorname{Jac}^{\perp} f(x) = \sup_{\sigma} \frac{\operatorname{vol}_b(d_x f(\sigma))}{\operatorname{vol}_b(\sigma)}$$

where σ ranges over all b-dimensional parallelotopes in $T_x M$.

Obviously, if x is a regular point of f, then the above supremum is attained at σ lying in the orthogonal complement of the fiber $f^{-1}(f(x))$, and $\operatorname{Jac}^{\perp} f(x) = 0$ if x is not a regular point. We will use the following coarea formula. Every smooth map $f\colon M^n \to M'^b$ satisfies the identity

$$\int_M \operatorname{Jac}^{\perp} f(x)\, d\operatorname{vol}_n(x) = \int_{M'} \operatorname{vol}_{n-b}(f^{-1}(y))\, d\operatorname{vol}_b(y). \tag{19.1.1}$$

We present a version of Proposition 18.4.1 and Proposition 18.4.3, which allows us to prove inequality (18.3.5) for the conformal systole, as well as to handle the case of equality when $n \geq b$.

PROPOSITION 19.1.2 (Bound on orthogonal Jacobian). *Let M^n be a closed Riemannian manifold of unit volume. Let V^b be a vector space, where $n \geq b$. Let $p \geq \max\{b, 2\}$. Let Γ be lattice in V, and let $\varphi : M \to V/\Gamma$ be a continuous map inducing an epimorphism of the fundamental groups. Consider the associated L^p-norm $\| \ \|_p$ on V, cf. Definition 18.3.3. Denote by $\| \ \|_E$ the Euclidean norm on V, defined by the John ellipsoid of $\| \ \|_p$. Then the homotopy class of φ contains a smooth map $f : M \to (V/\Gamma, \| \ \|_E)$ such that*

$$\int_M \operatorname{Jac}^{\perp} f \leq 1. \tag{19.1.2}$$

PROPOSITION 19.1.3. *In the hypotheses of Proposition 19.1.2, assume a strict inequality $p > \max\{b, 2\}$. If equality is attained in inequality* (19.1.2), *then the map f is a harmonic Riemannian submersion.*

In Section 19.2, we will generalize the BI construction of Section 18.7, and use it in Section 19.3 to prove the Propositions 19.1.2 and 19.1.3.

19.2. The BI construction in the case $n \geq b$

Applying Lemma 18.4.2 to the norm $\| \, \|_p$ yields a decomposition

$$\| \, \|_E^2 = \sum_{i=1}^N \lambda_i L_i^2,$$

where $\lambda_i > 0$, $\sum \lambda_i = b$, $L_i \in V^*$ and $\|L_i\|_p^* = 1$. Then a linear map $L : V \to \mathbb{R}^N$ defined by

$$L(x) = (\lambda_1^{1/2} L_1(x), \lambda_2^{1/2} L_2(x), \ldots, \lambda_N^{1/2} L_N(x))$$

is an isometry from $(V, \| \, \|_E)$ onto a subspace $L(V)$ of $\mathbb{R}^N$, equipped with the restriction of the standard coordinate metric of $\mathbb{R}^N$. (Note that the L_i are not the components of L.)

We identify the dual space V^* of V with $H^1(V/\Gamma; \mathbb{R})$. Then, we have a pull-back map $\varphi^* : V^* \to H^1(M; \mathbb{R})$. For every $\alpha \in V^*$, we have

$$\|\alpha\|_p^* = \|\varphi^*(\alpha)\|_p^* = \inf \left\{ |\omega|_p \;\middle|\; \omega \in \varphi^*(\alpha)_{\mathrm{dR}} \right\}.$$

DEFINITION 19.2.1. Fix a real $p > \max\{b, 2\}$ and let ω_i, $i = 1, \ldots, N$ denote the L^p-minimizer in the cohomology class $\varphi^*(L_i) \in H^1(M; \mathbb{R})$.

Since $\operatorname{vol}(M) = 1$, we have $|\omega_i|_p = \|L_i\|_p^* = 1$. The rest of the construction is the same as in Section 18.7. In particular, we define a map

$$F(x) = (\lambda_1^{1/2} f_1(x), \lambda_2^{1/2} f_2(x), \ldots, \lambda_N^{1/2} f_N(x)). \tag{19.2.1}$$

In conclusion, we obtain a map $\tilde{f} \colon \overline{M} \to V$ defined by

$$\tilde{f} = L^{-1} \circ \Pr_{L(V)} \circ F \tag{19.2.2}$$

where $\Pr_{L(V)} : \mathbb{R}^N \to L(V)$ is the orthogonal projection to $L(V)$. Clearly the map $\tilde{f}$ is Γ-equivariant, hence it is a lift of a map $f \colon M \to V/\Gamma$. (Note that the f_i are not the components of f.) We show in Section 19.3 that f satisfies inequality (19.1.2).

19.3. Proof of bound on orthogonal Jacobian

This Section contains the proof of Proposition 19.1.2 and Proposition 19.1.3. Consider the map f constructed in Section 19.2. Let $x \in M$. If x is a regular point of f, we let $\Sigma_x = (\ker d_x f)^\perp$. Otherwise let Σ_x be an arbitrary d-dimensional linear subspace of $T_x M$. Then

$$\operatorname{Jac}^\perp f(x) = \operatorname{Jac}((d_x f)|_{\Sigma_x}) \leq \operatorname{Jac}((d_x F)|_{\Sigma_x}) \tag{19.3.1}$$

since the projection in (19.2.2) is nonexpanding and L is an isometry. Let A be the linear map

$$A = (d_x F)|_{\Sigma_x} : \Sigma_x \to \mathbb{R}^N. \tag{19.3.2}$$

Then, by (19.2.1),

$$\operatorname{trace}(A^* A) = \sum \lambda_i \left| (d_x f_i)|_{\Sigma_x} \right|^2 \leq \sum \lambda_i |d_x f_i|^2 = \sum \lambda_i |\omega_i(x)|^2$$

By the inequality of geometric and arithmetic means, we have

$$\operatorname{Jac}((d_x F)|_{\Sigma_x}) = \det(A^* A)^{1/2} \leq \left(\tfrac{1}{b} \operatorname{trace}(A^* A) \right)^{b/2}. \tag{19.3.3}$$

Thus, the map f satisfies

$$\mathrm{Jac}^{\perp} f \leq \left(\sum \tfrac{\lambda_i}{b}|\omega_i|^2\right)^{b/2} \leq \left(\sum \tfrac{\lambda_i}{b}|\omega_i|^p\right)^{b/p}. \tag{19.3.4}$$

The last inequality follows from Jensen's inequality applied to the function $t \mapsto t^{p/2}$; recall that $p > 2$ and $\sum \lambda_i = b$. Integrating over M yields

$$\int_M \mathrm{Jac}^{\perp} f \leq \int_M \left(\sum \tfrac{\lambda_i}{b}|\omega_i|^p\right)^{b/p} \leq \left(\int_M \sum \tfrac{\lambda_i}{b}|\omega_i|^p\right)^{b/p}, \tag{19.3.5}$$

since $b/p < 1$ and $\mathrm{vol}(M) = 1$. Here the integrals are with respect to the volume form. Since $|\omega_i|_p \leq 1$, we have

$$\int_M \sum \tfrac{\lambda_i}{b}|\omega_i|^p = \sum \tfrac{\lambda_i}{b} \int_M |\omega_i|^p \leq \sum \tfrac{\lambda_i}{b} = 1, \tag{19.3.6}$$

and the desired inequality (19.1.2) follows.

REMARK 19.3.1. Let $f : M \to H_1(M,\mathbb{R})/H_1(M,\mathbb{Z})_{\mathbb{R}}$ be Lichnerowicz's harmonic map, *cf.* Section 17.2 and (17.2.1). Let $p = b = 2$. Let ω_1, ω_2 be an orthonormal basis for the harmonic 1-forms with respect to the L^2 inner product. Then one has $\lambda_1 = \lambda_2 = 1$, and the computations above can be simplified as follows:

$$\int_M \mathrm{Jac}^{\perp} f = \int_M |\omega_1 \wedge \omega_2| \leq \tfrac{1}{2} \int_M \left(|\omega_1|^2 + |\omega_2|^2\right) = 1,$$

where the integrals are with respect to the volume form. The desired inequality (19.1.2) follows, *cf.* Remark 18.3.7.

In the case of equality in (19.1.2), observe that all functions under integrals are continuous, hence all the inequalities throughout the argument turn to equalities.

LEMMA 19.3.2. *Equality in* (19.1.2) *implies that* $|\omega_i| = 1$ *everywhere.*

PROOF. Equality in (19.3.4) implies that $|\omega_i| = |\omega_j|$ for all i, j. The equality in (19.3.5) implies that the function

$$\sum_i \frac{\lambda_i}{b}|\omega_i|^p$$

is constant. These together imply that $|\omega_i|$ is constant. Then equality in (19.3.6) implies that $|\omega_i| = 1$. □

PROOF OF PROPOSITION 19.1.3. Since, by Lemma 19.3.2, we have $|\omega_i| = 1$ everywhere and ω_i is an L^p-minimizer, Proposition 18.6.1 implies that ω_i is harmonic for every i. Hence f is harmonic (and, in particular, smooth).

Equality in (19.3.3) is attained precisely when the map A of formula (19.3.2) is a conformal linear map. Then equality in (19.3.1) implies that $(df)|_{\Sigma_x}$ is conformal, too. Finally, equality in (19.3.4) implies that $\mathrm{Jac}^{\perp} f = 1$, thus f is a Riemannian submersion. □

19.4. Attaining the conformal BIG(n, b) bound

This Section contains the proof of Proposition 18.3.5 and of Theorem 18.3.8.

PROOF OF PROPOSITION 18.3.5. We apply Proposition 19.1.2 to the Abel-Jacobi map $\mathcal{A} : M \to \mathbb{T}^b = H_1(M;\mathbb{R})/H_1(M;\mathbb{Z})_{\mathbb{R}}$. We obtain a map f which is homotopic to $\mathcal{A}$ and satisfies inequality (19.1.2), where $\mathbb{T}^b$ is equipped with the flat Euclidean metric defined by the John ellipsoid of the unit ball of the norm $\| \ \|_p$, where the case $p = \infty$ corresponds to the stable norm. Then, by the coarea formula (19.1.1),

$$\int_M \operatorname{Jac}^{\perp} f(x)\, d\operatorname{vol}_n(x) = \int_{\mathbb{T}^b} \operatorname{vol}_{n-b}(f^{-1}(y))\, d\operatorname{vol}_b(y).$$

Observe that if $y \in \mathbb{T}^b$ is a regular value for f, then

$$\operatorname{vol}_{n-b}(f^{-1}(y)) \geq \deg(\mathcal{A}), \tag{19.4.1}$$

as in Definition 18.2.1. Hence we have

$$\deg(\mathcal{A}) \operatorname{vol}_b\left(\mathbb{T}^b, \| \ \|_E\right) \leq \operatorname{vol}_n(M).$$

by inequality (19.1.2). The proposition now results from the definition of the Hermite constant (5.3.1). □

PROOF OF THEOREM 18.3.8. The theorem follows from the combination of Proposition 18.3.5 and Proposition 18.6.1, by the argument of the proof of Theorem 18.1.3 in Section 18.9. Alternatively, one could argue as follows. In the case of equality, the equalities in (19.1.2) and (19.4.1) are attained. This implies the inequality (18.3.1), and the theorem follows from Lemma 19.3.2 characterizing the equality case in Proposition 19.1.2. Note that equality in (19.4.1) implies that $f^{-1}(y)$ is minimal. □

CHAPTER 20

Four-manifold systole asymptotics

20.1. Schottky problem and the surjectivity conjecture

The work of P. Buser and P. Sarnak [**BusS94**] on Riemann surfaces Σ_g in connection with the Schottky problem shows that the maximum, over the moduli space, of the least conformal length of a nonseparating loop in Σ_g behaves *logarithmically* in the genus g, *cf.* (20.6.1) below and also [**Gro83**, Theorem 5.5.C'].

We provide a lower bound which is *polynomial* in the second Betti number, for the analogous 2-dimensional conformal systolic invariant for a 4-manifold M with indefinite intersection form. For background on 4-manifolds, see [**FreU84, FriM94**]. The successive minima λ_k of a normed lattice were defined in Section 4.4, Definition 4.4.1. Note that the second successive minimum is exploited in Corollary 20.8.2 below.

THEOREM 20.1.1. *Let $n \in \mathbb{N}$ and consider the complex projective plane blown up at n points, $M_n = \mathbb{CP}^2 \# n\overline{\mathbb{CP}}^2$, where bar denotes reversal of orientation, while $\#$ is connected sum. Then*

$$C^{-1}\sqrt{n} < \sup_{\mathcal{G}} \left\{ \lambda_1 \left(H^2(M_n, \mathbb{Z}), |\,|_{L^2} \right) \right\}^2 < Cn, \quad \forall n > 0, \tag{20.1.1}$$

where $C > 0$ is a numerical constant, the supremum is over all smooth metrics $\mathcal{G}$ on M_n, and $|\,|_{L^2}$ is the norm (20.3.1) defined by $\mathcal{G}$.

Here the upper bound may be replaced by $\frac{2}{3}(n+1)$ in view of estimate (20.4.3), while the lower bound, by $\sqrt{k(n)}$, where $k(n)$ is asymptotic to $\frac{n}{2\pi e}$ as $n \to \infty$, *cf.* Theorem 20.2.2. The theorem is proved in Section 20.11. The desired metric is specified in formula (20.11.1) in terms of the inversion of the period map.

Formula (20.1.1) can be rewritten as follows in terms of the notation of Section 20.4:

$$C^{-1}\sqrt{n} < \sup_{\mathcal{G}} \operatorname{confsys}_2(M_n, \mathcal{G})^2 < Cn, \quad \forall n > 0. \tag{20.1.2}$$

The theorem results from the combination of two ingredients, namely the density of the image of the period map, *cf.* Remark 20.1.4, combined with the following lattice-theoretic result.

PROPOSITION 20.1.2. *Consider the Lorentzian lattice $\mathbb{Z}^{n,1}$, with inner product written as $x.y$ for $x, y \in \mathbb{Z}^{n,1}$. Given $v \in \mathbb{Z}^{n,1}$ with $v.v < 0$, apply a "sign reversal procedure" SR, to construct a positive definite lattice*

$$L_v = \mathrm{SR}(\mathbb{Z}^{n,1}, v)$$

of rank $n+1$, obtained by reversing the sign of the quadratic form in the v direction, cf. (20.7.1). Then

$$\sup\left\{\lambda_1(L_v) \,\middle|\, v \in \mathbb{Z}^{n,1},\ v.v < 0\right\} \geq \sqrt{\tfrac{n}{2\pi e}}(1 + o(1)), \quad \text{for } n \to \infty.$$

Let $(M, \mathcal{G})$ be a Riemannian 4-manifold. Let

$$* : H^2_{\mathrm{dR}}(M) \to H^2_{\mathrm{dR}}(M)$$

be the Hodge star operator in de Rham cohomology identified with the space $\mathcal{H}$ of harmonic 2-forms on M. Assume that $b^+(M) = 1$, so that the self-dual subspace (*i.e.* the $(+1)$-eigenspace of the operator $*$) is 1-dimensional, *cf.* [**LeB03**, p. 237]. Recall that the cup-product form in $H^2(M)$ is dual to the intersection form in $H_2(M)$.

CONJECTURE 20.1.3 (Surjectivity conjecture). *For every line V in the positive cone in $H^2_{\mathrm{dR}}(M)$ defined by the cup product form, there is a metric $\mathcal{G}$ on M whose self-dual subspace is exactly V.*

REMARK 20.1.4. When these results originally appeared in [**Katz03**], they were said to hold modulo the conjectured surjectivity of the period map. In fact, our bound does not depend on such surjectivity but only on the density of the image, proved in [**Li01**, Theorem 4]. An alternative proof of the density may be found in [**GayK04**, Theorem 1, item (8)]. Discussions with P. Biran, C. LeBrun, and S. Donaldson in 2003 suggested that one may be able to remove the dependence on the conjectured surjectivity of the period map, at least in the case of the blow-ups of the projective plane, by exploiting the action of the automorphism group of the intersection form, *cf.* Lemma 20.7.1 and Remark 20.7.2, and using metrics adapted to symplectic forms which represent classes from a suitable fundamental domain for the action, *cf.* [**Bir01**, Theorem 3.2] and Remark 20.9.1. Such a program was, in fact, carried out in [**Li01**, Theorem 4], *cf.* Appendix A.

A similar remark is made by M. Hamilton [**Hami06**], who, furthermore, generalizes Theorem 20.1.1 to all 4-manifolds with $b^+ = 1$.

QUESTION 20.1.5. Can one improve the lower bound in (20.1.1) to *linear* dependence on n? Here one could envision an averaging argument, using Siegel's formula as in [**MilnH73**, Theorem 9.5], over integral vectors satisfying $q_{n,1}(v) = -p$. Here one seeks a vector $v \in \mathbb{R}^{n,1}$ such that the integer lattice $\mathbb{Z}^{n,1} \subset \mathbb{R}^{n,1}$ has the Conway-Thompson behavior (20.2.1) with respect to the positive definite form $\mathrm{SR}(q_{n,1}, v)$.

QUESTION 20.1.6. Is there an *asymptotically infinite* lower bound similar to Theorem 20.1.1 for the stable 2-systole in place of the conformal 2-systole? This is related to understanding the discrepancy between the comass norm and the L^2 norm in 2-dimensional cohomology. Note that Remark 20.4.1, concerning the 1-systoles of surfaces, suggests that *a priori* there may exist, instead, an *asymptotically vanishing* upper bound for the stable 2-systole, *cf.* (20.5.2) in the definite case.

A number of systolic inequalities are now available in the literature. Nontrivial cup product relations lead to stable systolic inequalities, *cf.* Chapter 15 and inequality (20.4.3) below. Some of these inequalities are sharp, *cf.* Chapter 16 and [**NV04**]. Meanwhile, nontrivial Massey products also admit systolic repercussions, *cf.* Section 14.3.

The present chapter is organized as follows. Section 20.2 introduces the Conway-Thompson lattices and describes the idea of the proof. Section 20.3 defines the L^2-norm in cohomology and describes its relation to the intersection form. Section 20.4 defines the conformal and stable systoles. Section 20.5 discusses the definite case. Our main Theorem 20.1.1 can be thought of as a higher-dimensional analogue of the Buser-Sarnak theorem, presented in Section 20.6. Section 20.7 explains a useful sign reversal relation between definite and indefinite forms. Section 20.8 describes a Lorentzian construction of lattices inspired by a result of J. Conway and N. Sloane, and presents a lower bound for the second successive minimum. Section 20.9 presents the necessary linear algebraic ingredient. Section 20.10 deforms a lower bound for the second successive minimum, into a lower bound for the first successive minimum. The proof is completed by a successive minimum calculation in Section 20.11.

20.2. Conway-Thompson lattices CT_n and idea of proof

The density of the image of the period map furnishes a lot of latitude in prescribing the position of the integer lattice in middle-dimensional de Rham cohomology, with respect to the L^2-norm. In particular, we show that the least norm, $\lambda_1\left(H^p(M,\mathbb{Z})_{\mathbb{R}}, |\ |_{L^2}\right)$, of a nonzero lattice element, can be made arbitrarily large as the Betti number grows. Here one relies on the existence of Euclidean unimodular lattices L with arbitrarily high $\lambda_1(L)$, as well as on the (elementary) classification of indefinite odd unimodular forms, *cf.* (20.8.2). We acknowledge the influence on our approach of the Lorentzian construction (*i.e.* using indefinite forms) of the Leech lattice of J. Conway and N. Sloane [**ConS99**, Chapter 26], namely the following result.

THEOREM 20.2.1 (J. H. Conway, N. J. A. Sloane). *If*

$$t = (3, 5, 7, \ldots, 45, 47, 51|145)$$

is a vector with $q_{24,1}(t) = -1$ *in* $I_{24,1}$*, then* $t^{\perp} \cap I_{24,1}$ *is a copy of the Leech lattice.*

The *first* step of our approach can be described as adapting the Lorentzian construction by replacing the Leech lattice by the Conway-Thompson lattices. The latter are unimodular lattices which define packings of high asymptotic density. More precisely, we have the following result [**MilnH73**, Theorem 9.5].

THEOREM 20.2.2 (Conway, Thompson). *For every dimension* n*, there exists a positive definite inner product space over* $\mathbb{Z}$ *of odd type and rank* n *with*

$$\min\{x \cdot x \,|x \neq 0\} \geq k(n), \qquad (20.2.1)$$

where $k(n)$ *is asymptotic to* $\frac{n}{2\pi e}$ *as* $n \to \infty$.

DEFINITION 20.2.3. A family of unimodular lattices satisfying (20.2.1) will be denoted (CT_n). We will refer to a lattice CT_n from such a family as a *Conway-Thompson lattice.*

It should be pointed out that the lower bound proved by Conway and Thompson is a theoretical bound. One does not know how explicitly to construct lattices satisfying it. On the other hand, fairly explicit constructions exist of lattices with the property that $x \cdot x$ greater than $\sqrt{n}$, *cf.* [**BarW59, DT99**].

The *second* step of our approach is explained in Section 20.9.

20.3. Norms in cohomology

Let $(M, \mathcal{G})$ be a closed orientable $(2p)$-dimensional Riemannian manifold. Consider the lattice $H^p(M,\mathbb{Z})_{\mathbb{R}} \subset \mathcal{H} = H^p_{\mathrm{dR}}(M)$, defined as the image of $H^p(M,\mathbb{Z})$ in $H^p(M,\mathbb{R})$ under the inclusion $\mathbb{Z} \subset \mathbb{R}$ of coefficients, *i.e.* quotient by its torsion subgroup. We will sometimes delete the subscript $_{\mathbb{R}}$, by abuse of notation, when the torsion subgroup is trivial. Consider the L^2-norm $|\ |_{L^2}$ in $\mathcal{H}$, defined by

$$|\alpha|_{L^2}^2 = \int_M \alpha \wedge *\alpha \tag{20.3.1}$$

for each harmonic p-form $\alpha \in \mathcal{H}$, where $*$ is the Hodge operator for the metric $\mathcal{G}$. The following lemma is obvious, *cf.* [**FreU84**, Lemma 2.21], [**FriM94**].

LEMMA 20.3.1. *Let p be even. Then the L^2-norm in $H^p(M,\mathbb{R})$ is related to the intersection form*

$$q_M(\alpha,\beta) = \int_M \alpha \cup \beta$$

by means of the "sign reversal" formula

$$|\alpha|_{L^2}^2 = \langle \alpha, \alpha \rangle = q_M(\alpha^+,\alpha^+) - q_M(\alpha^-,\alpha^-) \tag{20.3.2}$$

where $\alpha = \alpha^+ + \alpha^-$ is the decomposition given by the splitting $\mathcal{H} = V^+ + V^-$ into the (± 1)-eigenspaces of the involution $*$.

Similarly to the notation of formula (20.7.1) below, we can restate the conclusion of Lemma 20.3.1 as follows:

$$\langle\ ,\ \rangle = \mathrm{SR}(q_M, V^-). \tag{20.3.3}$$

The lattice $H^p(M,\mathbb{Z})_{\mathbb{R}}$ is equipped with the L^2-norm defined by formula (20.3.1). The dual norm in the similarly defined lattice

$$H_p(M,\mathbb{Z})_{\mathbb{R}} \subset H_p(M,\mathbb{R})$$

will also be denoted $|\ |_{L^2}$.

20.4. Conformal length and systolic flavors

In this section, we define several flavors of systolic invariants of a Riemannian manifold $(M,\mathcal{G})$ of even dimension $2p$. The conformal p-systole of the metric $\mathcal{G}$ is denoted $\mathrm{confsys}_p(\mathcal{G})$, *cf.* Section 15.9. It is defined as the least norm of a nonzero element in the integer lattice in p-dimensional cohomology (or, equivalently, homology; *cf.* below Remark 20.4.2), with respect to the L^2-norm (20.3.1) defined by $\mathcal{G}$:

$$\begin{aligned}\mathrm{confsys}_p(\mathcal{G}) &= \lambda_1\left(H^p(M^{2p},\mathbb{Z})_{\mathbb{R}}, |\ |_{L^2}\right)\\ &= \min\left\{ |v|_{L^2} \;\middle|\; v \in H^p(M,\mathbb{Z})_{\mathbb{R}}, v \neq 0\right\}.\end{aligned}$$

Meanwhile, the *stable p-systole* is the quantity

$$\mathrm{stsys}_p(\mathcal{G}) = \lambda_1(H_p(M^{2p},\mathbb{Z})_{\mathbb{R}}, \|\ \|),$$

where $\|\ \|$ is the stable norm in homology, dual to the comass norm in cohomology, *cf.* [**Fed74**, 4.10], [**BaK03**]. The conformal systole is related to the stable systole as follows:

$$\mathrm{stsys}_p(\mathcal{G})\,\mathrm{vol}_{2p}(\mathcal{G})^{-\frac{1}{2}} \leq \binom{2p}{p}^{\frac{1}{2}} \mathrm{confsys}_p(\mathcal{G}). \tag{20.4.1}$$

Here the binomial coefficient appears due to the discrepancy between the linear comass norm and the natural Euclidean norm on the space of p-forms, *cf.* Section 15.8. In the case $p = 1$, the binomial coefficient may be replaced by 1.

REMARK 20.4.1 (1-systole asymptotics). It should be kept in mind that the asymptotic behavior of the (stable) 1-systole as a function of the genus is completely different from the conformal systole. Thus, M. Gromov [**Gro96**, 2.C] reveals the existence of a universal constant C such that we have an asymptotically *vanishing* upper bound $\frac{(\log g)^2}{g}$ for the systolic ratio:

$$\frac{\operatorname{sys}_1(\Sigma_g)^2}{\operatorname{area}(\Sigma_g)} \leq C\,\frac{(\log g)^2}{g},$$

for every orientable surface Σ_g of genus $g \geq 2$, with a Riemannian metric, see [**CrK03**, (2.9) and (2.10)] and Chapter 11 above for related bounds. In contrast, P. Buser and P. Sarnak [**BusS94**] provide an asymptotically *infinite* lower bound for the maximum of the conformal systole over the moduli space, *cf.* inequality (20.6.1).

REMARK 20.4.2 (Extremal length). The Poincare duality map induces an isometry

$$\mathrm{PD} : (H^p(M,\mathbb{Z})_{\mathbb{R}}, |\ |_{L^2}) \to (H_p(M,\mathbb{Z})_{\mathbb{R}}, |\ |_{L^2}), \tag{20.4.2}$$

proving that the integer lattice in middle dimension is *isodual* in the sense of [**ConS94, BeM89**]. Thus for $p = 1$, the invariant $\operatorname{confsys}_1$ is the extremal length (or conformal length) of the surface.

We have the following upper bound on conformal systole:

$$\lambda_1(H^p(M^{2p},\mathbb{Z})_{\mathbb{R}}, |\ |_{L^2})^2 \leq \gamma_b < \tfrac{2}{3}\, b_p(M^{2p}) \quad \text{for } b_p(M) \geq 2, \tag{20.4.3}$$

see Chapter 15 for stable systolic generalisations based on multiplicative relations in cohomology.

20.5. Systoles of definite intersection forms

While the main result of this chapter is Theorem 20.1.1 in the case of *indefinite* intersection form (which may be viewed as a higher dimensional analogue of the Buser-Sarnak theorem (20.6.1)), the *definite* case is also of interest. In this section, we briefly discuss the definite case. Consider the family of manifolds $n\mathbb{CP}^2$, defined as the connected sum of n copies of the complex projective plane with the standard orientation. Recall that these exhaust the smooth positive definite case by S. Donaldson's theorem, *cf.* [**Katz95b**]. In contrast to Theorem 20.1.1, the maximal conformal systole in the *definite* case is uniformly bounded as the second Betti number grows:

$$\begin{aligned} \lambda_1(H^2(n\mathbb{CP}^2,\mathbb{Z}), |\ |_{L^2}) &= \lambda_1(H^2(n\mathbb{CP}^2,\mathbb{Z}), \sqrt{q_M}) \\ &= \lambda_1(\mathbb{Z}^n) = 1, \end{aligned} \tag{20.5.1}$$

for every Riemannian $n\mathbb{CP}^2$, $n = 1, 2, \ldots$. This is immediate from formula (20.3.2) which identifies the L^2-norm and the intersection form q_M if the latter is positive definite. We thus obtain the following result, pointed out by C. Lebrun.

PROPOSITION 20.5.1. *Every Riemannian $n\mathbb{CP}^2$ satisfies the inequality*

$$\operatorname{stsys}_2\left(n\mathbb{CP}^2\right)^2 \leq 2\operatorname{vol}_4\left(n\mathbb{CP}^2\right). \tag{20.5.2}$$

PROOF. We will exploit the two conformal systolic invariants defined in Section 15.9. Since in dimension 4, every 2-form is either simple, or has length 2, *cf.* (13.1.3), we have an inequality

$$|x|^2 \leq 2\|x\|^2$$

between the pointwise Euclidean norm and the pointwise comass, for all tangent vectors $x \in T(n\mathbb{CP}^2)$. It follows that

$$\mathrm{Confsys}_2^2 \le 2\,\mathrm{confsys}_2^2 .$$

A metric of unit volume satisfies $\mathrm{stsys}_k \le \mathrm{Confsys}_k$, thus

$$\mathrm{stsys}_2^2 \le 2\,\mathrm{confsys}_2^2 .$$

Meanwhile, every metric on a connected sum $n\mathbb{CP}^2$ satisfies $\mathrm{confsys}_2 = 1$ because of the identification of the L^2 norm and the intersection form, *cf.* (20.5.1). We thus reprove Gromov's optimal inequality

$$\mathrm{stsys}_2^2 \le 2\,\mathrm{vol}_4,$$

but now it is valid for the connected sum of n copies of $\mathbb{CP}^2$. □

REMARK 20.5.2. In fact, the inequality can be stated in terms of the last successive minimum of the integer lattice in homology with respect to the stable norm $\|\ \|$. Namely, this minimum λ_n satisfies the inequality

$$\lambda_n\left(H_2(n\mathbb{CP}^2,\mathbb{Z}), \|\ \|\right)^2 < 2\,\mathrm{vol}_4(n\mathbb{CP}^2)$$

The proof is the same as before. This inequality is in fact optimal for all n.

QUESTION 20.5.3. What is the asymptotic behavior for the stable systole of $n\mathbb{CP}^2$ when $n \to \infty$? Can the constant in (20.5.2) be replaced by a function which tends to zero as $n \to \infty$?

20.6. Buser-Sarnak theorem

Our Theorem 20.1.1 may be viewed as a higher dimensional analogue of the following theorem of P. Buser and P. Sarnak [**BusS94**, formula (1.13)]. Let Σ_g be a closed orientable surface of genus g. Then the conformal 1-systole satisfies the bounds

$$C^{-1}\log g < \sup_{\mathcal{G}}\left\{\lambda_1\left(H^1(\Sigma_g,\mathbb{Z}), |\ |_{L^2}\right)\right\}^2 < C\log g,\ \forall g \ge 2 \tag{20.6.1}$$

where $C > 0$ is a numerical constant, while $|\ |_{L^2}$ is the norm (20.3.1) associated with $\mathcal{G}$. Here the supremum is over all metrics $\mathcal{G}$ on Σ_g. Since the L^2 norm is an invariant of the conformal class of the metric, the supremum in (20.6.1) can be thought of extending over the entire moduli space. The systolic quantity $\lambda_1\left(H^1(\Sigma_g,\mathbb{Z}), |\ |_{L^2}\right)$ may be viewed as the conformal (extremal) length of the surface, in view of the isomorphism of formula (20.4.2).

An explicit upper bound of $\frac{3}{\pi}\log(4g+3)$ is provided in [**BusS94**, formula (1.13)]. Note that a (weaker) upper bound of $C\sqrt{g}$ (in place of $C\log g$) results from R. Lazarsfeld's work [**Laz96**, p. 441, Proposition, part (i)].

20.7. Sign reversal procedure SR and $\mathrm{Aut}(I_{n,1})$-invariance

Let q be an indefinite quadratic form of index $+1$ (*i.e.* with a single negative direction) on a vector space E over $\mathbb{R}$, and let $v \in E$ be a vector satisfying $q(v) < 0$. Denote by $v^\perp \subset E$ the q-orthogonal complement of $v \in E$, or, more precisely, the Q-orthogonal complement, where $Q(u,w) = \frac{1}{4}(q(u+w) - q(u-w))$ is the polarisation of q. Thus, we have a decomposition $E = v^\perp \oplus \mathbb{R}v$. The *sign reversal*, $\mathrm{SR}(q,v)$, is the positive definite form on E obtained by reversing the sign of q in direction v, while keeping it fixed on $v^\perp \subset E$:

$$\mathrm{SR}(q,v)(x) = q(x^+) - q(x^-), \tag{20.7.1}$$

where $x = x^+ + x^-$ is the decomposition of $x \in E$ following the splitting $E = v^\perp \oplus \mathbb{R}v$, *cf.* formula (20.3.3). Let $\mathbb{R}^{p,q}$ denote the standard real vector space with quadratic form

$$q_{p,q}(x) = x_1^2 + \ldots + x_p^2 - x_{p+1}^2 - \cdots - x_{p+q}^2, \tag{20.7.2}$$

and let $I_{p,q} \subset \mathbb{R}^{p,q}$ denote its integer lattice. For the purposes of the proof of Theorem 20.1.1, it is convenient to reverse the orientation and work instead on the manifold $n\mathbb{CP}^2\#\overline{\mathbb{CP}}^2$, while hoping that such a step may not prove baffling to an algebraic geometer (the usual orientation is retained in Apendix A).

Recall that the intersection form on $n\mathbb{CP}^2\#\overline{\mathbb{CP}}^2$ is $q_{n,1}$, and the integer lattice in two-dimensional homology becomes a copy of $I_{n,1}$.

LEMMA 20.7.1. *The invariant* $\mathrm{confsys}_2\left(n\mathbb{CP}^2\#\overline{\mathbb{CP}}^2, \mathcal{G}\right)$ *only depends on the orbit of the anti-self-dual line of* $\mathcal{G}$ *in* $H^2_{\mathrm{dR}}\left(n\mathbb{CP}^2\#\overline{\mathbb{CP}}^2\right)$ *under the action of the automorphism group of* $I_{n,1}$.

PROOF. An endomorphism f of $H^2_{\mathrm{dR}}(n\mathbb{CP}^2\#\overline{\mathbb{CP}}^2)$ which is an automorphism of the indefinite lattice $I_{n,1}$, induces an isometry of the definite form $\mathrm{SR}(q_{n,1}, v)$, since f maps the subspace $v^{\perp_q}$ to the subspace $f(v)^{\perp_q}$, and hence

$$\mathrm{SR}(q_{n,1}, v)(x) = \mathrm{SR}(q_{n,1}, f(v))(f(x)).$$

Here if $x \in I_{n,1}$, then $f(x) \in I_{n,1}$ by the hypothesis that f preserves the integer lattice. Now the lemma follows from the formula

$$\mathrm{confsys}_2(\mathcal{G}) = \lambda_1\left(H^2\left(n\mathbb{CP}^2\#\overline{\mathbb{CP}}^2, \mathbb{Z}\right), \mathrm{SR}\left(q_{n,1}, V^-\right)^{\frac{1}{2}}\right),$$

where V^- is the anti-self-dual direction of $\mathcal{G}$. □

REMARK 20.7.2. Note that not all automorphisms of the intersection form can be realized by a diffeomorphism of the manifold, *cf.* [**Kot91**].

20.8. Lorentz construction of Leech lattice and line $\mathrm{CT}_n^\perp$

Let $\mathrm{CT}_n \subset \mathbb{R}^{n,0}$ be a Conway-Thompson lattice of Theorem 20.2.2, *i.e.* a unimodular lattice satisfying $\lambda_1(\mathrm{CT}_n)^2 \ge k(n)$. Then the lattice $\mathrm{CT}_n \oplus I_{0,1}$ is odd, indefinite, and unimodular, *cf.* (20.7.2) and notation there. Now we apply the classification of odd indefinite unimodular forms, following J. Milnor and D. Husemoller [**MilnH73**, Theorem 4.3, p. 22].

THEOREM 20.8.1. *Every indefinite odd inner product space over* $\mathbb{Z}$ *possesses an orthogonal basis, and hence is isomorphic to an orthogonal sum of copies of* $\langle 1\rangle$ *and* $\langle -1\rangle$.

We reproduce the short proof from [**MilnH73**] for the reader's convenience.

PROOF. The proof is by induction on the rank of the inner product space X. By the Hasse-Minkowski theorem, there exists a vector $x_1 \in X$ with $x_1 \cdot x_1 = 0$. Without loss of generality, we may assume that x_1 is prime. We complete x_1 to a basis $x_1, \ldots, x_n$ for the inner product space X. Let $y_1, \ldots, y_n$ be the dual basis. Then $x_1 \cdot y_1 = 1$.

By hypothesis, there exists a vector y such that $y \cdot y$ is odd. Hence one of the basis vectors y_k has the property that $y_k \cdot y_k$ is odd. Choose a subspace $X_0 \subset X$ as follows.

If $y_1 \cdot y_1$ is odd, let X_0 be spanned by x_1 and y_1.

If $y_1 \cdot y_1$ is even, let X_0 be spanned by x_1 and $y_1' = y_1 + y_k$, where k is chosen so that

$$(y_1 + y_k) \cdot (y_1 + y_k) \equiv y_k \cdot y_k \equiv 1 \pmod 2.$$

Note that $x_1 \cdot y_1' = 1$. For a suitable ℓ, the space X_0 has inner product matrix of the form

$$\begin{pmatrix} 0 & 1 \\ 1 & 2\ell + 1 \end{pmatrix}$$

so that the two vectors $x' = y_1' - \ell x_1$ and $y' = y_1' - (\ell+1)x_1$ are mutually orthogonal, with inner product matrix $\begin{pmatrix} 1 & 0 \\ 0 & -1 \end{pmatrix}$. Thus, we have $X_0 \cong \langle 1 \rangle \oplus \langle -1 \rangle$. Clearly, X splits as an orthogonal direct sum

$$X \cong X_0 \oplus X_0^{\perp} \cong \langle 1 \rangle \oplus \langle -1 \rangle \oplus X_0^{\perp}. \tag{20.8.1}$$

Choose $\epsilon \in \{1, -1\}$ so that the space $\langle \epsilon \rangle \oplus X_0^{\perp}$ is indefinite. Then the space $\langle \epsilon \rangle \oplus X_0^{\perp}$ admits an orthogonal basis by the inductive hypothesis, and therefore so does X by (20.8.1). □

We revert to the notation $I_{1,0} = \langle 1 \rangle$ and $I_{0,1} = \langle -1 \rangle$. Note that the lattice $\mathrm{CT}_n \oplus I_{0,1}$ is odd and unimodular. The classification implies that the lattice $I_{n,1}$ contains an isometric copy of CT_n such that the $q_{n,1}$-orthogonal complement of CT_n in $I_{n,1}$, is a copy of the line $I_{0,1}$. In formulas, there exists an isomorphism

$$\phi_n : \mathrm{CT}_n \oplus I_{0,1} \to I_{n,1} \tag{20.8.2}$$

preserving the bilinear forms. We will use the suggestive notation

$$\mathrm{CT}_n^{\perp} \subset I_{n,1} \tag{20.8.3}$$

for the line identified by the isomorphism of formula (20.8.2), namely the $q_{n,1}$-orthogonal complement of $\phi_n(\mathrm{CT}_n \times \{0\}) \subset I_{n,1}$.

COROLLARY 20.8.2. *Let $I_{n,1} \subset \mathbb{R}^{n,1}$ be the integer lattice. Choose a generator $v \in \mathrm{CT}_n^{\perp} \subset I_{n,1}$ as in (20.8.3). In other words, $v = \phi_n(0, e)$, where $e \in I_{0,1}$ is a generator, as in isomorphism (20.8.2). Consider the norm $\| \; \|_v$ defined for all $x \in \mathbb{R}^{n,1}$ by*

$$\|x\|_v = \left(\mathrm{SR}(q_{n,1}, v)(x)\right)^{1/2},$$

in the notation of formula (20.7.1). Then the integer lattice has successive minima $\lambda_1(I_{n,1}, \| \; \|_v) = +1$, and

$$\lambda_2(I_{n,1}, \| \; \|_v)^2 \geq k(n),$$

cf. Definition 4.4.1, where $k(n)$ is as in Theorem 20.2.2.

In other words, all integral vectors of square-norm smaller than $k(n)$ are proportional to each other.

PROOF. Given a lattice L with a positive definite form, we have the identity $\mathrm{SR}(L \oplus I_{0,1}, \iota(e)) = L \oplus I_{1,0}$, where ι is the inclusion of the second factor. In particular,

$$\mathrm{SR}(I_{n,1}, \phi_n(\iota(e))) = \mathrm{CT}_n \oplus I_{1,0}, \tag{20.8.4}$$

proving the corollary. □

As an indication of how nontrivial the isomorphism ϕ of (20.8.2) could be, consider Theorem 20.2.1, which exhibits an isomorphism $\Lambda_{24} \oplus I_{0,1} \to I_{24,1}$, where Λ_{24} is the Leech lattice.

Our eventual objective is the lower bound of Theorem 20.1.1. We first prove Proposition 20.8.3 below. Recall that the intersection form on $n\mathbb{CP}^2 \# \overline{\mathbb{CP}}^2$ is the diagonal form $q_{n,1}$, *cf.* formula (20.7.2). Let ϕ_n be the isomorphism (20.8.2).

PROPOSITION 20.8.3. *If $\mathcal{G}$ is a metric on $n\mathbb{CP}^2 \# \overline{\mathbb{CP}}^2$ whose anti-self-dual direction is the vector $\phi_n(0, e) \in I_{n,1}$, then all surfaces of "conformal area" smaller than $\sqrt{k(n)}$ with respect to $\mathcal{G}$ are homologous to multiples of one another.*

PROOF. The integer lattice in the self-dual subspace V^+ is isometric to the Conway-Thompson lattice:

$$V^+ \cap H^2(n\mathbb{CP}^2 \# \overline{\mathbb{CP}}^2, \mathbb{Z}) \simeq \mathrm{CT}_n.$$

Moreover, this copy of the Conway-Thompson lattice is a direct summand, where the second summand is isometric to $I_{0,1}$. The sign reversal formula (20.3.2) shows that the integer lattice

$$\left(H^2(n\mathbb{CP}^2 \# \overline{\mathbb{CP}}^2, \mathbb{Z}),\ \mathrm{SR}\left(q_M, \mathrm{CT}_n^\perp\right)^{\frac{1}{2}}\right),$$

is isometric to the positive definite lattice $\mathrm{CT}_n \oplus I_{1,0}$, where $I_{0,1}$ has been replaced by $I_{1,0}$ as in formula (20.8.4). Thus the proposition is a restatement of Corollary 20.8.2. □

20.9. Three quadratic forms in the plane

The main result of this section is Lemma 20.9.3 below on the interplay of three quadratic forms in the plane, an indefinite one, q, and a pair of definite ones, q_1 and q_s, where the parameter value s will be judiciously chosen in (20.10.7).

To go beyond Proposition 20.8.3 and prove our theorem, the lattice $\mathrm{CT}_n \oplus I_{1,0}$ is not sufficient, as it contains vectors of unit norm in the second summand $I_{1,0}$, so that the quantity $\lambda_1(\mathrm{CT}_n \oplus I_{1,0}) = 1$ is too small. In other words, we need to replace a lower bound for the successive minimum λ_2 of the integer lattice, by a lower bound for the successive minimum λ_1 for the same lattice, but with respect to a new norm. The idea is to deform appropriately the choice of the negative definite direction $v = \phi_n(0, e)$, responsible for the Conway-Thompson behavior of its complement.

Thus, to prove Theorem 20.1.1, we will apply the surjectivity of the period map, not to the line

$$\mathrm{CT}_n^\perp \subset H^2_{\mathrm{dR}}\left(n\mathbb{CP}^2 \# \overline{\mathbb{CP}}^2\right),$$

but rather to the image of $\mathrm{CT}_n^\perp$ under a suitable "Lorentz deformation", as in formula (20.11.1).

REMARK 20.9.1. Since the quantity λ_1 (as in Definition 4.4.1) is continuous as a function on the space of positive definite lattices, while the form $\mathrm{SR}(q_M, V)$ is continuous in both parameters, and V^- depends continuously on the metric, it follows that for our purposes, Conjecture 20.1.3 can be relaxed to assume the density of the image in place of surjectivity.

The argument relies on a rather crude bound on the operator norm of the deformation. The deformation needs to be sufficient to eliminate short vectors, but with operator norm controlled so as not to negate entirely the Conway-Thompson effect.

Sign reversal on the line $\mathrm{CT}_n^\perp \subset I_{n,1}$ produces a quadratic form with respect to which most vectors are suitably long, except for a single direction. To weed out the remaining short vector, we apply a sutiable deformation, whose linear algebraic content is presented in Lemma 20.9.3 below.

Let π be the xy-plane. Let e_1, e_2 be the standard basis and x, y the standard coordinates. Consider the indefinite form $q = dxdy$, and let $s > 0$ be a real parameter.

DEFINITION 20.9.2. Our "Lorentz transformation" $A_s : \pi \to \pi$ is defined by the matrix

$$A_s = \begin{pmatrix} s & 0 \\ 0 & s^{-1} \end{pmatrix}$$

with respect to the standard basis, and we set

$$u_s = A_s(e_1 + e_2) = se_1 + \tfrac{1}{s}e_2$$

and $v_s = A_s(e_1 - e_2) = se_1 - \frac{1}{s}e_2$.

LEMMA 20.9.3. *Consider the positive definite quadratic form*

$$q_s = \mathrm{SR}(q, v_s)$$

on π, obtained from q by reversing the sign in the direction v_s, as in formula (20.7.1). *Then the map $A_s : (\pi, q_1) \to (\pi, q_s)$ is an isometry.*

PROOF. Since the "Lorentz transformation" A_s preserves q and sends v_1 to v_s, it is clear that it also sends q_1 to q_s, but we will give a short explicit calculation. We have $q(u_s, v_s) = 0$. Let (x', y') be the coordinates with respect to the basis $\{u_s, v_s\}$ of the plane π. Then the two pairs of coordinates are related by $x = s(x' + y'), y = \frac{1}{s}(x' - y')$. Now we have

$$q = dxdy = s(dx' + dy')\,\frac{1}{s}(dx' - dy') = dx'^2 - dy'^2.$$

Therefore by definition, $q_s = \mathrm{SR}(q, v_s) = dx'^2 + dy'^2$. Thus $q_s(u_s, v_s) = 0$, while $q_s(u_s) = q_s(v_s) = 1$. Similarly, the vectors $u = e_1 + e_2$ and $v = e_1 - e_2$ form an orthonormal basis for q_1, proving the Lemma. □

20.10. Replacing λ_1 by the geometric mean $(\lambda_1\lambda_2)^{1/2}$

The transformation described in Section 20.9 allows one to replace the first two successive minima of a lattice, by their geometric mean. Let $L = I_{n,1} \subset (\mathbb{R}^{n,1}, q_{n,1})$ be the integer lattice. Let $v \in L$ be a vector satisfying $q_{n,1}(v) = -1$ and

$$L = \mathbb{Z}v \oplus v^\perp \simeq I_{0,1} \oplus v^\perp, \tag{20.10.1}$$

where the sublattice $\left(v^\perp, (q_{n,1}|_{v^\perp})^{\frac{1}{2}}\right)$ is positive definite. Let $\mathrm{SR}(q_{n,1}, v)$ be the positive definite form obtained by sign reversal. Let

$$\lambda_i = \lambda_i\left(L, \mathrm{SR}(q_{n,1}, v)^{\frac{1}{2}}\right)$$

be the successive minima with respect to the new form. We have $\lambda_1 = 1$ but we will ignore this in the statement of the proposition below, so as to emphasize the geometric mean inherent in the proof. Note that we have

$$\lambda_2 = \lambda_1\left(v^\perp, \left(q_{n,1}\left|_{v^\perp}\right.\right)^{\frac{1}{2}}\right). \tag{20.10.2}$$

PROPOSITION 20.10.1. *There is a $q_{n,1}$-preserving transformation A of $\mathbb{R}^{n,1}$ such that $\lambda_1\left(L, \mathrm{SR}\left(q_{n,1}, Av\right)^{\frac{1}{2}}\right) \geq \sqrt{\lambda_1\lambda_2}$.*

PROOF. Let $\pi \subset \mathbb{R}^{n,1}$ be any 2-plane containing the vector v as in (20.10.1). We choose coordinates (x, y) in π with the following three properties:

(1) the union of the x-axis and the y-axis in π is the intersection of the isotropic cone of $q_{n,1}$ with π;
(2) the restriction of $q_{n,1}$ to π is the form q of Lemma 20.9.3;
(3) with respect to the standard basis e_1, e_2 in π, we have $v = e_1 - e_2$.

Now let $s \in \mathbb{R}$, and set $v_s = se_1 - \frac{1}{s}e_2$. Let q_s be the positive definite quadratic form obtained by sign reversal $q_s = \mathrm{SR}(q_{n,1}, v_s)$. Thus, for $s = 1$, replacing q by q_1 has the effect of replacing $I_{0,1}$ by $I_{1,0}$ in the decomposition (20.10.1). Hence we have the following isometry of lattices:

$$(L, q_1) \simeq I_{1,0} \oplus v^\perp. \tag{20.10.3}$$

We wish to understand the position of the integer lattice L with respect to the definite form q_s "deforming" q_1. By Lemma 20.9.3, the map

$$A_s \oplus Id_{\pi^\perp}, \tag{20.10.4}$$

also denoted A_s, is an isometry from q_1 to q_s. Thus the pullback lattice $\left(A_s^{-1}(L), q_1\right)$ is isometric to (L, q_s). We have $A_s^{-1}(v) = \frac{1}{s}e_1 - se_2$, and hence

$$q_s(v) = q_1\left(A_s^{-1}v\right) = q_1\left(\tfrac{1}{s}e_1 - se_2\right) = \tfrac{1}{s^2} + s^2 \geq s^2. \tag{20.10.5}$$

Now consider an element $x \in L = \mathbb{Z}v \oplus v^\perp$ which is not proportional to the generator v of the first summand. By the Pythagorean theorem applied to formula (20.10.3), the element x satisfies

$$q_1(x)^{\frac{1}{2}} \geq \lambda_1\left(v^\perp\right) = \lambda_2\left(L, \sqrt{q_1}\right),$$

by formula (20.10.2). Meanwhile, we have the following bound on the operator norm with respect to the form q_1: $\|A_s\| = \|A_s^{-1}\| \leq s$, and therefore

$$q_s(x) = q_1(A_s^{-1}x) \geq \frac{\lambda_2^2}{s^2}. \tag{20.10.6}$$

Combining (20.10.5) and (20.10.6), we obtain the lower bound

$$\lambda_1\left(L, \sqrt{q_s}\right) \geq \min\left\{s, \tfrac{\lambda_2}{s}\right\}.$$

Choosing the parameter value

$$s = \sqrt{\tfrac{\lambda_2}{\lambda_1}} = \sqrt{\lambda_2}, \tag{20.10.7}$$

we complete the proof of the proposition. □

COROLLARY 20.10.2. *Let* $\mathbb{Z}v = \mathrm{CT}_n^\perp \subset L = \mathbb{Z}^{n,1}$, *as in Theorem 20.2.2 and isomorphism* (20.8.2). *Then there is a transformation* $A = A_{k(n)^{\frac{1}{4}}}$ *of* $\mathbb{R}^{n,1}$ *such that* $\lambda_1\left(L, \mathrm{SR}(q_{n,1}, Av)^{\frac{1}{2}}\right) \geq k(n)^{\frac{1}{4}}$.

20.11. Period map and proof of main theorem

We are now in a position to prove Theorem 20.1.1. The above inequality (20.4.3) proves the upper bound of estimate (20.1.1), insofar as $b_2(n\mathbb{CP}^2 \# \overline{\mathbb{CP}^2}) = n+1$. Let us write down a formula, (20.11.1), for a metric $\mathcal{G}_n$ satisfying the lower bound. Let $M_n = n\mathbb{CP}^2 \# \overline{\mathbb{CP}^2}$, so that $H^2(M_n, \mathbb{Z}) = I_{n,1}$, with intersection form $q_M = q_{n,1}$. Recall that the L^2-norm in $H^2(M_n, \mathbb{R})$ is related to the intersection form $q_M(\alpha, \beta) = \int_{M_n} \alpha \cup \beta$ by means of the "sign reversal" formula

$$|\alpha|_{L^2}^2 = \langle \alpha, \alpha \rangle = q_M(\alpha^+, \alpha^+) - q_M(\alpha^-, \alpha^-),$$

where $\alpha = \alpha^+ + \alpha^-$ is induced by the splitting $H^2(M_n, \mathbb{R}) = V^+ + V^-$ into the (± 1)-eigenspaces of the Hodge involution $*$. It is convenient to introduce the notation SR, for the "sign reversal" procedure, whose effect is to replace an indefinite $(n, 1)$ form by a positive definite form: $\langle\ ,\ \rangle = \mathrm{SR}(q_M, V^-)$, *cf.* formula (20.3.3).

By the Conway-Thompson theorem [**MilnH73**, Theorem 9.5], there exist positive definite unimodular lattices CT_n of rank n satisfying $\lambda_1(\mathrm{CT}_n)^2 \geq k(n)$, where $k(n)$ is asymptotic to $\frac{n}{2\pi e}$ as $n \to \infty$. Here λ_1 is the least length of a nonzero lattice element, *cf.* Definition 4.4.1. By the classification of odd indefinite unimodular forms [**MilnH73**, p. 22], there exists a vector $v \in I_{n,1}$ with $q_{n,1}(v) = -1$, whose orthogonal complement with respect to the polarisation of $q_{n,1}$ is the lattice CT_n. Denote by $\mathrm{CT}_n^\perp \subset H^2(M_n, \mathbb{R})$ the negative definite line $\mathbb{R}v$.

In Proposition 20.10.1, we presented a Lorentzian endomorphism A_s of $\mathbb{R}^{n,1}$, with the following property. Let λ_1 and λ_2 be the first two successive minima of the integer lattice, *cf.* Definition 4.4.1, with respect to the definite quadratic form $\mathrm{SR}(q_M, v)$. The endomorphism A_s replaces λ_1 and λ_2 by their geometric mean, when one passes to the new definite form $\mathrm{SR}(q_M, A_s v)$.

Let $\mathcal{M}(M_n)$ be the space of all Riemannian metrics on M_n, and let $\mathcal{N}$ be the projectivisation of the negative cone of the form q_M. Let

$$\mathrm{P} : \mathcal{M}(M_n) \to \mathcal{N}$$

be the map assigning to each metric, its anti-self-dual direction. Exploiting the surjectivity of P, we would like to set

$$\mathcal{G}_n = \mathrm{P}^{-1} A_{k(n)^{\frac{1}{4}}} \left(\mathrm{CT}_n^\perp\right), \tag{20.11.1}$$

where P^{-1} is a right inverse. Since the full surjectivity is unavailable but rather only dense image [**Li01**, Theorem 4], we perturb the line $A_{k(n)^{\frac{1}{4}}}\left(\mathrm{CT}_n^\perp\right)$ into the image, clearly without significantly affecting the relevant lower bounds for the L^2 norm of integral classes.

Finally, the lower bound results from the following calculation:

$$\begin{aligned}
\operatorname{confsys}_2(\mathcal{G}_n) &= \lambda_1(H^2(M_n,\mathbb{Z}),|\ |_{L^2}) \\
&= \lambda_1\left(H^2\left(n\mathbb{CP}^2\#\overline{\mathbb{CP}^2},\mathbb{Z}\right),\ \mathrm{SR}\left(q_{n,1}, A_{k(n)^{\frac{1}{4}}}\left(\mathrm{CT}_n^{\perp}\right)\right)^{\frac{1}{2}}\right) \\
&\geq \sqrt{\lambda_2\left(H^2(n\mathbb{CP}^2\#\overline{\mathbb{CP}^2},\mathbb{Z}),\ \mathrm{SR}\left(q_{n,1},\mathrm{CT}_n^{\perp}\right)^{\frac{1}{2}}\right)} \\
&= \sqrt{\lambda_1(\mathrm{CT}_n)} \\
&\geq k(n)^{\frac{1}{4}}.
\end{aligned}$$

A proof of the density of the image is presented in the next chapter.

APPENDIX A

Period map image density (by Jake Solomon)

A.1. Introduction and outline of proof

We present a proof of the density of the image of the period map for the case of the n-fold blow-up of projective plane, $\mathbb{CP}^2\#n\overline{\mathbb{CP}^2}$. This result was first obtained by T. J. Li and A. Liu [**Li01**]. An alternative proof of the density may be found in [**GayK04**, Theorem 1, item (8)]. Our proof, similarly to that of [**Li01**], relies on the connection between Seiberg-Witten invariants and Gromov-Witten invariants, due to C. Taubes [**Ta95**], as further developed by D. McDuff [**Mc96**] and P. Biran [**Bir97**]. Our goal is to provide a non-technical exposition, accessible to a wide range of geometers and topologists.

More precisely, let M be a closed oriented four-manifold with intersection form

$$q : H^2(M,\mathbb{Z}) \otimes H^2(M,\mathbb{Z}) \to \mathbb{Z}.$$

By definition, $b_2^+(M)$ is the maximal rank of a submodule of $H^2(M,\mathbb{Z})$ on which the restriction of q is positive definite. We assume throughout that $b_2^+(M) = 1$. Given a vector space V, we denote by $\mathbb{P}(V)$ its projectivisation.

DEFINITION A.1.1. The positive cone $\mathcal{P} = \mathcal{P}(M)$ of M is

$$\mathcal{P} := \left\{ [\alpha] \in \mathbb{P}\left(H^2(M,\mathbb{R})\right) \middle| \, q(\alpha,\alpha) > 0 \right\}.$$

Given a Riemannian metric $\mathcal{G}$ on M, its Hodge star operator $*$ defines an involution on the space of harmonic two-forms $\mathcal{H}^2(M)$ and hence, by the Hodge theorem, on $H^2(M,\mathbb{R})$. We denote by

$$\mathcal{P}_{\mathcal{G}} \subset H^2(M,\mathbb{R})$$

the $+1$ eigenspace of the operator $*$, also known as the space of self-dual harmonic two-forms on M. If $\alpha \in \mathcal{P}_{\mathcal{G}}$, $\alpha \neq 0$, is a self-dual harmonic 2-form, we have

$$q(\alpha,\alpha) = \int_M \alpha \wedge \alpha = \int_M \alpha \wedge *\alpha = \|\alpha\|_{L^2} > 0.$$

Our hypothesis $b_2^+(M) = 1$ therefore implies $\dim \mathcal{P}_{\mathcal{G}} = 1$. Let $\mathcal{M}$ denote the space of Riemannian metrics on M.

DEFINITION A.1.2. The period map

$$\mathrm{P} : \mathcal{M} \longrightarrow \mathcal{P}(M)$$

is defined by $\mathcal{G} \mapsto [\mathcal{P}_{\mathcal{G}}]$.

The goal of this appendix will be to prove the following theorem in the special case when $M = \mathbb{CP}^2\#n\overline{\mathbb{CP}^2}$:

THEOREM A.1.3. *The image of the period map is dense in $\mathcal{P}$.*

This result was first obtained by T. J. Li and A. Liu in [**Li01**], where it is proved for arbitrary symplectic four-manifolds with $b_2^+ = 1$. Nonetheless, the result is perhaps most difficult precisely in the case we treat.

In general outline, the proof proceeds as follows. First, we show that it suffices, for $x = [\alpha]$ belonging to a dense subset of $\mathcal{P}$, to construct a symplectic form

$$\omega \in \Omega^2(M),\ d\omega = 0,\ \omega^2 \neq 0,$$

such that $[\omega] = x$. We note that the diffeomorphism group $\mathrm{Diff}(M)$ acts by pullback both on the set of symplectic forms on M and on the positive cone $\mathcal{P}$ of M. Thus, it suffices to construct ω such that $[\omega] = x$ for x in a dense subset of a fundamental domain for the action of $\mathrm{Diff}(M)$ on $\mathcal{P}$. In fact, we will specify a subset $\mathcal{S} \subset \mathcal{P}$ that contains a fundamental domain for a suitable subgroup

$$G_{(-1,-2)} \subset \mathrm{Diff}(M),$$

cf. (A.4.6), such that the image of the map P is dense even in $\mathcal{S}$. We explicitly construct a finite set of diffeomorphisms of M which generate $G_{(-1,-2)}$, and then prove that $\mathcal{S}$ contains a fundamental domain for the action of $G_{(-1,-2)}$ on $\mathcal{P}$ by techniques of hyperbolic geometry.

On the other hand, to prove the density of the image of the map P in $\mathcal{S} \subset \mathcal{P}$, we employ powerful techniques combining gauge theory and the theory of holomorphic curves on M.

We wish to thank P. Biran for several helpful conversations, and R. Borcherds for pointing out that $\mathcal{S}$ indeed contains a fundamental domain for $G_{(-1,-2)}$.

A.2. Symplectic forms and the self-dual line

The proposition below highlights the role of symplectic structures in understanding the image of the period map.

Let M be a $2p$ dimensional smooth manifold. By definition, an almost complex structure on M is an automorphism J of the tangent bundle TM satisfying $J^2 = -1$. Now, suppose M is equipped with a symplectic form $\omega \in \Omega^2(M)$, *cf.* Section 14.2. Thus, we have $d\omega = 0, \omega^p \neq 0$.

DEFINITION A.2.1. An almost complex structure J on M is called ω-*tame* if the $(0,2)$-tensor $\omega(\cdot, J\cdot)$ defines a positive definite bilinear form on each tangent space of M. It is called ω-*compatible* if, in addition, the bilinear form is symmetric.

It is not hard to show that the space of all ω-compatible almost complex structures on M is non-empty and contractible [**McS98**].

PROPOSITION A.2.2. *Assume* $\dim M = 4$ *and* $b_2^+(M) = 1$. *An element of* $\mathcal{P}(M)$ *that can be represented by a symplectic form, necessarily lies in the image of the period map.*

PROOF. Let ω be a symplectic form on a four-manifold M. We first choose any ω-compatible almost complex structure J. Consider the metric $\mathcal{G} := \omega(\cdot, J\cdot)$. Let $* : \Omega^*(M) \to \Omega^*(M)$ denote the Hodge-star operator associated to $\mathcal{G}$. It follows from linear algebra that

$$*\omega = \omega.$$

Since $d * \omega = d\omega = 0$ by the definition of a symplectic form, we conclude that ω is harmonic with respect to the $\mathcal{G}$-Laplacian on 2-forms on M. Since ω is self-dual, it represents the line $\mathcal{P}_{\mathcal{G}}$. □

A.3. A lemma from hyperbolic geometry

Let $\langle\cdot,\cdot\rangle$ denote the standard Lorentzian inner product on $\mathbb{R}^{n+1}$ given by a diagonal matrix with entries $-1,\ldots,-1,1$, on the diagonal.

Recall that the isometry group of n-dimensional hyperbolic space $\mathcal{H}^n$ is the group $O(1,n;\mathbb{R})$ of the linear transformations of $\mathbb{R}^{n+1}$ preserving $\langle\cdot,\cdot\rangle$. The action of $O(1,n)$ on $\mathcal{H}^n$ is given by restricting the action on $\mathbb{R}^{n+1}$ to the subset $\mathcal{H}^n_+ \subset \mathbb{R}^{n+1}$ defined by

$$\mathcal{H}^n_+ = \left\{ x = (x_1,\ldots,x_n;x_{n+1}) \in \mathbb{R}^{n+1} \middle| \langle x,x\rangle = +1, \quad x_{n+1} > 0 \right\}.$$

The space $\mathcal{H}^n_+$, equipped with pull back metric from $(\mathbb{R}^{n+1},\langle\cdot,\cdot\rangle)$, is one of the standard models for $\mathcal{H}^n$.

DEFINITION A.3.1. An element $r \in O(1,n)$ defines a *reflection* if for all $v \in \mathbb{R}^{n+1}$, we have

$$r(v) = v - 2\frac{\langle e,v\rangle}{\langle e,e\rangle}e$$

for a suitable $e \in \mathbb{R}^{n+1}$ with $\langle e,e\rangle < 0$.

Let $\Gamma \subset O(1,n;\mathbb{Z})$ be a subgroup generated by reflections. To each reflection $r \in \Gamma$ (not necessarily one of the generators), there corresponds a totally geodesic hyperplane $\mu_r \subset \mathcal{H}^n$ called the *mirror* of r, which is the locus of the fixed points of r. Consider the union of all mirrors of reflections in Γ. We call the closure of a connected component of the complement in $\mathcal{H}^n$ of this union a cell. We employ the notation $\{C_\iota \mid \iota \in \mathcal{I}\}$ for the collection of all such cells. The following result was stated, for example, in [**Vin75**, p. 324].

PROPOSITION A.3.2. *Each cell C_ι is a fundamental domain for the action of Γ on $\mathcal{H}^n$.*

PROOF. Fix a cell C_ι. Any point $x \in \mathcal{H}^n$ lies in a suitable cell C_β, $\beta \in \mathcal{I}$. It suffices to find an element of Γ carrying C_β and hence also x to C_ι.

LEMMA A.3.3. *Only a finite number of mirrors of reflections in Γ can meet any compact subset $K \subset \mathcal{H}^n$.*

PROOF. Without loss of generality, by replacing K with a ball containing it, we may assume $K = B_r(p)$, the ball of radius r centered at some point $p \in \mathcal{H}^n$. Note the following one to one correspondence between points of $B_r(p)$ and hyperplanes intersecting $B_r(p)$: By the convexity of the distance function on $\mathcal{H}^n$, any totally geodesic hyperplane $\mu \subset \mathcal{H}^n$ contains a unique point q closest to p. Clearly, if μ intersects $B_r(p)$, then $q \in B_r(p)$. On the other hand, we may recover μ from q as the totally geodesic hyperplane perpendicular to the unique length minimizing geodesic connecting q to p. Now, suppose $\{\mu_i\}$ is an infinite sequence of distinct mirrors of reflections in Γ each of which meet $B_r(p)$ and let q_i be the point of μ_i closest to p. Since $B_r(p)$ is compact, by passage to a subsequence, we may assume the sequence q_i converges. It follows that the corresponding hyperplanes μ_i converge, and so do the corresponding reflections. This contradicts the discreteness of $O(1,n;\mathbb{Z})$. □

In particular, a path connecting C_β with C_ι intersects only a finite number of mirrors. So, we may choose a finite sequence of adjacent cells

$$C_\beta = C_{\beta_1}, C_{\beta_2}, \ldots, C_{\beta_m} = C_\iota.$$

Adjacency means that each pair of cells $(C_{\beta_i}, C_{\beta_{i+1}})$ shares a unique mirror μ_i, where $i = 1, \ldots, n-1$. Reflecting sequentially in the mirrors μ_i moves the cell C_β to the cell C_ι by an isometry belonging to Γ. In particular, this isometry moves x into the closure of C_ι, as required. □

A.4. Diffeomorphism group of blow-up of projective plane

Let $M_n := \mathbb{CP}^2 \# n\overline{\mathbb{CP}^2}$. Thus, M_n is $\mathbb{CP}^2$ blown up at n points. We specify a basis of $H_2(M_n, \mathbb{Z})$ as follows. Let $L \in H_2(M_n, \mathbb{Z})$ be the homology class of a line in $\mathbb{CP}^2$. Let E_i denote the homology class of the ith exceptional divisor. Denote by $\mathrm{PD} : H^2(M_n, \mathbb{Z}) \to H_2(M_n, \mathbb{Z})$ the Poincaré duality isomorphism. Consider the basis in cohomology given by

$$\lambda = \mathrm{PD}^{-1}(L) \text{ and } \epsilon_i = \mathrm{PD}^{-1}(E_i) \in H^2(M_n, \mathbb{Z}).$$

We denote the intersection form on $H_2(M_n, \mathbb{Z})$ by Q or "$\cdot$", and we write $q = Q \circ \mathrm{PD}$ for the intersection form on $H^2(M_n, \mathbb{Z})$. An easy calculation yields

$$L \cdot L = 1, \qquad E_i \cdot L = 0, \qquad E_i \cdot E_j = -\delta_{ij}. \tag{A.4.1}$$

In other words, the intersection form is a Lorentzian inner product on $H_2(M_n, \mathbb{Z})$.

LEMMA A.4.1. *There is an element $r_i \in \mathrm{Diff}(M_n)$ that is the connect sum of the diffeomorphism given by complex conjugation on the ith copy of $\overline{\mathbb{CP}^2}$ with the identity diffeomorphism on the remaining $M_{n-1} = \mathbb{CP}^2 \#(n-1)\overline{\mathbb{CP}^2}$.*

PROOF. We must show that the two diffeomorphisms can be connected, to obtain a global diffeomorphism of $M_{n-1} \# \overline{\mathbb{CP}^2}$. That is, we must show that the identity map of $S^3 \subset \mathbb{C}^2$ can be smoothly isotoped through diffeomorphisms to the map arising from complex conjugation restricted to $S^3 \subset \mathbb{C}^2$. Indeed, both maps arise from matrices belonging to the special orthogonal group $SO(4)$. Since the Lie group $SO(4)$ is connected, we may connect these two matrices by a smooth path of matrices in $SO(4)$, each of which induces a diffeomorphism on S^3. □

We note that r_i^* acts on cohomology by the formula

$$r_i^*(\alpha) = \alpha - 2\frac{q(\alpha, \epsilon_i)}{q(\epsilon_i, \epsilon_i)}\epsilon_i, \qquad \alpha \in H^2(M_n, \mathbb{Z}). \tag{A.4.2}$$

Thus, r_i^* is the reflection of the Lorentzian lattice $(H^2(M_n, \mathbb{Z}), q)$, which corresponds to the vector ϵ_i. An element β of self-intersection -2 defines a reflection by a similar formula

$$r_\beta^*(\alpha) = \alpha - 2\frac{q(\alpha, \beta)}{q(\beta, \beta)}\beta. \tag{A.4.3}$$

The following lemma may be found in [**FriM94**].

LEMMA A.4.2. *Let M be a four-manifold. Let $\beta \in H^2(M, \mathbb{Z})$ be Poincaré dual to an embedded 2-sphere of self-intersection -2. Then there exists a suitable diffeomorphism $r_\beta \in \mathrm{Diff}(M)$ such that r_β^* acts on $H^2(M, \mathbb{Z})$ by reflection corresponding to β in the sense of* (A.4.3).

Now let $M = M_n$, and let $i, j, k = 1, \ldots, n$, assumed pairwise distinct. We set

$$\beta = \epsilon_{ijk} := \lambda - \epsilon_i - \epsilon_j - \epsilon_k. \tag{A.4.4}$$

Let r_β denote the diffeomorphism provided by Lemma A.4.2. In fact, in this particular case, we may choose the diffeomorphism r_β to be a biholomorphism. Using algebraic geometry, these biholomorphisms can be constructed explicitly and are known as Cremona transformations [**GriH78**]. We define r_{ijk} to be the Cremona transformation associated to ϵ_{ijk}. We set

$$\mathcal{S}_n := \left\{ [\alpha] \in \mathcal{P}(M_n \ ) \middle| \, q(\alpha, \epsilon_i) > 0, \, q(\alpha, \epsilon_{ijk}) > 0 \quad \forall i, j, k \right\}. \tag{A.4.5}$$

Furthermore, we let

$$G_{(-1,-2)} \subset \mathrm{Diff}(M_n) \tag{A.4.6}$$

be the subgroup generated by the n diffeomorphisms r_i and the $\binom{n}{3}$ diffeomorphisms r_{ijk}. Here the indices $(-1,-2)$ help recall the nature of the generators of the group $G_{(-1,-2)}$.

PROPOSITION A.4.3. *The set $\mathcal{S} = \mathcal{S}_n$ contains a fundamental domain of the action of $G_{(-1,-2)}$ on $\mathcal{P}(M_n)$.*

PROOF. Let $\rho : \mathrm{Diff}(M_n) \to \mathrm{End}(H^2(M_n), q)$ be the natural homomorphism, and set

$$\tilde{G}_{(-1,-2)} := \rho\left(G_{(-1,-2)}\right).$$

Clearly, it suffices to show that $\mathcal{S}$ contains a fundamental domain of $\tilde{G}_{(-1,-2)}$. Let

$$\pi : H^2(M_n) \setminus \{0\} \to \mathbb{P}(H^2(M_n))$$

denote the natural projection, and let

$$H^+_{M_n} = \{\alpha \in H^2(M_n) | q(\alpha, \alpha) = 1\}.$$

Clearly, the restriction

$$\pi|_{H^+_{M_n}} : H^+_{M_n} \to \mathcal{P}$$

is a $\tilde{G}_{(-1,-2)}$-equivariant diffeomorphism. We now apply Proposition A.3.2 above with $\Gamma = \tilde{G}_{(-1,-2)}$, noting that since the inequalities (A.4.5) defining $\mathcal{S}$ arise from the mirrors of reflections in $\tilde{G}_{(-1,-2)}$, $\mathcal{S}$ must contain a cell C_ι. $\square$

A.5. Background material from symplectic geometry

We will exploit P. Biran's criterion [**Bir97, Bir01**] for second cohomology classes of a symplectic four-manifold to be approximately represented by symplectic forms. Let (M, ω) be a closed symplectic four-manifold. A submanifold $N \subset M$ is called symplectic if the restriction of ω to N is a symplectic form. By analogy with algebraic geometry, we will call a class $E \in H_2(M)$ *exceptional* if it can be represented by an embedded symplectic 2-sphere $\Sigma \subset M$ such that $\Sigma \cdot \Sigma = -1$.

DEFINITION A.5.1. We will denote by $\mathrm{E}_\omega \subset H_2(M)$ the set of all ω-exceptional classes.

Before stating Biran's theorem, we mention that four-manifolds whose Seiberg-Witten invariants vanish when the Seiberg-Witten moduli space has non-zero dimension are said to have Seiberg-Witten simple type. For our purposes, all we need to know is that closed simply connected symplectic four-manifolds with $b_2^+ = 1$ do *not* have Seiberg-Witten simple type. As before, we denote by $q(\cdot, \cdot)$ the intersection form on $H^2(M)$.

THEOREM A.5.2 (P. Biran). *Let (M,ω) be a closed symplectic four-manifold. Assume M does not have Seiberg-Witten simple type. Let $\alpha \in H^2(M)$ belong to the positive cone of M, i.e. $q(\alpha,\alpha) > 0$. Suppose that α also satisfies*

(1) $q(\alpha, [\omega]) > 0$,
(2) $\alpha(E) > 0$ *for all* $E \in \mathrm{E}_\omega$.

Then there exist symplectic forms on M representing cohomology classes arbitrarily close to α.

This result was proved in [**Bir97**] using Taubes's Seiberg-Witten Gromov-Witten correspondence [**Ta95**], as further developed by McDuff [**Mc96**], combined with the technique of symplectic inflation due to F. Lalonde and D. McDuff [**LaM96**].

REMARK A.5.3. Although in general, when $b_2^+ = 1$, the Seiberg-Witten invariants depend on the choice of a chamber, there is a canonical choice of chamber for symplectic manifolds. So, in the case under consideration, we may refer to the Seiberg-Witten invariants unambiguously.

REMARK A.5.4. The main idea behind the proof Theorem A.5.2 is to use the non-vanishing of the Seiberg-Witten invariants to construct a symplectically embedded surface representing a multiple of the Poincare dual of the chosen class α. Taubes did this in a remarkable string of papers outlined in [**Ta95**]. Symplectic inflation uses the embedded symplectic surface to construct a symplectic form representing a cohomology class arbitrarily close to α.

In very broad outline, Taubes's argument runs as follows: A solution to the Seiberg-Witten equations consists of a section of a certain vector bundle as well as a connection on that bundle. The non-vanishing of the Seiberg-Witten invariant implies the existence of such a solution to the equations. Moreover, this solution persists under a large variety of perturbations. In the limit of a specially chosen large perturbation, the vanishing sets of the sections become symplectically embedded surfaces.

In order to apply Theorem A.5.2, we will need to analyze which classes in $H_2(M)$ can be ω-exceptional. This will rely heavily on the theory of J-holomorphic curves, which we now outline. Let M be a smooth manifold admitting an almost complex structure J and let Σ be a Riemann surface with complex structure j.

DEFINITION A.5.5. A J-holomorphic Σ-curve in M is a map $u : \Sigma \to M$ satisfying the Cauchy-Riemann equation

$$J \circ du - du \circ j = 0.$$

Here, we will be interested exclusively in J-holomorphic spheres, *i.e.* in the case $\Sigma \simeq S^2$. Now we formulate the connection between ω-exceptional classes, on the one hand, and J-holomorphic curves, on the other. If (M,ω) is a symplectic manifold, we denote by $\mathcal{J}_\omega(M)$ the set of ω-tame complex structures on M, *cf.* Definition A.2.1. The following result was proved by McDuff [**Mc90**].

THEOREM A.5.6. *For J belonging to a residual subset of $\mathcal{J}_\omega(M)$, there exists a unique embedded J-holomorphic sphere representing each element $E \in \mathrm{E}_\omega$.*

Our main tool will be the property of positivity of intersections for J-holomorphic curves, first stated by M. Gromov [**Gro85**]. For a detailed history and exposition of the proof of this theorem, see [**McS04**, Appendix E].

A J-holomorphic curve $u : \Sigma \to M$ is called *simple* if it does not factor as the composition $u' \circ \phi$ of a map of Riemann surfaces $\phi : \Sigma \to \Sigma'$ with a J-holomorphic curve $u' : \Sigma' \to M$ such that the degree of ϕ is is greater than one.

THEOREM A.5.7. *Each pair A_0, $A_1 \in H_2(M)$ of homology classes represented by distinct, connected, simple J-holomorphic curves, satisfies $A_0 \cdot A_1 \geq 0$.*

Since embedded J-holomorphic spheres are always simple, we can combine Theorem A.5.6 with Theorem A.5.7 to deduce constraints on E_ω.

Finally, we note that if (M, ω) is a symplectic manifold, then each almost complex structure $J \in \mathcal{J}_\omega(M)$ makes TM into a complex vector bundle. Recall that the Chern classes of a complex vector bundle belong to integral cohomology. Since $\mathcal{J}_\omega(M)$ is contractible, it follows that the Chern classes of TM do not depend on the choice of $J \in \mathcal{J}_\omega(M)$. We will use the following topological lemma relating intersection numbers in a symplectic four-manifold (M, ω) with the first Chern class of its tangent bundle $c_1(TM)$.

LEMMA A.5.8 (The adjunction formula). *Let (M, ω) be a symplectic four-manifold and let $\Sigma \subset M$ be a symplectic submanifold of* $\dim \Sigma = 2$. *Then*

$$c_1(TM)([\Sigma]) = [\Sigma] \cdot [\Sigma] + \chi(\Sigma).$$

PROOF. Choosing $J \in \mathcal{J}_\omega(M)$, we may consider TM as a complex vector bundle. Since Σ is embedded, choosing a Hermitian metric on TM, we have the splitting

$$TM|_\Sigma \simeq T\Sigma \oplus \nu_\Sigma,$$

where ν_Σ denotes the normal bundle of Σ. Since the first Chern class of a line bundle is its Euler class, we have

$$c_1(\nu_\Sigma) = \Sigma \cdot \Sigma, \qquad c_1(T\Sigma) = \chi(\Sigma).$$

The lemma follows immediately from the Whitney sum formula. $\square$

A.6. Proof of density of image of period map

As before, let

$$\pi : H^2(M_n) \setminus \{0\} \to \mathbb{P}(H^2(M_n))$$

denote the natural projection. Define

$$\tilde{\mathcal{S}} := \{\alpha \in \pi^{-1}(\mathcal{S}) \mid q(\alpha, \lambda) > 0\}.$$

By Proposition A.4.3, it suffices to prove that if $\alpha \in \tilde{\mathcal{S}}$, we can find cohomology classes arbitrarily close to α represented by symplectic forms. To this end, we will need to apply Theorem A.5.2.

Theorem A.5.2 starts with a symplectic manifold (M, ω). So, to apply Theorem A.5.2 in the case $M = M_n$, we first need to construct a symplectic form ω on M. In fact, in order to facilitate the study of E_ω, we will work in the more rigid Kähler category. Recall that a Kähler form on M is a symplectic form ω such that $\mathcal{J}_\omega$ contains an integrable complex structure induced by holomorphic coordinate charts on M. We use the fact that M_n arises as the complex analytic blowup of $\mathbb{CP}^2$ at n points. As such, it admits a canonical complex structure, which we denote by J_n.

LEMMA A.6.1. *There exists a Kähler form ω_δ on M_n representing the class*

$$\lambda - \delta(\epsilon_1 + \ldots + \epsilon_k) \in H^2(M_n)$$

for all sufficiently small $\delta > 0$. The canonical complex structure J_n on M_n arising from the complex analytic blow-up construction is compatible with ω_δ for all δ.

PROOF. The Fubini-Study form on $\mathbb{CP}^2$ is well known to be Kähler [**GriH78**]. The lemma then follows from the general construction of Kähler forms on blow-ups [**GriH78**, p. 192]. We note that if we were not interested in the existence of a compatible integrable complex structure, this argument could be carried through entirely in the symplectic category using McDuff's symplectic blow-up construction [**McS98**]. □

It is important to emphasize that in order to apply Theorem A.5.2, we need first to *fix* a symplectic structure ω on M. Theorem A.5.2 then allows us to construct new symplectic forms representing cohomology classes α that satisfy conditions (1) and (2). However, Lemma A.6.1 allows us some freedom as to which symplectic structure we fix. So, for any fixed class $\alpha \in \tilde{\mathcal{S}}$, we can choose a symplectic form ω such that condition (1) holds. Indeed, since by assumption $q(\alpha, \lambda) > 0$, for δ_0 sufficiently small we have $q(\alpha, \omega_{\delta_0}) > 0$. At this point, we fix $\omega = \omega_{\delta_0}$.

In order to verify condition (2) of Theorem A.5.2, we need to derive constraints on the set of exceptional classes E_ω. The following lemma, which shows that certain classes actually are exceptional, via the combination of Theorems A.5.6 and A.5.7 will help constrain which other classes might be exceptional. We define

$$E_{ij} := L - E_i - E_j$$

for $i \neq j$.

LEMMA A.6.2. *The classes E_i and E_{ij} are exceptional for any symplectic form ω with respect to which J_n is compatible. In particular, we have $E_i, E_{ij} \in \mathrm{E}_\omega$ for $\omega = \omega_{\delta_0}$.*

PROOF. It suffices to show that E_i, E_{ij}, can be represented by spheres holomorpically embedded with respect to J_n. Then since J_n is ω compatible, such spheres will automatically be symplectic.

Indeed, let $\Sigma \subset M$ be a holomorphically embedded sphere, let $x \in \Sigma$ and let $X \in T_x\Sigma$. Then the vectors X, J_nX, form a basis for $T_x\Sigma$, and by compatibility $\omega(X, J_nX) > 0$. Hence, $\omega|_{T\Sigma}$ is non-degenerate, *i.e.* a symplectic form.

In particular, E_i is represented by the exceptional divisor $\mathfrak{E}_i$ of the ith blowup. This is well known to be an embedded holomorphic sphere with respect to J_n, *cf.* [**GriH78**, p. 182]. Furthermore, using the fact that the Cremona transformation r_{ijk} is a biholomorphism, and noting that

$$r_{ijk}^*(\epsilon_i) = \lambda - \epsilon_j - \epsilon_k,$$

we see that E_{jk} is represented by the embedded holomorphic sphere $r_{ijk}(\mathfrak{E}_i)$. □

Given $E \in \mathrm{E}_\omega$, we write $E = e_0L - \sum_i e_iE_i$. Also, we use the abbreviated notation $c_1 = c_1(TM) \in H^2(M)$.

LEMMA A.6.3. *Every exceptional class E satisfies the following* 4 *conditions:*

(i) $E \cdot E = -1$, *i.e.* $e_0^2 - \sum_i e_i^2 = -1$.
(ii) $c_1(E) = 1$, *i.e.* $3e_0 - \sum_i e_i = 1$.

(iii) *Either* $E = E_i$, *or* $e_0 > 0$ *and* $e_i \geq 0$, $\forall i$.
(iv) *Either* $E = E_{ij}$ *for some* $i \neq j$, *or* $E \cdot E_{ij} \geq 0$, *i.e.* $e_0 - e_i - e_j \geq 0$ *for all* $i \neq j$.

PROOF. Item (i) is part of the definition of an exceptional class. To prove (ii), we apply Lemma A.5.8 to obtain

$$c_1(E) = -1 + 2 = 1, \quad c_1(L) = 1 + 2 = 3, \quad c_1(E_i) = -1 + 2 = 1.$$

To prove item (iii), assume $E \neq E_i$. Combining Lemma A.6.2 with Theorems A.5.6 and A.5.7, we deduce that $E \cdot E_i \geq 0$. Hence $e_i \geq 0$ for all $i \geq 1$. It then follows from (ii) that $e_0 > 0$. Finally, (iv) follows immediately from Lemma A.6.2 combined with Theorems A.5.6 and A.5.7, as before. □

We now apply the preceding lemma to verify condition (2). By definition, if $\alpha \in \tilde{\mathcal{S}}$, we have $\alpha(E_i) = q(\alpha, \epsilon_i) > 0$. In particular, condition (2) is satisfied when $E = E_i$. To deal with the case $E \neq E_i$, we introduce the following lemma.

LEMMA A.6.4. *Let* $\alpha \in \tilde{\mathcal{S}}$. *Write* $\alpha = a_0\lambda - \sum_i a_i\epsilon_i$. *Let* $a_{i_1} \geq a_{i_2} \geq a_{i_3}$ *be the three largest coefficients among the* $a_1, \ldots, a_n$. *Let*

$$E = e_0 L - \sum_i e_i E_i$$

be an exceptional class, and assume $E \neq E_i$. *Then*

$$\alpha(E) \geq e_0 q(\alpha, \epsilon_{i_1 i_2 i_3}), \tag{A.6.1}$$

where $\epsilon_{i_1 i_2 i_3}$ *is the class defining the appropriate Cremona map as in* (A.4.4).

PROOF. First we prove that if $E \neq E_i$ then

$$0 \leq e_i \leq e_0 \tag{A.6.2}$$

for all i. Indeed, if $E = E_{ij}$ then inequality (A.6.2) is immediate. Otherwise, by rearranging the inequality of (iv), we obtain

$$e_i \leq e_0 - e_j.$$

Then we use the fact that by (iii) we have $e_j \geq 0$ and $e_i \geq 0$ to conclude equation (A.6.2).

Combining (A.6.2) and (ii), we know that the coefficients e_i of any exceptional divisor $E \neq E_i \in \mathrm{E}_\omega$ must satisfy the inequalities

$$0 \leq e_i \leq e_0 \text{ and } \sum_i e_i = 3e_0 - 1. \tag{A.6.3}$$

Now, by the formula for the intersection form,

$$q(\alpha, E) = a_0 e_0 - \sum_i a_i e_i. \tag{A.6.4}$$

We think of (A.6.1) as a lower bound for expression (A.6.4) as we hold α fixed and allow E to vary over all exceptional classes with fixed non-zero e_0. By (A.6.3), for a fixed value of e_0, the number of possible choices for the coefficients e_i are finite. So, for one particular choice of the e_i, the expression (A.6.4) must be minimized. Clearly it suffices to prove the lower bound (A.6.1) for this minimizing choice. It is not important whether this choice of the e_i can actually be realized by an exceptional class.

Recall that i_1, i_2, i_3, are the indices such that $a_{i_1} \geq a_{i_2} \geq a_{i_3}$ are the largest of the coefficients a_i of α. We assert that the choice of the coefficients e_i that minimizes the expression (A.6.4) is given by

$$e_{i_1} = e_{i_2} = e_0, \qquad e_{i_3} = e_0 - 1, \qquad e_i = 0,\ i \neq i_1,\ i_2,\ i_3. \tag{A.6.5}$$

Indeed, we prove by induction that for an arbitrary choice of the coefficients e_i the value of expression (A.6.4) cannot be less than the value given the choice (A.6.5). Indeed, starting with an arbitrary choice of the coefficients e_i satisfying (A.6.3), we execute the following algorithm repeatedly until it terminates:

(1) If e_{i_1} or $e_{i_2} < e_0$ and $e_j > 0$ for some $j \neq i_1, i_2, i_3$, then increment e_{i_1} or e_{i_2} by 1 and decrement e_j by 1.
(2) Otherwise, if e_{i_1} or $e_{i_2} < e_0$ and $e_{i_3} = e_0$ then increment e_{i_1} or e_{i_2} by 1 and decrement e_{i_3} by 1.
(3) Otherwise, if $e_{i_3} < e_0 - 1$ and $e_j > 0$ for some $j \neq i_1, i_2, i_3$, increment e_{i_3} by 1 and decrement e_j by 1.
(4) Otherwise, terminate.

It is clear that executing this algorithm can only decrease the expression (A.6.4). On the other hand, using equations (A.6.3), it is easy to see that this algorithm terminates only when e_i are as in (A.6.5). Finally, since e_{i_1}, e_{i_2}, e_{i_3}, are bounded by e_0, the process must terminate after a finite number of steps. Indeed, the value of one of the coefficients e_{i_1}, e_{i_2}, e_{i_3}, must increase at each step. Furthermore, the value of e_{i_3} can only decrease once, wherease the values of e_{i_1} and e_{i_2} can never decrease.

Now, substituting choice (A.6.5) into expression (A.6.4), we obtain

$$\begin{aligned} \alpha(E) &= a_0 e_0 - \sum_i a_i e_i \\ &\geq a_{i_3} + a_0 e_0 - (a_{i_1} + a_{i_2} + a_{i_3}) e_0 \\ &= a_{i_3} + e_0 q(\alpha, \epsilon_{123}), \end{aligned}$$

Since $\alpha \in \widetilde{S}$ we know that $q(\alpha, \epsilon_{i_3}) > 0$ and hence $a_{i_3} > 0$. Equation (A.6.1) follows immediately. $\square$

Finally, to check condition (2) of Theorem A.5.2, we apply Lemma A.6.4 to conclude $\alpha(E) \geq e_0 q(\alpha, \epsilon_{i_1 i_2 i_3}) > 0$ by the defining equation of the fundamental domain which corresponds to the appropriate Cremona transformation.

APPENDIX B

Open problems

We conclude with a list of open problems in systolic topology, geometry, and arithmetic.

B.1. Topology

B.1.1. In Subsection 1.3.3 above, we described an application of a map between classifying spaces to the (still open) problem of the determination of the optimal systolic ratio of projective spaces, see [**BaKSS06**]. The argument in [**BaKSS06**] exploits the inclusion of classifying spaces, $BS^1 \to BS^3$, resulting from the Lie group inclusion $S^1 \to S^3$. There is a natural way to try to generalize such an argument.

Since the group S^3 is the standard building block in Lie group theory, in principle the field is wide open for generalisation, by exploiting the inclusion of classifying spaces corresponding to the subgroup inclusion $S^3 \hookrightarrow G$, where G is a compact Lie group. It remains to be seen whether geometrically meaningful results can be obtained in this fashion.

More fundamentally, while the link of homotopy theory to systolic geometry tends to involve integral homotopy theory, many of the arguments, *e.g.* for general systolic freedom [**Katz95a, CrK03**], and the above results on quaternionic projective space, tend to be rational (see, *e.g.* [**Su74**]). It would be desirable to forge a direct connection between rational homotopy theory and systolic geometry. This would constitute a considerable simplification, and requires clarification.

B.1.2. The equality of the systolic category and the Lusternik-Schnirelmann category for 3-manifolds stems from the fact both are determined via the simple "free *versus* non-free" dichotomy for the fundamental group. In this dimension, the only intermediate value of either category is 2, and the result for this value follows from the appropriate classification in terms of sphere bundles over the circle. The main mathematical content on the systolic side is provided by Gromov's 1-systolic inequality for essential manifolds, which characterizes the maximal value of the category.

The situation is more complex for 4-manifolds, where there is a pair intermediate values, namely, 2 and 3. Thus, in the non-essential case (not covered by Gromov's inequality), there is a potential for a deeper invariant. The distinction between the two intermediate values may depend on a study of Massey products in 4-manifolds.

Given a (non-essential) 4-manifold M with $b_1(M) \geq 1$, we note the following. If the cup product on $H^1(M, \mathbb{R})$ does not vanish, both categories must be at least 3 (by comparison with real cup-length). Otherwise, one expects the existence of a nontrivial Massey product in this case to be the deciding factor in distinguishing between the two intermediate values of systolic category, similarly to the Lescop invariant for 3-manifolds. The issue is related to the surjectivity of the Abel-Jacobi map $M \to J_1(M)$ to the Jacobi torus $J_1(M) = T^{b_1(M)}$.

This surjectivity can be studied at the level of lifts to suitable covers, such as the universal free abelian cover. Here the nontriviality of the homology fiber of the lift at this level, is a sufficient condition.

Alternatively, one can study the map from M (with $b_1(M) = 2$) to a more general space, *e.g.* exploiting the step 2 nilpotent completion and the map to the corresponding 3-manifold, instead of the map to the 2-torus, *cf.* Remark 12.7.5.

B.2. Geometry

B.2.1. It was shown in [**BaKSS06**] that, contrary to the *complex* case clarified by M. Gromov, the symmetric metric on the *quaternionic* projective plane turns out not to be its systolically optimal metric, contrary to expectation expressed in a number of publications in Riemannian geometry, see Subsection 1.3.3 above.

The technique used in [**BaKSS06**] involves the exceptional Lie algebra E_7. Can this technique be refined to determine the precise value of the optimal systolic ratio of the two-point homogeneous spaces of non-complex type?

B.2.2. It was proved in [**KatzS06a**] that a genus 2 surface satisfies the Loewner inequality, and in [**KatzS05**], that surfaces of genus at least 20 are Loewner. The cases $3 \leq g \leq 19$ are open. One approach would be to use the capacity of annuli, *cf.* Remark 6.2.2.

Can efficient pair of pants decompositions for hyperbolic surfaces, combined with capacity of cylinders, be used to obtain efficient lower bounds for systoles of surfaces? By using the results by P. Buser and M. Seppälä [**BusSe02**] on "short" pair of pant decompositions of hyperbolic surfaces, one may be able to obtain suitable bounds for the capacity of the corresponding cylinders (*i.e.* annuli obtained by cutting open each pair of pants), and derive inequalities for higher genus surfaces.

B.3. Arithmetic

B.3.1. Consider the Weeks 3-manifold M_W. It is well known that its orbifold fundamental group $\pi \subset \mathrm{PSL}(2,\mathbb{C})$ is arithmetic. Consider principal congruence subgroups of π and the corresponding covers $M(p)$ of M_W. Do such covers satisfy a bound similar to the 4/3-bound (1.2.3) for the principal congruence tower of Hurwitz surfaces, *i.e.* $\mathrm{sys}\pi_1(M(p)) \geq \frac{2}{3}\log\|M(p)\|$ (without an additive constant)?

B.3.2. For hyperbolic 3-manifolds, the orthogonal group $O(3,1)$ is identified with the complex 2 by 2 matrices, hence admits trace estimates for congruence towers similar to the $\mathrm{SL}(2,\mathbb{R})$ case, dealt with in [**KatzSV07**]. For n-manifolds, no such identification is available for $n \geq 4$. What kind of lower bounds can one obtain for congruence towers of arithmetic hyperbolic n-manifolds?

B.3.3. The most "symmetric" genus 2 surface, namely the Bolza surface, is not a Hurwitz surface (which is, by definition, a $(2,3,7)$ triangle surface), but rather a (2,3,8) triangle surface, *cf.* Section 10.2. Do (2,3,8) triangle surfaces satisfy a bound similar to the bound (1.2.3) in the Hurwitz case?

B.3.4. In the case of arithmetic Fuchsian groups defined over $\mathbb{Q}$, P. Buser and P. Sarnak [**BusS94**] use the existence of a non-congruence subgroup of index 2, to prove a lower bound for the conformal systole via an eigenvalue estimate. Does this technique extend to arbitrary arithmetic groups? Does it extend to Kleinian groups?

Bibliography

[Ab81] W. Abikoff, *The uniformization theorem.* Amer. Math. Monthly **88** (1981), no. 8, 574–592.

[Ab86] ———, *Remark on the size of a Riemann surface.* J. Analyse Math. **46** (1986), 11–12.

[Ac60] R. Accola, *Differentials and extremal length on Riemann surfaces.* Proc. Nat. Acad. Sci. USA **46** (1960), 540–543.

[Ad98] C. Adams, *Maximal cusps, collars, and systoles in hyperbolic surfaces.* Indiana Univ. Math. J. **47** (1998), no. 2, 419–437.

[Ad98] ———, *Detecting incompressibility of boundary in 3-manifolds.* Geom. Dedicata **99** (2003), 47–60.

[AM99] C. Adams and F. Morgan, *Isoperimetric curves on hyperbolic surfaces.* Proc. Amer. Math. Soc. **127** (1999), no. 5, 1347–1356.

[AdR00] C. Adams and A. Reid, *Systoles of hyperbolic* 3*-manifolds.* Math. Proc. Cambridge Philos. Soc. **128** (2000), no. 1, 103–110.

[AlAR90] *Interview with L. Bers,* More mathematical people. Contemporary conversations. Edited by D. Albers, G. Alexanderson, and C. Reid, Harcourt Brace Jovanovich, Boston, MA, 1990, pp. 3–21.

[AlR87] D. Albers and C. Reid, *An interview with Lipman Bers.* College Math. J. **18** (1987), no. 4, 266–290.

[Alt99] H. Alt, *Lineare Funktionalanalysis,* 3 Auflage, Springer-Verlag, Berlin, 1999.

[AmK00] L. Ambrosio and B. Kirchheim, *Currents in metric spaces.* Acta Math. **185** (2000), no. 1, 1–80.

[Amm04] B. Ammann, *Dirac eigenvalue estimates on two-tori.* J. Geom. Phys. **51** (2004), no. 3, 372–386.

[Ar83] M. Armstrong, *Basic topology.* Corrected reprint of the 1979 original. Undergraduate Texts in Math., Springer-Verlag, New York–Berlin, 1983.

[Bab93] I. Babenko, *Asymptotic invariants of smooth manifolds.* Russian Acad. Sci. Izv. Math. **41** (1993), 1–38.

[Bab02a] ———, *Forte souplesse intersystolique de variétés fermées et de polyèdres.* Annales Inst. Fourier **52**, (2002), 1259–1284.

[Bab02b] ———, *Loewner's conjecture, the Besicovich's barrel, and relative systolic geometry.* Mat. Sb. **193** (2002), no. 4, 3–16; English transl., Sb. Math. **193** (2002), no. 4, 473–486.

[Bab04] ———, *Géométrie systolique des variétés de groupe fondamental* $\mathbb{Z}_2$, Sémin. Théor. Spectr. Géom. Grenoble **22** (2004), 25–52.

[Ba06] ———, *Topologie des systoles unidimensionelles.* L'Enseignement Math. (2) **52** (2006), 109–142.

[BabB05] I. Babenko and F. Balacheff, *Géométrie systolique des sommes connexes et des revêtements cycliques.* Math. Ann. **333** (2005), no. 1, 157–180.

[BabK98] I. Babenko and M. Katz, *Systolic freedom of orientable manifolds.* Ann. Sci. Ecole Norm. Sup. (Paris) **31** (1998), 787–809.

[BaiW03] I. Baird and J. Wood, *Harmonic morphisms between Riemannian manifolds*, London Math. Soc. Monographs. New Series, vol. 29, The Clarendon Press, Oxford University Press, Oxford, 2003.

[Bal04] F. Balacheff, *Sur des problèmes de la géométrie systolique.* Sémin. Théor. Spectr. Géom. Grenoble **22** (2004), 71-82.

[Bana93] W. Banaszczyk, *New bounds in some transference theorems in the geometry of numbers.* Math. Ann. **296** (1993), 625–635.

[Bana96] ———, *Inequalities for convex bodies and polar reciprocal lattices in* $\mathbb{R}^n$ II: *Application of K-convexity.* Discrete Comput. Geom. **16** (1996), 305–311.

[BaCIK05] V. Bangert, C. Croke, S. Ivanov, and M. Katz, *Filling area conjecture and ovalless real hyperelliptic surfaces.* Geom. Funct. Anal. **15** (2005) no. 3, 577–597. See `arXiv:math.DG/0405583`

[BaCIK07] ———, *Boundary case of equality in optimal Loewner-type inequalities.* Trans. Amer. Math. Soc. **359** (2007), no. 1, 1–17. See `arXiv:math.DG/0406008`

[BaK03] V. Bangert and M. Katz, *Stable systolic inequalities and cohomology products.* Comm. Pure Appl. Math. **56** (2003), 979–997. See `arXiv:math.DG/0204181`

[BaK04] ———, *An optimal Loewner-type systolic inequality and harmonic one-forms of constant norm.* Comm. Anal. Geom. **12** (2004), no. 3, 703-732. `arXiv:math.DG/0304494`

[BaKSS06] V. Bangert, M. Katz, S. Shnider, and S. Weinberger, E_7*, Wirtinger inequalities, Cayley 4-form, and homotopy.* See `arXiv:math.DG/0608006`

[Bar57] E. Barnes, *On a theorem of Voronoi.* Proc. Cambridge Phil. Soc. **53** (1957), 537–539.

[BarW59] E. Barnes and G. Wall,*Some extreme forms defined in terms of Abelian groups.* J. Austral. Math. Soc. **1** (1959/1961), 47–63.

[Bav86] C. Bavard, *Inégalité isosystolique pour la bouteille de Klein.* Math. Ann. **274** (1986), no. 3, 439–441.

[Bav88] ———, *Inégalités isosystoliques conformes pour la bouteille de Klein.* Geom. Dedicata **27** (1988), 349–355.

[Bav92] ———, *La systole des surfaces hyperelliptiques,* Prepubl. Ec. Norm. Sup. Lyon **71** (1992).

[Bav06] ———, *Une remarque sur la géométrie systolique de la bouteille de Klein.* Arch. Math. (Basel) **87** (2006), no. 1, 72–74.

[BeM89] A.-M. Bergé and J. Martinet, *Sur un problème de dualité lié aux sphères en géométrie des nombres.* J. Number Theory **32** (1989), 14–42.

[Berg65] M. Berger, *Lectures on geodesics in Riemannian geometry. Tata Institute Fundamental Research Lectures on Mathematics,* no. 33, Tata Inst. Fundamental Research, Bombay 1965.

[Berg70] ———, *Quelques problèmes de géométrie riemannienne ou deux variations sur les espaces symétriques compacts de rang un.* L'Enseignement Math. (2) **16** (1970) 73–96.

[Berg72a] ———, *Du côté de chez Pu.* Ann. Sci. Ecole Norm. Sup. (4) **5** (1972), 1–44.

[Berg72b] ———, *A l'ombre de Loewner.* Ann. Scient. Ec. Norm. Sup. (4) **5** (1972), 241–260.

[Berg93] ———, *Systoles et applications selon Gromov.* Séminaire N. Bourbaki, exposé 771, Astérisque **216** (1993), 279–310.

[Berg03] ———, *A panoramic view of Riemannian geometry.* Springer-Verlag, Berlin, 2003.

[Bers76] I. Berstein, *On the Lusternik–Schnirelmann category of Grassmannians,* Proc. Camb. Phil. Soc. **79** (1976), 129-134.

[Bes87] A. Besse, *Einstein manifolds.* Springer-Verlag, Berlin, 1987.

[BesCG03] G. Besson, G. Courtois, and S. Gallot, *Hyperbolic manifolds, amalgamated products and critical exponents.* C. R. Math. Acad. Sci. Paris **336** (2003), no. 3, 257–261.

[Bie88] K. R. Biermann, *Die Mathematik und ihre Dozenten an der Berliner Universität 1810–1933.* Akademie-Verlag, Berlin, 1988.

[Bir97] P. Biran, *Symplectic packing in dimension* 4. Geom. Funct. Anal. **7** (1997), no. 3, 420–437.

[Bir01] ———, *From symplectic packing to algebraic geometry and back.* Proc. Third European Congress Math., Birkhäuser, Basel, 2001, pp. 506–524.

[Bl61a] C. Blatter, *Über Extremallängen auf geschlossenen Flächen. Comment. Math. Helv.* **35** (1961), 153–168.

[Bl61b] ———, *Zur Riemannschen Geometrie im Grossen auf dem Möbiusband.* Compositio Math. **15** (1961), 88–107.

[BlW99] J. Block and Sh. Weinberger, *Arithmetic manifolds of positive scalar curvature.* J. Differential Geom. **52** (1999), no. 2, 375–406.

[Bol78] B. Bollobás, *Extremal graph theory.* Reprint of the 1978 original. Dover, Mineola, NY, 2004.

[Bor94] J. Borzellino, *Pinching theorems for teardrops and footballs of revolution.* Bull. Austral. Math. Soc. **49** (1994), no. 3, 353–364.

[BoT82] R. Bott and L. Tu, *Differential forms in algebraic topology.* Graduate Texts in Math., vol. 82, Springer-Verlag, New York–Berlin, 1982.

[BouG76] A. Bousfield, and V. Gugenheim, *On* PL *de Rham theory and rational homotopy type.* Mem. Amer. Math. Soc. **8** (1976), no. 179.

[Bra85] L. de Branges, *A proof of the Bieberbach conjecture.* Acta Math. **154** (1985), no. 1-2, 137–152.

[BrH99] M. Bridson and A. Haefliger, *Metric spaces of non-positive curvature.* Grundlehren der Mathematischen Wissenschaften, vol. 319. Springer-Verlag, Berlin, 1999.

[Bro99] R. Brooks, *Platonic surfaces.* Comment. Math. Helv. **74** (1999), no. 1, 156–170.

[BroBS01] R. Brooks, R. Brown, and H. Schirmer, *The absolute degree and the Nielsen root number of compositions and Cartesian products of maps.* Theory of Fixed Points and its Applications (São Paulo, 1999), Topology Appl. **116** (2001), no. 1. Elsevier, Amsterdam, 2001. pp. 5–27.

[BroM04] R. Brooks and E. Makover, *Random construction of Riemann surfaces.* J. Differential Geom. **68** (2004), 121–157.

[BrowHK84] R. Brown, P. Heath, and K. Kamps, *Coverings of groupoids and Mayer–Vietoris type sequences.* Categorical topology (Toledo, Ohio, 1983), 147–162, Sigma Ser. Pure Math., vol. 5, Heldermann, Berlin, 1984. pp. 147–162.

[Bry96] R. Bryant, *On extremals with prescribed Lagrangian densities.* Manifolds and Geometry (Pisa, 1993), Sympos. Math., vol. 36, Cambridge Univ. Press, Cambridge, 1996, pp. 86–111.

[BuraI94] D. Burago and S. Ivanov, *Riemannian tori without conjugate points are flat.* Geom. Funct. Anal. **4** (1994), no. 3, 259–269.

[BuraI95] ———, *On asymptotic volume of tori.* Geom. Funct. Anal. **5** (1995), no. 5, 800–808.

[BuraZ80] Y. Burago and V. Zalgaller, *Geometric inequalities.* "Nauka", Moscow, 1980; English transl., Grundlehren der Mathematischen Wissenschaften, vol. 285, Springer-Verlag, Berlin, 1988.

[BurgM02] M. Burger, T. Gelander, A. Lubotzky, and S. Mozes, *Counting hyperbolic manifolds.* Geom. Funct. Anal. **12** (2002), no. 6, 1161–1173.

[Bus78] P. Buser, *Riemannsche Flächen mit grosser Kragenweite.* Comment. Math. Helv. **53** (1978), no. 3, 395–407.

[BusK81] P. Buser and H. Karcher, *Gromov's almost flat manifolds.* Astérisque, vol. 81, Soc. Math. France, Paris, 1981.

[BusS94] P. Buser and P. Sarnak, *On the period matrix of a Riemann surface of large genus.* With an appendix by J. H. Conway and N. J. A. Sloane. Invent. Math. **117** (1994), no. 1, 27–56.

[BusSe02] P. Buser and M. Seppälä, *Short homology bases and partitions of Riemann surfaces.* Topology **41** (2002), no. 5, 863–871.

[Cal96] E. Calabi, *Extremal isosystolic metrics for compact surfaces.* Actes de la table ronde de geometrie differentielle, Séminares et Congrès, vol. 1, Soc. Math. France, Paris, 1996, pp. 167–204.

[Car76] M. do Carmo, *Differential geometry of curves and surfaces.* Prentice-Hall, Englewood Cliffs, NJ, 1976.

[Car92] ———, *Riemannian geometry.* Birkhäuser, Boston, 1992.

[Cas71] J. Cassels, *An introduction to the geometry of numbers.* Grundlehren der mathematischen Wissenschaften, vol. 99, Springer-Verlag, Berlin–Heidelberg–New York, 1971.

[CW06] S. Chang and S. Weinberger, *Topological nonrigidity of nonuniform lattices.* Comm. Pure Appl. Math. **59** (2006).

[Cha72] I. Chavel, *Extremal length in real projective spaces.* Symposia Mathematica, Vol. X (Convegno di Geometria Differenziale, INDAM, Rome, 1971), Academic Press, London, 1972, pp. 159–167.

[Cha06] ———, *Riemannian geometry. A modern introduction.* Second edition. Cambridge Studies in Advanced Mathematics, vol. 98, Cambridge Univ. Press, Cambridge, 2006.

[ChE75] J. Cheeger and D. Ebin, *Comparison theorems in Riemannian geometry.* North-Holland, Amsterdam–Oxford, 1975.

[CM05] T. Cochran and J. Masters, *The growth rate of the first Betti number in abelian covers of 3-manifolds.* Math. Proc. Cambridge Philos. Soc. (to appear). See `arXiv:math.GT/0508294`

[ConS94] J. H. Conway and N. J. A. Sloane, *On lattices equivalent to their duals.* J. Number Theory, **48** (1994), no. 3, 373–382.

[ConS99] ———, *Sphere packings, lattices and groups.* Third edition, Springer-Verlag, New York, 1999.

[CorT03] O. Cornea, G. Lupton, J. Oprea, and D. Tanré, *Lusternik-Schnirelmann category.* Math. Surveys and Monographs, vol. 103, Amer. Math. Soc., Providence, RI, 2003.

[Cr88] C. Croke, *Area and the length of the shortest closed geodesic.* J. Differential Geom. **27** (1988), no. 1, 1–21.

[CrK03] C. Croke and M. Katz, *Universal volume bounds in Riemannian manifolds.* Surveys in Differential Geometry, vol. 8, Lectures on Geometry and Topology held in honor of Calabi, Lawson, Siu, and Uhlenbeck at Harvard University, May 3–5, 2002, edited by S. T. Yau, International Press, Somerville, MA, 2003, pp. 109–137. See `arXiv:math.DG/0302248`

[DaHM88] J. Dadok, R. Harvey, and F. Morgan, *Calibrations on* $\mathbb{R}^8$. Trans. Amer. Math. Soc. **307** (1988), no. 1, 1–40.

[DaS98] V. Danilov and V. Shokurov, *Algebraic curves, algebraic manifolds and schemes.* Itogi Nauki i Tekhiniki, Fund. Napravleniya, vol. 23, VINITI, Moscow, 1988; English transl., Algebraic geometry. I, Encyclopaedia Math. Sci., vol. 23, Springer-Verlag, Berlin, 1994, 1998.

[Dav89] R. Davitt, *William Marshall Bullitt and his amazing mathematical collection.* Math. Intelligencer **11** (1989), no. 4, 26–33.

[Do79] P. Dombrowski, *150 years after Gauss' "Disquisitiones generales circa superficies curvas".* With the original text of Gauss. Astérisque, **62**, Soc. Math. France, Paris, 1979.

[DR03] A. Dranishnikov and Y. Rudyak, *Examples of non-formal closed* $(k-1)$*-connected manifolds of dimensions* $\geq 4k-1$. Proc. Amer. Math. Soc. **133** (2005), no. 5, 1557–1561. See `arXiv:math.AT/0306299`.

[DT99] N. Dummigan and P. Tiep, *Lower bounds for the minima of certain symplectic and unitary group lattices.* Amer. J. Math. **121** (1999), no. 4, 889–918.

[EeS64] J. Eells and J. Sampson, *Harmonic mappings of Riemannian manifolds.* Amer. J. Math. **86** (1964) 109–160.

[Ei13] *Entwurf einer verallgemeinerten Relativitätstheorie und einer Theorie der Gravitation.* I. Physikalischer Teil von Albert Einstein in Zürich II. Mathematischer Teil von Marcel Grossmann in Zürich, Teubner, Leipzig und Berlin, 1913.

[Ell88] G. Ellis, *Homotopy classification the J. H. C. Whitehead way.* Exposition. Math. **6** (1988), no. 2, 97–110.

[FarbW05] B. Farb and Sh. Weinberger, *Hidden symmetries and arithmetic manifolds.* Geometry, spectral theory, groups, and dynamics, Contemp. Math., vol. 387, Amer. Math. Soc., Providence, RI, 2005, pp. 111–119.

[FarK92] H. Farkas and I. Kra, *Riemann surfaces.* Second edition. Graduate Texts in Math., vol. 71, Springer-Verlag, New York, 1992.

[FaS04] A. Fathi and A. Siconolfi, *Existence of* C^1 *critical subsolutions of the Hamilton-Jacobi equation.* Invent. Math. **155** (2004), no. 2, 363–388.

[Fed69] H. Federer, *Geometric measure theory.* Grundlehren der mathematischen Wissenschaften, vol. 153, Springer-Verlag, Berlin, 1969.

[Fed74] ———, *Real flat chains, cochains, and variational problems.* Indiana Univ. Math. J. **24** (1974), 351–407.

[FedF60] H. Federer and W. Fleming, *Normal and integral currents.* Ann. of Math. (2) **72** (1960), 458–520.

[FelHL98] Y. Felix, S. Halperin, and J.-M. Lemaire, *The rational LS category of products and of Poincaré duality complexes.* Topology **37** (1998), no. 4, 749–756.

[FelHT01] Y. Felix, S. Halperin, and J.-C. Thomas, *Rational homotopy theory.* Graduate Texts in Math., vol. 205, Springer-Verlag, New York, 2001.

[Fer77] J. Ferrand, *Sur la regularite des applications conformes.* C. R. Acad. Sci. Paris **284** (1977), no. 1, A77–A79.

[FreU84] D. Freed and K. Uhlenbeck, *Instantons and four-manifolds.* Math. Sci. Res. Inst. Publ., vol. 1, Springer-Verlag, New York, 1984.

[Fre99] M. Freedman, *$\mathbb{Z}_2$-Systolic-Freedom.* Proc. of the Kirbyfest (J. Hass and M. Scharlemann, eds.), Geometry and Topology Monographs, vol. 2, Geometry & Topology, Coventry, 1999, pp. 113–123.

[FriM94] R. Friedman and J. W. Morgan, *Smooth four-manifolds and complex surfaces.* Ergebnisse der Mathematik und ihrer Grenzgebiete (3), vol. 27, Springer-Verlag, Berlin, 1994.

[Fu16] P. Funk, *Uber eine geometrische Anwendung der Abelschen Integralgleichung.* Math. Ann. **77** (1916), 129-135.

[GaHL04] S. Gallot, D. Hulin, and J. Lafontaine, *Riemannian geometry.* Third edition. Springer-Verlag, Berlin, 2004.

[Gan71] T. Ganea, *Some problems on numerical homotopy invariants.* Symposium on Algebraic Topology (Battelle Seattle Res. Center, Seattle, WA, 1971), Lecture Notes Math., vol. 249, Springer-Verlag, Berlin, 1971, pp. 23–30.

[Gar06] A. Garsia, http://math.ucsd.edu/ garsia/cv.html#students

[GayK04] D. Gay and R. Kirby, *Constructing symplectic forms on 4-manifolds which vanish on circles.* Geom. Topol. **8** (2004), 743–777.

[Gela04] T. Gelander, *Homotopy type and volume of locally symmetric manifolds.* Duke Math. J. **124** (2004), no. 3, 459–515.

[Gelb53] A. Gelbart, *Testimony of Abe Gelbart.* Hearings before the Committee on Un-American Activities. Communist Methods of Infiltration (Education–Part 5). United States Government Printing Office. Washington, 1953, pp. 1565–1584.

[GeP69] I. Gelfand, M. Graev, I. Pyatetskii-Shapiro, *Representation theory and automorphic functions.* Generalized Functions, 6. Academic Press, Boston, MA, 1990.

[GG92] J. Gómez-Larrañaga and F. González-Acuña, *Lusternik-Schnirel'mann category of* 3-*manifolds.* Topology **31** (1992), no. 4, 791–800.

[Go83] A. Goodman, *Univalent functions.* Mariner, Tampa, 1983.

[Gre67] M. Greenberg, *Lectures on algebraic topology.* Benjamin, New York–Amsterdam, 1967.

[GreHV76] W. Greub, S. Halperin, and R. Vanstone, *Connections, curvature, and cohomology.* III*: Cohomology of principal bundles and homogeneous spaces.* Pure and Applied Math., vol. 47, Academic Press, New York–London, 1976.

[GriH78] P. Griffiths and J. Harris, *Principles of algebraic geometry.* Wiley, New York, 1978.

[Gro78] M. Gromov, *Almost flat manifolds.* J. Differential Geom. **13** (1978), no. 2, 231–241.

[Gro81a] ———, *Structures métriques pour les variétés riemanniennes.* (J. Lafontaine and P. Pansu, eds.) Textes Mathématiques, vol. 1. Cedic, Paris, 1981.

[Gro81b] ———, *Volume and bounded cohomology.* IHES Publ. Math. **56** (1981), 213–307.

[Gro83] *Filling Riemannian manifolds.* J. Diff. Geom. **18** (1983), 1-147.

[Gro85] *Pseudoholomorphic curves in symplectic manifolds.* Invent. Math. **82** (1985), no. 2, 307–347.

[Gro96] ———, *Systoles and intersystolic inequalities.* Actes de la Table Ronde de Géométrie Différentielle (Luminy, 1992), Sémin. et Congr., vol. 1, Soc. Math. France, Paris, 1996, pp. 291–362. `www.emis.de/journals/SC/1996/1/ps/smf_sem-cong_1_291-362.ps.gz`

[Gro99] *Metric structures for Riemannian and non-Riemannian spaces.* Birkhäuser, Boston, 1999.

[GroH81] B. Gross and J. Harris, *Real algebraic curves.* Ann. Sci. Ecole Norm. Sup. (4) **14** (1981), no. 2, 157–182.

[GruL87] P. Gruber and Lekkerkerker, *Geometry of numbers.* North-Holland, Amsterdam, 1987.

[Grun78] H. Grunsky, *Lectures on theory of functions in multiply connected domains.* Studia Mathematica, Skript 4. Vandenhoeck & Ruprecht, Göttingen, 1978.

[Hal83] S. Halperin, *Lectures on minimal models.* Mém. Soc. Math. France (N.S.) No. 9-10 (1983), 1–261.

[Ham92] C. Hamburger, *Regularity of differential forms minimizing degenerate elliptic functionals.* J. Reine Angew. Math. **431** (1992), 7–64.

[Hami06] M. Hamilton, *On the conformal systoles of four-manifolds.* Manuscripta Math. **121** (2006), 417-424. See `arXiv:math.DG/0512127`

[HarL82] R. Harvey and H. B. Lawson, *Calibrated geometries.* Acta Math. **148** (1982), 47–157.

[Hat02] A. Hatcher, *Algebraic topology.* Cambridge Univ. Press, Cambridge, 2002.

[Heb81] J. Hebda, *Some lower bounds for the area of surfaces.* Invent. Math. **65** (1981/82), no. 3, 485–490.

[Heb86] ———, *The collars of a Riemannian manifold and stable isosystolic inequalities.* Pacific J. Math. **121** (1986), 339–356.

[Hel99] S. Helgason, *The Radon transform.* 2nd edition, Progr. Math., vol. 5. Birkhäuser, Boston, 1999.

[Hem76] J. Hempel, *3-Manifolds.* Ann. of Math. Studies, vol. 86, Princeton Univ. Press, Princeton, NJ, 1976.

[Her55] J. Hersch, *Longueurs extrémales et géométrie globale.* Ann. Sci. Ecole Norm. Sup. (3) **72** (1955), 401–414.

[Hi03] J. Hillman, *Tits alternatives and low dimensional topology.* J. Math. Soc. Japan **55** (2003), no. 2, 365–383.

[Hi04] ———, *An indecomposable PD_3-complex*: II. Algebraic Geometric Topology **4** (2004), 1103–1109.

[Hu00] J. Hu, *Gromov–Witten invariants of blow-ups along points and curves.* Math. Z. **233** (2000), no. 4, 709–739.

[Hub59] H. Huber, *Zur analytischen Theorie hyperbolischen Raumformen und Bewegungsgruppen.* Math. Ann. **138** (1959), 1–26.

[Iv02] S. Ivanov, *On two-dimensional minimal fillings.* St. Petersburg Math. J. **13** (2002), no. 1, 17–25.

[IvK04] S. Ivanov and M. Katz, *Generalized degree and optimal Loewner-type inequalities.* Israel J. Math. **141** (2004), 221–233. See `arXiv:math.DG/0405019`

[Iw02] N. Iwase, *A_∞-method in Lusternik-Schnirelmann category.* Topology **41** (2002), no. 4, 695–723.

[JaP04] D. Jakobson, M. Levitin, N. Nadirashvili, and I. Polterovich, *Spectral problems with mixed Dirichlet-Neumann boundary conditions: isospectrality and beyond.* See `arXiv:math.SP/0409154`

[JaP05] D. Jakobson, M. Levitin, N. Nadirashvili, N. Nigam, and I. Polterovich, *How large can the first eigenvalue be on a surface of genus two?* Intern. Math. Res. Not. **2005**, no. 63, 3967–3985.

[Je84] F. Jenni, *Über den ersten Eigenwert des Laplace-Operators auf ausgewählten Beispielen kompakter Riemannscher Flächen.* Comment. Math. Helv. **59** (1984), no. 2, 193–203.

[Jo48] F. John, *Extremum problems with inequalities as subsidiary conditions.* Studies and Essays Presented to R. Courant on his 60th Birthday, January 8, 1948, Interscience, New York, 1948, pp. 187–204.

[KamT82] F. Kamber and P. Tondeur, *Harmonic foliations.* Harmonic maps (R. J. Knill, M. Kalka, and H. C. J. Sealey, eds.), Lecture Notes Math., vol. 949, Springer-Verlag, Berlin–Heidelberg, 1982, pp. 87–121.

[Kat83] A. Katok, *Entropy and closed geodesics.* Ergodic Th. Dynam. Sys. **2** (1983), 339–365.

[KatH95] A. Katok and B. Hasselblatt, *Introduction to the modern theory of dynamical systems. With a supplementary chapter by A. Katok and L. Mendoza.* Encyclopedia of Mathematics and its Applications, vol. 54, Cambridge Univ. Press, Cambridge, 1995.

[Kato92] S. Katok, *Fuchsian groups.* Chicago Lectures in Mathematics, Univ. Chicago Press, Chicago, IL, 1992.

[Katz83] M. Katz, *The filling radius of two-point homogeneous spaces.* J. Diff. Geom. **18** (1983), 505–511.

[Katz88] ———, *The first diameter of 3-manifolds of positive scalar curvature.* Proc. Amer. Math. Soc. **104** (1988), no. 2, 591–595.

[Katz95a] ———, *Counterexamples to isosystolic inequalities.* Geom. Dedicata **57** (1995), 195–206.

[Katz95b] ———, *A proof via the Seiberg–Witten moduli space of Donaldson's theorem on smooth 4-manifolds with definite intersection forms.* R.C.P. 25, vol. 47 (Strasbourg, 1993–1995), Prépubl. Inst. Rech. Math. Av., 1995/24, Univ. Louis Pasteur, Strasbourg, 1995, pp. 269–274.

[Katz02] ———, *Local calibration of mass and systolic geometry.* Geom. Funct. Anal. **12** (2002), no. 3, 598–621. See `arXiv:math.DG/0204182`

[Katz03] ———, *Four-manifold systoles and surjectivity of period map.* Comment. Math. Helv. **78** (2003), 772–786. See arXiv:math.DG/0302306

[Katz07] ———, *Systolic inequalities and Massey products in simply-connected manifolds.* Israel J. Math. (2007). See arXiv:math.DG/0604012

[KatzL05] M. Katz and C. Lescop, *Filling area conjecture, optimal systolic inequalities, and the fiber class in abelian covers.* Geometry, spectral theory, groups, and dynamics, Contemp. Math., vol. 387, Amer. Math. Soc., Providence, RI, 2005, pp. 181–200. See arXiv:math.DG/0412011

[KatzR06] M. Katz and Y. Rudyak, *Lusternik-Schnirelmann category and systolic category of low dimensional manifolds.* Comm. Pure Appl. Math. **59** (2006), no. 10, 1433–1456. See arXiv:math.DG/0410456

[KatzR07] ———, *Bounding volume by systoles of* 3*-manifolds.* See arXiv:math.DG/0504008

[KatzRS06] M. Katz, Y. Rudyak, and S. Sabourau, *Systoles of 2-complexes, Reeb graph, and Grushko decomposition.* Intern. Math. Research Notices, 2006 (2006). Art. ID 54936, pp. 1–30. See arXiv:math.DG/0602009

[KatzS05] M. Katz and S. Sabourau, *Entropy of systolically extremal surfaces and asymptotic bounds.* Ergodic Th. Dynam. Sys. **25** (2005), no. 4, 1209–1220. See arXiv:math.DG/0410312

[KatzS06a] ———, *Hyperelliptic surfaces are Loewner.* Proc. Amer. Math. Soc. **134** (2006), no. 4, 1189–1195. See arXiv:math.DG/0407009

[KatzS06b] ———, *An optimal systolic inequality for CAT(0) metrics in genus two.* Pacific J. Math. **227** (2006), no. 1, 95–107. See arXiv:math.DG/0501017

[KatzSV07] M. Katz, M. Schaps, and U. Vishne, *Logarithmic growth of systole of arithmetic Riemann surfaces along congruence subgroups.* J. Diff. Geom. (to appear). See arXiv:math.DG/0505007

[Kod87] S. Kodani, *On two-dimensional isosystolic inequalities.* Kodai Math. J. **10** (1987), no. 3, 314–327.

[Kon03] J. Kong, *Seshadri constants on Jacobian of curves.* Trans. Amer. Math. Soc. **355** (2003), no. 8, 3175–3180.

[Kor86] J. Korevaar, *Ludwig Bieberbach's conjecture and its proof by Louis de Branges.* Amer. Math. Monthly **93** (1986), no. 7, 505–514.

[Kot91] D. Kotschick, SO(3)*-invariants for* 4*-manifolds with* $b_2^+ = 1$. Proc. London Math. Soc. (3) **63** (1991), no. 2, 426–448.

[Kot01] ———, *On products of harmonic forms.* Duke Math. J. **107** (2001), no. 3, 521–531.

[Kr66] V. Kraines, *Topology of quaternionic manifolds.* Trans. Amer. Math. Soc. **122** (1966), 357–367.

[Ku78] H. T. Kung and C. E. Leiserson, *Systolic arrays (for VLSI).* Sparse Matrix Proceedings 1978 (Sympos. Sparse Matrix Comput., Knoxville, TN, 1978), SIAM, Philadelphia, PA, 1979, pp. 256–282.

[KuN95] T. Kuusalo and M. Näätänen, *Geometric uniformization in genus* 2. Ann. Acad. Sci. Fenn. Ser. A Math. **20** (1995), no. 2, 401–418.

[Laf79] J. Lafontaine, *Une caractérisation des tores riemanniens plats.* C. R. Acad. Sci. Paris Sér. A-B **288** (1979), no. 4, A291–A293.

[LaLS90] J. Lagarias, H. Lenstra, and C. Schnorr, *Bounds for Korkin-Zolotarev reduced bases and successive minima of a lattice and its reciprocal lattice.* Combinatorica **10** (1990), 343–358.

[LaM96] F. Lalonde and D. McDuff, *The classification of ruled symplectic* 4*-manifolds.* Math. Res. Lett. **3** (1996), no. 6, 769–778.

[Law75] H. B. Lawson, *The stable homology of a flat torus.* Math. Scand. **36** (1975), 49–73.

[LawM89] H. B. Lawson and M. Michelsohn, *Spin geometry.* Princeton Math. Series, vol. 38. Princeton Univ. Press, Princeton, NJ, 1989.

[Laz96] R. Lazarsfeld, *Lengths of periods and Seshadri constants of abelian varieties.* Math. Res. Lett. **3** (1996), no. 4, 439–447.

[Lê93] Vân Lê Hông, *Curvature estimate for the volume growth of globally minimal submanifolds.* Math. Ann. **296** (1993), no. 1, 103–118.

[LeB03] C. LeBrun, *Einstein metrics, four-manifolds, and differential topology.* Surveys in differential geometry, vol. 8,, International Press, Somerville, MA, 2003, pp. 235–255.

[LeiR02] C. Leininger an A. Reid, *The co-rank conjecture for 3-manifold groups.* Algebr. Geom. Topol. **2** (2002), 37–50.

[Les96] C. Lescop, *Global surgery formula for the Casson-Walker invariant.* Annals Math. Studies, vol. 140, Princeton Univ. Press, Princeton, NJ, 1996.

[Li01] T.-J. Li and A.-K. Liu, *Uniqueness of symplectic canonical class, surface cone and symplectic cone of 4-manifolds with* $B^+ = 1$. J. Differential Geom. **58** (2001), no. 2, 331–370.

[Lich69] A. Lichnerowicz, *Applications harmoniques dans un tore.* C. R. Acad. Sci., Sér. A **269** (1969), 912–916.

[Liu05] L. Liu, *The mapping properties of filling radius and packing radius and their applications.* Differential Geom. Appl. **22** (2005), no. 1, 69–79.

[Lo23] K. Löwner, *Untersuchungen über schlichte konforme Abbildungen des Einheitskreises,* I. Math. Ann. **89** (1923), 103–121.

[Lo71] C. Loewner, *Theory of continuous groups.* Notes by H. Flanders and M. Protter. Mathematicians of Our Time. vol. 1, The MIT Press, Cambridge, MA–London, 1971.

[Lo88] ———, *Collected papers.* Edited and with an introduction by Lipman Bers, Birkhäuser, Boston, MA, 1988.

[Lu90] W. Lu, Z. Tang, H. Xiong, and S. Bai, *In memory of Bao Ming Pu.* Adv. in Math. (China) **19** (1990), no. 2, 239–240. (Chinese)

[Lub94] A. Lubotzky, *Discrete groups, expanding graphs and invariant measures.* With an appendix by Jonathan D. Rogawski. Progress in Mathematics, vol. 125, Birkhäuser, Basel, 1994.

[LuS34] L. Lusternik and L. Schnirelmann, *Méthodes topologiques dans les problèmes variationnels.* Hermann, Paris, 1934.

[MacR03] C. Maclachlan and A. Reid, *The arithmetic of hyperbolic* 3*-manifolds.* Graduate Texts Math., vol. 219. Springer-Verlag, New York, 2003.

[MaM05] E. Makover and J. McGowan, *The length of closed geodesics on random Riemann surfaces.* See `arXiv:math.DG/0504175`

[Man79] A. Manning, *Topological entropy for geodesic flows.* Ann. of Math. (2) **110** (1979), 567–573.

[May69] J. May, *Matric Massey products.* J. Algebra **12** (1969) 533–568.

[Mc90] D. McDuff, *The structure of rational and ruled symplectic* 4*-manifolds.* J. Amer. Math. Soc. **3** (1990), no. 3, 679–712.

[Mc96] ———, *From symplectic deformation to isotopy.* Topics in symplectic 4-manifolds (Irvine, CA, 1996), First Int. Press Lect. Ser., I, International Press, Cambridge, MA, 1998, pp. 85–99.

[McS98] D. McDuff, and D. Salamon, *Introduction to symplectic topology.* Second edition. Oxford Mathematical Monographs. The Clarendon Press, Oxford Univ. Press, New York, 1998.

[McS04] ———, *J-holomorphic curves and symplectic topology.* Colloquium Publications, vol. 52, Amer. Math. Soc., Providence, RI, 2004.

[MilmS86] V. Milman and G. Schechtman, *Asymptotic theory of finite-dimensional normed spaces.* With an appendix by M. Gromov. Lecture Notes Math., vol. 1200. Springer-Verlag, Berlin, 1986.

[MilnH73] J. Milnor and D. Husemoller, *Symmetric bilinear forms.* Springer-Verlag, Heidelberg, 1973.

[Mir95] R. Miranda, *Algebraic curves and Riemann surfaces.* Graduate Studies Math., vol. 5, Amer. Math. Soc., Providence, RI, 1995.

[Mo65] J. Moser, *On the volume elements on a manifold.* Trans. Amer. Math. Soc. **120** (1965), 286–294.

[Mo86] ———, *Minimal solutions of variational problems on a torus.* Ann. Inst. H. Poincaré–Analyse non linéaire **3** (1986), 229–272.

[MS92] J. Moser and M. Struwe, *On a Liouville-type theorem for linear and nonlinear elliptic differential equations on a torus.* Bol. Soc. Brasil. Mat. **23** (1992), 1–20.

[NabW00] A. Nabutovsky and Sh. Weinberger, *Variational problems for Riemannian functionals and arithmetic groups.* Inst. Hautes Etudes Sci. Publ. Math. **92** (2000), 5–62 (2001).

[Nag06] P.-A. Nagy, *On length and product of harmonic forms in Kähler geometry.* Math. Z. **254** (2006), no. 1, 199–218. See `arXiv:math.DG/0406341`

[NV04] P.-A. Nagy and C. Vernicos, *The length of harmonic forms on a compact Riemannian manifold.* Trans. Amer. Math. Soc. **356** (2004), no. 6, 2501–2513. See `arXiv:math.DG/0301369`

[OR01] J. Oprea and Y. Rudyak, *Detecting elements and Lusternik-Schnirelmann category of 3-manifolds.* Lusternik-Schnirelmann category and related topics (South Hadley, MA, 2001), Contemp. Math., vol. 316, Amer. Math. Soc., Providence, RI, 2002, pp. 181–191. See `arXiv:math.AT/0111181`

[Par04] H. Parlier, *Fixed-point free involutions on Riemann surfaces.* See `arXiv:math.DG/0504109`

[Pat01] G. P. Paternain, *Schrödinger operators with magnetic fields and minimal action functionals.* Israel J. Math. **123** (2001), 1–27.

[PatP03] G. P. Paternain and J. Petean, *Minimal entropy and collapsing with curvature bounded from below.* Invent. Math. **151** (2003), no. 2, 415–450.

[Pe02] G. Perelman, *The entropy formula for the Ricci flow and its geometric applications.* See `arXiv:math.DG/0211159`

[PlR94] V. Platonov and A. Rapinchuk, *Algebraic Groups and Number Theory.* Academic Press, Boston, MA, 1994.

[Po93] R. Pomerantz, *The world in flames.* A short history of the Holocaust. CIS Publishers, New York, 1993.

[Pu52] P. M. Pu, *Some inequalities in certain nonorientable Riemannian manifolds.* Pacific J. Math. **2** (1952), 55–71.

[Re86] C. Reid, *Hilbert–Courant.* Springer-Verlag, New York, 1986.

[Rei75] I. Reiner, *Maximal orders.* Academic Press, London, 1975.

[Ri1854] B. Riemann, *Hypothèses qui servent de fondement à la géométrie.* Oeuvres mathématiques de Riemann. J. Gabay, Sceaux, 1990.

[RT90] M. Rosenbloom and M. Tsfasman, *Multiplicative lattices in global fields.* Invent. Math. **101** (1990), no. 3, 687–696.

[Ru87] W. Rudin, *Real and complex analysis.* Third edition. McGraw-Hill, New York, 1987.

[Rud99] Y. Rudyak, *On category weight and its applications.* Topology **38** (1999), no. 1, 37–55.

[RudS06] Y. Rudyak and S. Sabourau, *Systolic invariants of groups and 2-complexes via Grushko decomposition.* See `arXiv:math.DG/0609379`

[RudT00] Y. Rudyak and A. Tralle, *On Thom spaces, Massey products, and nonformal symplectic manifolds.* Internat. Math. Res. Notices **10** (2000), 495–513.

[Rum79] H. Rummler, *Quelques notions simples en géométrie riemannienne et leurs applications aux feuilletages compacts.* Comment. Math. Helv. **54** (1979), 224–239

[Sab04] S. Sabourau, *Systoles des surfaces plates singulières de genre deux.* Math. Z. **247** (2004), no. 4, 693–709.

[Sab06] ———, *Systolic volume and minimal entropy of aspherical manifolds.* J. Differential Geom. **74** (2006), no. 1, 155–176. See `arXiv:math.DG/0603695`

[Sak88] T. Sakai, *A proof of the isosystolic inequality for the Klein bottle.* Proc. Amer. Math. Soc. **104** (1988), no. 2, 589–590.

[Sc93] P. Schmutz, *Riemann surfaces with shortest geodesic of maximal length.* Geom. Funct. Anal. **3** (1993), no. 6, 564–631.

[Sc94a] ———, *Systoles on Riemann surfaces.* Manuscripta Math. **85** (1994), no. 3-4, 429–447.

[Sc94b] ———, *Congruence subgroups and maximal Riemann surfaces.* J. Geom. Anal. **4** (1994), no. 2, 207–218.

[Sc96a] ———, *Arithmetic Fuchsian groups and the number of systoles.* Math. Z. **223** (1996), no. 1, 13–25.

[Sc96b] ———, *Compact Riemann surfaces with many systoles.* Duke Math. J. **84** (1996), no. 1, 191–198.

[Sc96c] ———, *Systoles of arithmetic surfaces and the Markoff spectrum.* Math. Ann. **305** (1996), no. 1, 191–203.

[Sc98] P. Schmutz Schaller, *Geometry of Riemann surfaces based on closed geodesics.* Bull. Amer. Math. Soc. (N.S.) **35** (1998), no. 3, 193–214.

[Sen91a] W. Senn, *Strikte Konvexität für Variationsprobleme auf dem n-dimensionalen Torus.* Manuscripta Math. **71** (1991), 45–65.

[Sen91b] ———, *Über Mosers regularisiertes Variationsproblem für minimale Blätterungen des n-dimensionalen Torus.* Z. Appl. Math. Physics (ZAMP) **42** (1991), 527–546.

[Serg88] V. Sergiescu, *Basic cohomology and tautness of Riemannian foliations.* Appendix B in P. Molino, Riemannian foliations, Progr. Math., vol. 73, Birkhäuser, Boston–Basel, 1988, pp. 235–248.

[Serr73] J-P. Serre, *A course in arithmetic.* Graduate Texts in Math., vol. 7, Springer-Verlag, New York–Heidelberg, 1973.

[Sib70] L. Sibner and R. Sibner, *A non-linear Hodge–de-Rham theorem.* Acta Math. **125** (1970), 57–73.

[Sik05] J. Sikorav, *Bounds on primitives of differential forms and cofilling inequalities.* See arXiv:math.DG/0501089

[Sin78] W. Singhof, *Generalized higher order cohomology operations induced by the diagonal mapping.* Math. Z. **162** (1978), no. 1, 7–26.

[Sm62] S. Smale, *On the structure of* 5*-manifolds.* Ann. of Math. (2) **75** (1962), 38–46.

[Sp79] M. Spivak, *A comprehensive introduction to differential geometry.* Vol. 4. Second edition. Publish or Perish, Wilmington, DE, 1979.

[St00] M. Streit, *Field of definition and Galois orbits for the Macbeath-Hurwitz curves.* Arch. Math. (Basel) **74** (2000), no. 5, 342–349.

[Su74] D. Sullivan, *Genetics of homotopy theory and the Adams conjecture.* Ann. of Math. **100** (1974), 1–79.

[Su79] ———, *A homological characterization of foliations consisting of minimal surfaces.* Comment. Math. Helv. **54** (1979), 218–223.

[Ta95] C. Taubes, *The Seiberg-Witten and Gromov invariants.* Math. Res. Lett. **2** (1995), no. 2, 221–238.

[To84] P. Tolksdorf, *Regularity for a more general class of quasilinear elliptic equations.* J. Differential Equations **51** (1984), no. 1, 126–150.

[Tra93] A. Tralle, *On compact homogeneous spaces with nonvanishing Massey products.* Differential geometry and its applications (Opava, 1992), Math. Publ., vol. 1, Silesian Univ. Opava, 1993, pp. 47–50.

[Tro90] M. Troyanov, *Coordonnées polaires sur les surfaces riemanniennes singulières.* Ann. Inst. Fourier (Grenoble) **40** (1990), no. 4, 913–937 (1991).

[Tu47] W. Tutte, *A family of cubical graphs.* Proc. Cambridge Philos. Soc. **43** (1947), 459–474.

[UD54] P. Ungar and B. Descartes, *Advanced Problems and Solutions: Solutions: 4526.* Amer. Math. Monthly **61** (1954), no. 5, 352–353.

[Vig80] M.-F. Vignéras, *Variétés riemanniennes isospectrales et non isométriques.* Ann. of Math. (2) **112** (1980), no. 1, 21–32.

[Vin75] E. B. Vinberg, *Some arithmetical discrete groups in Lobavcevskiĭ spaces.* Discrete subgroups of Lie groups and applications to moduli (Internat. Colloq., Bombay, 1973), Oxford Univ. Press, Bombay, 1975, pp. 323–348.

[Vis07] U. Vishne, *Traces in congruence subgroups of arithmetic groups, and girth of Riemannian manifolds.* In preparation.

[Vo04] R. Vogeler, *Combinatorics of curves on Hurwitz surfaces.* Dissertation, University of Helsinki, Helsinki, 2004; Ann. Acad. Sci. Fenn. Math. Diss. no. 137, (2004), 40 pp.

[Wal65] C. Wall, *Finiteness conditions for* CW*-complexes.* Ann. of Math. (2) **81** (1965), 56–69.

[War83] F. Warner, *Foundations of Differentiable Manifolds and Lie Groups.* Graduate Texts in Math., vol. 94, Springer-Verlag, New York–Heidelberg–Berlin, 1983.

[Wei05] Sh. Weinberger, *Computers, rigidity, and moduli. The large-scale fractal geometry of Riemannian moduli space.* M. B. Porter Lectures. Princeton Univ. Press, Princeton, NJ, 2005.

[Wen05] S. Wenger, *Isoperimetric inequalities of Euclidean type in metric spaces.* Geom. Funct. Anal. **15** (2005), no. 2, 534–554.

[Wh83] B. White, *Regularity of area-minimizing hypersurfaces at boundaries with multiplicity.* Seminar on minimal submanifolds (E. Bombieri, ed.), Annals Math. Studies, vol. 103, Princeton Univ. Press, Princeton NJ, 1983, pp. 293–301.

[Wh99] ———, *The deformation theorem for flat chains. Acta Math.* **183** (1999), no. 2, 255–271.

[Whi49] J. H. C. Whitehead,tiCombinatorial homotopy. II. Bull. Amer. Math. Soc. **55** (1949), 453–496.

[Wr74] A. Wright, *Monotone mappings and degree one mappings between PL manifolds.* Geometric topology (Proc. Conf., Park City, Utah, 1974), Lecture Notes Math., vol. 438, Springer-Verlag, Berlin, 1975, pp. 441–459.

Index

Titles in This Series

137 **Mikhail G. Katz,** Systolic geometry and topology, 2007

136 **Jean-Michel Coron,** Control and nonlinearity, 2007

135 **Bennett Chow, Sun-Chin Chu, David Glickenstein, Christine Guenther, James Isenberg, Tom Ivey, Dan Knopf, Peng Lu, Feng Luo, and Lei Ni,** The Ricci flow: Techniques and applications, Part I: Geometric aspects, 2007

134 **Dana P. Williams,** Crossed products of C^*-algebras, 2007

133 **Andrew Knightly and Charles Li,** Traces of Hecke operators, 2006

132 **J. P. May and J. Sigurdsson,** Parametrized homotopy theory, 2006

131 **Jin Feng and Thomas G. Kurtz,** Large deviations for stochastic processes, 2006

130 **Qing Han and Jia-Xing Hong,** Isometric embedding of Riemannian manifolds in Euclidean spaces, 2006

129 **William M. Singer,** Steenrod squares in spectral sequences, 2006

128 **Athanassios S. Fokas, Alexander R. Its, Andrei A. Kapaev, and Victor Yu. Novokshenov,** Painlevé transcendents, 2006

127 **Nikolai Chernov and Roberto Markarian,** Chaotic billiards, 2006

126 **Sen-Zhong Huang,** Gradient inequalities, 2006

125 **Joseph A. Cima, Alec L. Matheson, and William T. Ross,** The Cauchy Transform, 2006

124 **Ido Efrat, Editor,** Valuations, orderings, and Milnor K-Theory, 2006

123 **Barbara Fantechi, Lothar Göttsche, Luc Illusie, Steven L. Kleiman, Nitin Nitsure, and Angelo Vistoli,** Fundamental algebraic geometry: Grothendieck's FGA explained, 2005

122 **Antonio Giambruno and Mikhail Zaicev, Editors,** Polynomial identities and asymptotic methods, 2005

121 **Anton Zettl,** Sturm-Liouville theory, 2005

120 **Barry Simon,** Trace ideals and their applications, 2005

119 **Tian Ma and Shouhong Wang,** Geometric theory of incompressible flows with applications to fluid dynamics, 2005

118 **Alexandru Buium,** Arithmetic differential equations, 2005

117 **Volodymyr Nekrashevych,** Self-similar groups, 2005

116 **Alexander Koldobsky,** Fourier analysis in convex geometry, 2005

115 **Carlos Julio Moreno,** Advanced analytic number theory: L-functions, 2005

114 **Gregory F. Lawler,** Conformally invariant processes in the plane, 2005

113 **William G. Dwyer, Philip S. Hirschhorn, Daniel M. Kan, and Jeffrey H. Smith,** Homotopy limit functors on model categories and homotopical categories, 2004

112 **Michael Aschbacher and Stephen D. Smith,** The classification of quasithin groups II. Main theorems: The classification of simple QTKE-groups, 2004

111 **Michael Aschbacher and Stephen D. Smith,** The classification of quasithin groups I. Structure of strongly quasithin K-groups, 2004

110 **Bennett Chow and Dan Knopf,** The Ricci flow: An introduction, 2004

109 **Goro Shimura,** Arithmetic and analytic theories of quadratic forms and Clifford groups, 2004

108 **Michael Farber,** Topology of closed one-forms, 2004

107 **Jens Carsten Jantzen,** Representations of algebraic groups, 2003

106 **Hiroyuki Yoshida,** Absolute CM-periods, 2003

105 **Charalambos D. Aliprantis and Owen Burkinshaw,** Locally solid Riesz spaces with applications to economics, second edition, 2003

104 **Graham Everest, Alf van der Poorten, Igor Shparlinski, and Thomas Ward,** Recurrence sequences, 2003

103 **Octav Cornea, Gregory Lupton, John Oprea, and Daniel Tanré,** Lusternik-Schnirelmann category, 2003

102 **Linda Rass and John Radcliffe,** Spatial deterministic epidemics, 2003

101 **Eli Glasner,** Ergodic theory via joinings, 2003

100 **Peter Duren and Alexander Schuster,** Bergman spaces, 2004

99 **Philip S. Hirschhorn,** Model categories and their localizations, 2003

98 **Victor Guillemin, Viktor Ginzburg, and Yael Karshon,** Moment maps, cobordisms, and Hamiltonian group actions, 2002

97 **V. A. Vassiliev,** Applied Picard-Lefschetz theory, 2002

96 **Martin Markl, Steve Shnider, and Jim Stasheff,** Operads in algebra, topology and physics, 2002

95 **Seiichi Kamada,** Braid and knot theory in dimension four, 2002

94 **Mara D. Neusel and Larry Smith,** Invariant theory of finite groups, 2002

93 **Nikolai K. Nikolski,** Operators, functions, and systems: An easy reading. Volume 2: Model operators and systems, 2002

92 **Nikolai K. Nikolski,** Operators, functions, and systems: An easy reading. Volume 1: Hardy, Hankel, and Toeplitz, 2002

91 **Richard Montgomery,** A tour of subriemannian geometries, their geodesics and applications, 2002

90 **Christian Gérard and Izabella Łaba,** Multiparticle quantum scattering in constant magnetic fields, 2002

89 **Michel Ledoux,** The concentration of measure phenomenon, 2001

88 **Edward Frenkel and David Ben-Zvi,** Vertex algebras and algebraic curves, second edition, 2004

87 **Bruno Poizat,** Stable groups, 2001

86 **Stanley N. Burris,** Number theoretic density and logical limit laws, 2001

85 **V. A. Kozlov, V. G. Maz′ya, and J. Rossmann,** Spectral problems associated with corner singularities of solutions to elliptic equations, 2001

84 **László Fuchs and Luigi Salce,** Modules over non-Noetherian domains, 2001

83 **Sigurdur Helgason,** Groups and geometric analysis: Integral geometry, invariant differential operators, and spherical functions, 2000

82 **Goro Shimura,** Arithmeticity in the theory of automorphic forms, 2000

81 **Michael E. Taylor,** Tools for PDE: Pseudodifferential operators, paradifferential operators, and layer potentials, 2000

80 **Lindsay N. Childs,** Taming wild extensions: Hopf algebras and local Galois module theory, 2000

79 **Joseph A. Cima and William T. Ross,** The backward shift on the Hardy space, 2000

78 **Boris A. Kupershmidt,** KP or mKP: Noncommutative mathematics of Lagrangian, Hamiltonian, and integrable systems, 2000

77 **Fumio Hiai and Dénes Petz,** The semicircle law, free random variables and entropy, 2000

76 **Frederick P. Gardiner and Nikola Lakic,** Quasiconformal Teichmüller theory, 2000

75 **Greg Hjorth,** Classification and orbit equivalence relations, 2000

For a complete list of titles in this series, visit the
AMS Bookstore at **www.ams.org/bookstore/**.